D0217944

Integral Mechanical Attachment

Integral Mechanical Attachment

A Resurgence of the Oldest
Method of Joining

Robert W. Messler, Jr.

AMSTERDAM • BOSTON • HEIDELBERG • LONDON
NEW YORK • OXFORD • PARIS • SAN DIEGO
SAN FRANCISCO • SINGAPORE • SYDNEY • TOKYO
Butterworth-Heinemann is an imprint of Elsevier

Butterworth–Heinemann is an imprint of Elsevier
30 Corporate Drive, Suite 400, Burlington, MA 01803, USA
Linacre House, Jordan Hill, Oxford OX2 8DP, UK

∞ Recognizing the importance of preserving what has been written, Elsevier prints its
books on acid-free paper whenever possible.

Library of Congress Cataloging-in-Publication Data
Application submitted

British Library Cataloguing-in-Publication Data
A catalogue record for this book is available from the British Library.

ISBN 13: 978-0-7506-7965-7
ISBN 10: 0-7506-7965-4

For information on all Butterworth–Heinemann publications
visit our Web site at www.books.elsevier.com

Printed in the United States of America
06 07 08 09 10   11 10 9 8 7 6 5 4 3 2 1

I dedicate this book to my daughters,
Vicki and Kerri,
who both have given me nothing
but joy in my life!

CONTENTS

3

Rigid Integral Mechanical Attachments or Interlocks

4

Elastic Integral Mechanical Attachments or Interlocks

5

Plastic (Formed-In) Integral Mechanical Attachments or Interlocks

6

Integral Mechanical Attachment Classification Revisited

7

Metal Attachment Schemes and Attachments

8

Polymer Attachment Schemes and Attachments

9

Ceramics and Glass Attachments Schemes and Attachments

10

Concrete and Masonry-Unit Attachments Schemes and Attachments

11

Wood Attachment Schemes and Attachments

12

Mechanical Electrical Connections

13

The Future of Integral Mechanical Attachment: Where From Here?

PREFACE

As long as there have been humans, there has been a need for and practice of joining; that is, intentionally bringing two or more physical objects together to form a larger and more complex object or structure of greater utility. And, without question, we can be certain the very first joining was accomplished when our fore-bearers (fully human or not) took two objects that fit together because of their natural shape and made them into a more complex object. Whether the first decision to join two or more simple objects was to produce a crude tool, such as a hammer, or to create a shelter, two or more objects that fit together naturally were fitted together without anything but their shape and the resulting fit to cause them to nest one with the other and, at least to some degree, interlock to produce an assembly with some structural integrity. A hammer was made by wedging a flat, oval-shaped stone into the crotch of a forked stick. A shelter was made by carefully arranging sticks or stones one with the other, such that each newly-added stick or stone fit into or around and interlocked with previously assembled sticks or stones.

From this simple, but profound start, the practice of mechanically attaching or interlocking parts to create assemblies or structures using geometric features that were integral to the parts began, and the life of humans has never been the same since. Humans have the most highly developed (although not unique) ability to alter their world at will by using their minds to create what they need from what they have. In fact, the newfound ability to join objects by using their natural shapes must have quickly advanced (also not by humans alone) to include the shaping of natural objects so that they could be fitted together to produce a more complex assembly or structure. The proud producer of a hammer only had to have the stone hammer-head cleave into two halves with sharp edges to realize the potential to produce an ax by wedging the now-sharp stone into a suitably forked stick. From here, with small but steady advances, stones were intentionally shaped to make sharp edges for ax heads or sharp points for the tips of spears. And, similarly, sticks were intentionally split to receive and better grip a specially-shaped stone.

The rate of advance may have been slow from the Stone Age[1] to the Bronze Age[2] to the Iron Age[3] and beyond, but it continued to occur at an ever-accelerating rate that has not stopped—and with ever-greater sophistication. And while new methods of joining emerged along the way, including the use of fasteners to accomplish mechanical joining without relying on geometric features of the parts comprising the assembly, natural and, then, synthetic adhesives to bond one part to another, and welding using first simple hammering together of hot metal parts and, then, increasingly more intense sources of energy to join ever-higher melting materials into single entities, the very first method by which parts were joined has not gone away. In fact it, like other things in history, is experiencing a resurgence. *Integral mechanical attachment* has once again, as it has many times before, become not simply a fall-back option for joining but, in many instances, a preferred option. And there is nothing to suggest this pattern of resurgence will end with the current proliferation of snap-fit assembly used so widely for the simple assembly of detail parts molded from polymers.

This book looks at this fascinating method for joining (i.e., integral mechanical attachment), and it does so as never before in terms of thoughtfulness or thoroughness. In doing so, it creates a resource never before available, surely not in one convenient source. Oh, there have been—and are—some wonderful references (normally in unreadable handbook or encyclopedic formats) for the mechanical attachment of individual materials including Vallory H. Laughner and Augustus D. Hargan's unprecedented (but long out-of-print and nearly impossible to find) 1956 masterwork *Handbook of Fastening and Joining of Metal* Parts (McGraw-Hill), Robert O. Parmley's *Standard Handbook of Fastening & Joining* (McGraw-Hill, 2nd ed., 1989), Wolfram Graubner's *Encyclopedia of Wood Joints* (Taunton Press, 1992), and E.G. Nawy's *Concrete Construction Engineering Handbook* (CRC Press, 1997). But, there has never been a single reference that not only compiles knowledge on integral mechanical attachment methods but also, equally or more importantly, explains the underlying principles of operation and range of utility.

---

[1] Stone Age: The earliest known period of human culture; characterized by the use of stone tools. The Stone Age, like the Bronze Age and Iron Age, began and ended at different times in different parts of the world and in different cultures in those various parts. Most anthropologists believe that the Stone Age began about 100,000 B.C. and ended about 4,000 B.C.

[2] Bronze Age: A non-specific period of time between the Stone Age and the Iron Age, characterized by weapons and implements made of bronze, an alloy of copper and tin. In Europe, the period began around 1,800 B.C. and ran into the Christian era, although iron began to appear well before the end of the Bronze Age. In other parts of the world, notably in the Middle East (e.g., Egypt, Syria, Persia [Iran], Mesopotamia [Iraq], and Turkey) and in China and India, the Bronze Age clearly began much earlier, perhaps between 4,000 and 3,000 B.C.

[3] Iron Age: The generally prehistoric period succeeding the Bronze Age, characterized by the introduction of iron metallurgy; in Europe, beginning around the 8th century B.C.

The book is intended as a life-long, desk-top reference for designers of all kinds, from the most sophisticated design engineers to the most practical do-it-yourselfers. It is also intended to open the reader's eyes to the wealth of possibilities for this deceptively simple looking method of joining.

This book consists of 13 chapters, organized by the technology first (in Chapters 1 through 6) and the materials or application second (in Chapters 7 through 12). The last chapter (Chapter 13) tries to look into the future for the technology.

Chapter 1 introduces the technology of integral mechanical attachment, differentiates it from the much more familiar technology of mechanical fastening (both within the over-arching process technology of mechanical joining), and presents the relative advantages and potential shortcomings of the integral attachment approach. It also makes quite clear there really may be "nothing new under the sun," if one opens one's eyes! Chapter 2 proposes a classification scheme for various methods by which integral mechanical attachment can be accomplished and places integral attachment within the context of all methods by which materials and structures can be joined. It also looks at the underlying mechanism responsible for all integral mechanical attachment methods. Chapter 3, 4, and 5 look at each of the three major classifications, based on whether the attachments are intended to operate by remaining rigid (i.e., rigid interlocks), by deflecting elastically (i.e., elastic, snap-fit interlocks), or by being formed plastically (i.e., plastic, formed-in interlocks), respectively. The detailed forms and ways by which each type of attachment can be accomplished are presented type-by-type. Chapter 6 revisits the classification scheme to present a taxonomy for the method and compares methods.

The latter chapters of the book present the various ways by which various materials can be joined by integral mechanical attachment. Methods for joining metal (in all forms; sheet, castings, extrusions, etc.), plastics or polymers, wood, cement and concrete, ceramics and glass, and for specific application in electronic applications (including also microelectronic and high-voltage/high-current electric power applications) are presented in Chapters 7 through 12, respectively. What results is a compilation of methods within integral attachment that is unprecedented!

Finally, Chapter 13 considers the future for integral mechanical attachment, including how it is likely to evolve for higher performance applications, to smaller and smaller scales (including the nano-scale), into biotechnology, and into inhospitable (hostile) environments.

The text, hopefully more like a narrative than a treatise, is richly augmented by schematic illustrations, plots, photographs, and tables.

The book ends with a comprehensive index that lists and cross-lists key words to be as user-friendly as possible in a written format. It is formatted like "keywords" are for computer-based searches.

A book like this doesn't happen without inspiration and help from others. So, it is only proper to acknowledge all those who did one or the other or both.

I am most grateful to Vallory H. Laughner and Augustus D. Hargan, who I never had the pleasure of meeting, but with whom I feel a particular connection (no pun intended!). Their book *Handbook of Fastening & Joining of Metal Parts* published in 1956 by McGraw-Hill, and long out of print, is a masterwork. The book is an absolute wonder for the thousands of ideas for joining and for the marvelous illustrations. Its treatment of formed metal tabs in sheet-metal did more to cause me to think about integral mechanical attachment than any other experience I had except my involvement in the multi-client–sponsored Integral Fastening Program conducted at Rensselaer Polytechnic Institute between about 1993 and 1999. I am grateful for the intellectually stimulating interaction with Professor Gary A. Gabriele, Program Director, and all the many students who worked in the program. I am particularly grateful to Dean Q. Lewis for his assistance in this book with plastic snapfits. I am also grateful to Robert O. Parmley for his invaluable book *Standard Handbook of Fastening & Joining,* 2nd ed. (McGraw-Hill, 1989). Bob did a great job of covering the topic with superb tables and illustrations, but, regrettably, not much discussion.

For photographs and previously published illustrations, or data plots or tables, I thank the following, in alphabetical order: Airbus S.A.S (Franciose Maenhart), Blagnac, France; Aluminum Extrusion Council, Wauconda, IL; American Cement Institute (Dr. Shuab Ahmad), Farmington Hills, MI; American Welding Society (Andrew Davis), Miami, FL; Analog Devices, Inc., Cambridge, MA; Anatomical Charts Co., Skokie, IL; Bayer Materials Science (Liz Gage), Pittsburgh, PA; Birkhauser-Verlag, Basel, Switzerland; BTM Corporation (Ryan Bastick), Marysville, MI; Sam Chiappone, Rensselaer Polytechnic Institute, Troy, NY (numerous photos); Christian Prilhofer Consulting/New Building Systems, Freilassing, Germany; Concrete Steel Reinforcing Institute, Schaumburg, IL; Corning, Inc. (Kristine Gable), Corning, NY; Delta Structures, Wood Dale, IL; E.O. "Ned" Eddins (photo of a beaver dam); Elsevier Ltd. (Helen Gainford), London, England; Fisher-Price (Jack Woodworth), East Aurora, NY; Professor Ramanath Ganapathiraman, Rensselaer Polytechnic Institute, Troy, NY (photomicrographs of nano-materials); GE Lighting (Gerald Duffy), Cleveland, OH; General Motors Corporation (Dr. Roland Menassa, Dr. Wayne Cai, and Scott Abbate), Warren, MI; Hanser Gardner (Dr. Christine Strohm), Cincinnati, OH; Haas Automation, Inc. (Scott Rathburn), Oxnard, CA; Intel, Santa Clara, CA; KNex Industries, Inc. (Bob Glickman), Hatfield, PA; KraftMaid Cabinetry (Pat Depner), Middlefield, OH; LEGO Systems, Inc. (Janice E. Favreau), Enfield, CT; Marathon Ashland Petroleum LLC (Roy Whitt and Vicki Moolaw), Findlay, OH; McGraw-Hill (Cynthia Aquilera), New York, NY; Angus McIntyre (photo Machu Picchu); Metalock Engineering, Inc. (David Fowler), Exhall, Coventry, England; MEMX, Santa Clara, CA, and Albuquerque, NM; Nelson Stud Welding (Harry A. Chambers), Elyria, OH; Rob Planty, Rensselaer Polytechnic Institute, Troy, NY (photo of crimped leads); Portland Cement Association (Bill Burns), Skokie, IL; Karl Ernst Roehl (photo of Three Gorges Dam); Schott North America, Inc. (Brian Lynch), Elmsford, NY; Steelcase Corporation (Kurt Heidmann), Grand

Rapids, MI; Rick Sternbach (photo of the Asteroid Tug); Stryker Corporation (Stephen Brown), Mahwah, NJ; Thomasville Furniture Industries, Inc. (David Burkhart and Paul Bobb), Thomasville, NC; 3M, St. Paul, MN; University of California at Berkeley (Professor Roger Howe and Dr. Elliot Hui), Berkeley, CA; U.S. Forest Products Laboratory (Sandra L. Morgan), Madison, WI; Don VanSteele, Rensselaer Polytechnic Institute, Troy, NY (photos of extrusion); Paul Vlaar (photo of the Via Appia, Rome); Wood Truss Council of America (Emily Patterson), Madison, WI; Professor Dr. Klaus Zwerger, Institute for Artistic Design at Vienna Technical University, Vienna, Austria (photos of wood joints); and Carmela Zmiric (photo of Baska, Croatia).

I thank Joel Stein, my editor at Elsevier, for encouraging me with this project, and his assistant, Shelley Burke, for her invaluable assistance in production of the book. I also thank Nancy Fowler-Beatty, my secretary at Rensselaer Polytechnic Institute, for her endless patience with me for making excuses for me when I didn't show up where I should have because I was squirreled-away writing the manuscript, and for her assistance whenever I asked.

Most of all, I want to thank two people for their creative talents: first, Richard R. Rodney, an undergraduate at RPI, who I hired to create the hundred-plus schematic illustrations that appear throughout this book. I am impressed by his talent and professionalism, but even more so by his wonderful nature. Second, my son-in-law, Avram N. Kaufman, for his tremendously creative cover art and his equally creative and endearing depictions of evolving cavemen and their tools and weapons in Figure 1.1. God is really generous in doling out talent and passion to some people, and these two got quite a share of talent and passion!

Last, but surely not least, I thank my best friend, my greatest supporter, and my endlessly patient wife. I love you Joanie!

There have been—and are—books on mechanical fastening and fasteners, on adhesive bonding and adhesives, and on welding, brazing, soldering, and thermal spraying, but there has never been a book devoted to the oldest and most time-resistant method of all for joining—integral mechanical attachment. This book brings that fact to a long-overdue end!

I hope you enjoy it, as much as learn from it!

Robert W. Messler, Jr., Ph.D.
September 21, 2005

1   INTRODUCTION TO INTEGRAL MECHANICAL ATTACHMENT

1.1   THE OLDEST METHOD OF JOINING: USING NATURAL SHAPES AND FORMS

Our earliest ancestors doubtless lived off the land and were drawn by their necessity and driven by their resourcefulness to use what Nature provided. At first, however, they were also compelled by their ignorance to use whatever was provided in whatever form and shape it was provided. Surely they would have gripped a naturally-rounded stone in their hand to use as a hammer (Figure 1.1a). They probably even used this crude hammer as a weapon for killing animals for food and then used it to tenderize the meat and prepare the hind for use as clothing. With time, they learned they could hammer more efficiently by wedging such a round stone into the crotch of a three-pronged stick (see Figure 1.1b), gaining the mechanical advantage of a lever. And, thus, the earliest tool was born from the brain of the greatest tool-making species.

The need to follow roving grazing animals for food and to find new, more fertile land and the edible plants that grew there also drew our early ancestors out of the natural caves that surely served as shelters for at least some, forcing them to improvise shelters when natural shelters could not be found. Stone hammers were used to pound sticks into the ground to serve as the supporting framework for other sticks that were simply intertwined and interlocked to provide a roof and walls. Others probably built shelters from stones they gathered and stacked and nested to create an interlocking wall as shelter from the wind and predators. Again, roofs were probably made from sticks laid on top of the walls and intertwined and interlocked to be secure and protective against the weather, eventually being further insulated with bark, grass, moss, or animal pelts.

FIGURE 1.1 An artist's concept of how early humans may have evolved the design of a hammer and, then, an ax as they, themselves, evolved. A naturally-round stone held in the hand served as a hammer (a); a naturally-round stone wedged into a naturally-pronged stick created a more efficient hammer or club (b); a naturally-sharp or sharpened stone wedged into a split in a stick created an ax for use as a tool or a weapon (c); a sharpened stone wedged into a split in a stick was lashed into place for added durability (d); and a stick force-fit into a hole bored into a sharpened stone created an ax that is close to those used today (e). (Courtesy of Avram Kaufman and used with permission.)

The common links among all of these earliest fabricated devices and structures are two. First, all of the parts, regardless of form (e.g., stones or sticks), were selected and used for their natural shape and, thus, utility. Second, these parts were assembled into more elaborate tools or structures simply by attaching one naturally-occurring part to another by creating mechanical interference and interlocking to provide structural integrity. (The round stone in Figure 1.1a was grasped by wrapping one's fingers around it or, later, by wedging it into the supporting crotch of a three-pronged stick, as in Figure 1.1b. In both examples, the stone is locked into the hand or the stick by purely mechanical forces.) The technology of joining, by which either small things can be brought together to make bigger things or simple things can be brought together to make more geometrically and functionally complex things, was thus born. And, most notable for this book, the very first joining used purely mechanical forces that arose from the physical interference of one part of the assembly with another to keep those parts in proper proximity, orientation, and alignment; with resistance to separation or accidental disassembly being achieved when properly shaped parts actually interlocked one with the other.

The idea of using the geometric features of an object to accomplish mechanical joining was—and is—the basis for what is properly known as *integral mechanical attachment;* the subject of this book.

1.2   THE PROCESS EVOLVES, BUT NOT MUCH!

Who knows whether in using an early hand-held or stick-mounted stone hammer the stone split and the idea of an ax was born, or whether early man simply selected naturally-broken stones for their sharp edges and cutting ability, and used them as-is. In either case, a small but significant evolution occurred. More efficient tools, such as axes and knives and spears, could be made by wedging such sharp stones into wood stick handles (see Figure 1.1c). Also, doubtless with time, more functionally sophisticated tools and, inevitably (it seems!), weapons could be made by using stones as hammers to shape other stones for other purposes, such as cutting or piercing in a tool or weapon or for fitting better one with another to erect a shelter.

With these inevitable advances in part design and fabrication, assemblies inevitably became more sophisticated, more functional, and more diverse, allowing the creation of stone roads and pathways (Figure 1.2), hewn-stone buildings (Figure 1.3), and natural- and hewn-stone bridges (Figure 1.4), for example. However, one common link continued: parts were joined mechanically, using only mechanical forces arising from the physical interference of one part with another and with the possibility of interlocking to provide structural integrity and prevent accidental disassembly.

Interestingly, *Homo sapiens* were not and are not alone in their use of the integral mechanical attachment of naturally-occurring items to create useful

FIGURE 1.2 Combinations of naturally-shaped and hewn-to-fit stones were used to build roads throughout the ages. While the Romans set hewn stones into mortar and added smaller stones to lock the larger ones in place, such as on the Via Appia (left), one of the major consular roads out of ancient Rome, the Croates simply set naturally-shaped stones into dirt to create this seaside road and walkway (right). (The photograph of the Via Appia, as seen on www.encyclopedia.thefreedictionary.com, was taken by Paul Vlaar and is used with his kind permission. The photograph of the road and walkway in Baska, Croatia, as seen on www.trekearth.com/gallery/Europe/Croatia, was taken by Carmela Zmiric and is used with her kind permission.)

FIGURE 1.3 The ancient Incas built their marvelous cities, such as Machu Picchu, in the Andes of Peru, using hand-hewn stones. Their constructions included roads, temples, dwellings, walls, and, as shown here (left), terraces for agriculture. The fit of even very large stones was often extraordinarily precise, as shown in this wall with openings (right), so no mortar was needed or used. (Photographs by Angus McIntyre, were obtained from www.raingod.com. Despite numerous attempts, no contact could be made with Mr. McIntyre for permission to use what may be a copyrighted photograph.)

FIGURE 1.4 The Simahui (Sima Regret) Bridge, in the southeastern region of the county of Xin Chang, was built during the Tang Dynasty and is one of the oldest in all of China, where there are many hundreds of examples remaining. No mortar was used in this ancient bridge, as naturally-shaped and hewn stones were simply fitted to form the arch and bridge. (Photograph taken from http://library.sx.zj.cn/e-page/ancientbridge/ggq.htm, for which, despite numerous attempts, no contact could be made for permission.)

assemblies from naturally-shaped parts. Many birds build their nests by intertwining twigs, reeds, grass, and vines. Some even modestly alter the small pieces they employ so they fit and hold together better. Fine examples of such nests are created by the spotted-backed weaver. Beavers, as Mother Nature's "engineers," construct elaborate and (as anyone who has had to try to disassemble them knows) quite strong and durable dams and lodges composed of intertwined and interlocked sticks (Figure 1.5), many of which were purposely altered in their shape so that they worked better. Those most familiar with beavers claim to have watched them roll round stones, larger than themselves, into place to set the foundation for dams they plan to build in fast-moving water.

Inevitably, early humans were awakened to the possibility of integral mechanical attachment as a joining process by seeing what was being done in Nature by other creatures.

With time, another often more efficient method for interlocking was discovered: *mechanical fastening,* as is described in Section 1.4. Surely, an early example was when our prehistoric ancestors lashed pointed stones into split sticks to create spears and knives with greater durability (see Figure 1.1d). With this advancement, the human species began its incessant advancement over its fellow animal species.

FIGURE 1.5 Dams made by beavers, such as the dam shown here near Grand Teton, Wyoming, are created by interlacing and interlocking chewed down and shaped sticks among carefully positioned rocks. (Courtesy of E.O. "Ned" Eddins, with permission.)

However, neither mechanical fastening nor any other method of joining, for that matter, has ever completely replaced or made irrelevant the utility, no less the simple elegance and special capability, of integral mechanical attachment. As will be seen as one proceeds through this book, not only has integral mechanical attachment never been completely replaced, but in fact, as it has at various times throughout history, is experiencing resurgence.

1.3 INTEGRAL ATTACHMENT: A FORM OF MECHANICAL JOINING

Mechanical joining is, at once, the oldest, most widely used, and most diverse of all processes for joining parts into devices, assemblies, and structures. It gets its name from the fact that only forces of a mechanical origin allow and enable joining. The materials comprising the pieces being joined do not form any chemical bonds at the atomic or molecular level, as occurs in the sister-joining processes of adhesive bonding and welding (Brandon and Kaplan, 1997). What occurs to allow and enable joining and provide structural integrity is that one part of the intended assembly physically interferes with another part (or parts) to prevent motion in some direction(s), thereby allowing loads to be resisted in that(those) direction(s).

Applied loads or forces give rise to two types of stresses within a solid body: normal stresses and shear stresses. *Normal stresses* develop on planes at right angles to the direction of loading, whether tensile or compressive (Figure 1.6). When a body is fixed in a state of mechanical equilibrium, equal and opposite normal stresses develop on opposite faces of the normal planes of volume elements of the body. For tensile loading, normal forces try to pull the body (actually, the material of which the body is made) apart by exceeding the cohesive strength of the atomic-level bonding across adjacent layers of atoms

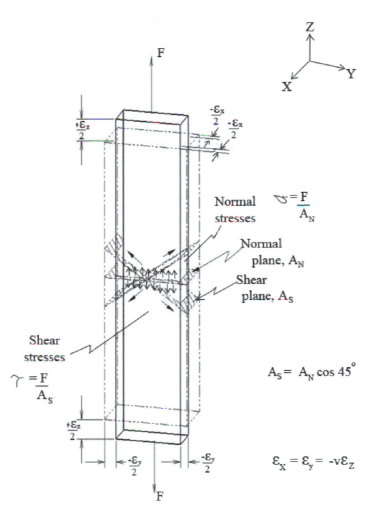

FIGURE 1.6 A schematic showing the normal *and* shear stresses that develop within a solid body; here, a thin rectangular strip under an applied force F. Even ignoring that real materials are composed of atoms, normal stresses develop on planes oriented 90-degrees to the direction of load application, while shear stresses develop on planes oriented at ±45 degrees from the direction of primary load application. In real materials, the strain caused to occur in the direction of loading, here ε_z, is offset by strains of the opposite sign in the two orthogonal directions, here $-\varepsilon_x$ or $-\varepsilon_y$, related through Poisson's ratio, v, by ε_x (or ε_y) $= -v\varepsilon_z$.

making up the material. The planes across which applied tensile forces try to pull a material apart are actually the closest packed planes of atoms. These planes are usually oriented at an angle θ other than 90 degrees to the applied forces, so those applied forces are resolved into normal and shear components on these planes through the appropriate trigonometric functions of the angle θ.

While all loading can be along one axis (i.e., uniaxial), in which case cohesive fracture occurs perpendicular to that axis due to the normal stresses that develop, strain along one axis or in one direction in a material always gives rise to strains along the other two orthogonal axes or in the other two orthogonal directions (see Figure 1.6). The relationship between the tensile strain in one direction (say in the z-direction of a Cartesian coordinate system) and in transverse orthogonal directions is given by Poisson's ratio, in which $\nu = -\varepsilon_x$ (or $\varepsilon_y)/\varepsilon_z$. Simplistically, Poisson's effect occurs because solid materials attempt to maintain a constant volume.[1] As a result of Poisson's ratio (which is a physical property of a material), even a uniaxial load gives rise to normal strains and stresses on all three orthogonal planes.

Besides normal stresses developing in response to any and all applied loads, shear stresses also develop (see Figure 1.6). *Shear stresses* develop on planes at ±45 degrees to the normal plane. Once again, in real materials, they occur in equal and opposite pairs across the line between two adjacent planes that are most densely packed with atoms. Because of Poisson's effect, shear stresses develop on all three orthogonal planes and lie at ±45 degrees between the three orthogonal normal planes in a three-dimensional (3-D) body, and these too, occur in equal and opposite pairs or couples.

In mechanically joined parts using either a fastener, such as a bolt, or integral geometric features of the mating parts in the joint, one part transfers loads applied to it to the other part either by tension (or compression), by shear, or both, as shown in Figure 1.7a and b. In fact, mechanical joints, whether created using integral attachment features or supplemental fasteners (described in Section 1.4, later in this chapter), transfer loads from one part to another either by shear, in which case one part (or the fastener) produces a bearing stress in the mating part, or by tension (or compression) (Messler, 2004).

The process of mechanical joining is deceptively simple for what it allows.[2] First and foremost, different parts can be joined to create larger and/or more complex-shaped assemblies or structures that can provide functions not obtainable in the individual parts. Second, the parts that are being joined can be of virtually any material, as they only need to interact based on their shape, not on

---

[1] In fact, while solids (as well as liquids) try (and tend) to be incompressible (i.e., maintain constant volume), most do not do so perfectly and some (e.g., many polymers) do not do so very well.

[2] In engineering, the juxtaposition of simplicity and impressive capability is often referred to as *elegance!*

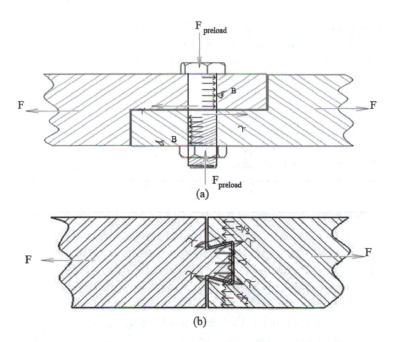

FIGURE 1.7 Applied stresses in a joint are transferred from one part to another either using integral geometric features of the mating parts or using a supplemental part known as a *fastener*. In either case, applied stresses resolve into both normal and shear stresses that can be resisted by parts bearing against one another, for example. Here, a bolt-and-nut method of mechanical fastening (a) and a dovetail-and-groove method of integral mechanical (b) are shown. The applied force F produces a shear stress τ in the bolt shank, as the bolt shank produces a bearing stress σ_B in the joint elements. As it always should, the bolt shown here has a preload in it from tightening the nut. Tensile stresses develop at the root cross-sections of the dovetail features, as shear stresses develop along their abutting faces.

their chemical makeup or nature. Third, and often under-appreciated, parts that are intentionally assembled mechanically can be intentionally disassembled. This allows pre-fabrication setup (as is done in modular homes, modern buildings, and modern bridges), portability, and ultimate disposal at the end of useful or needed service life. (In other words, mechanical joining is recycling's best friend![3]) Fourth, and surely not last nor least, only mechanical joining permits parts to be assembled while allowing relative motion, if wanted and where needed.

Like all joining processes, mechanical joining also allows large and/or complex-shaped structures or assemblies to be created from smaller and simpler-shaped detail parts. This is important because all processes for fabri-

---

[3] In Europe, beginning around 2000, automobile manufacturers were required to demonstrate that in excess of 90% of the parts of the vehicle (by weight) can be recycled, which means vehicles must be readily able to be disassembled. This puts a premium on mechanical methods of joining/assembly.

cating parts (casting, forging, molding, machining, etc.) have some limit on the size and shapes of parts that they can create.

Mechanical joining, in general, and integral mechanical attachment, specifically, enables the following:

- Provides needed contact or physical proximity between and among parts
- Maintains the relative position and/or specific orientation of parts
- Maintains the required precision of alignment of abutting parts
- Prevents unwanted movement in directions where it is not wanted
- Provides a retaining force against unwanted separation

As described in Section 1.1, the use of geometric features of pieces or parts to allow and enable them to be joined into functional devices, assemblies, or structures absolutely preceded any other method by which such pieces or parts could be joined. In the integral mechanical attachment or joining that results, the geometric features needed could exist naturally in the shape of the piece or part, or they might need to be and could be processed into the piece or part. Since time immemorial, the beaver has used its protruding front teeth to cut down trees, trim off unwanted branches, and shape the resulting part to fit its needs in creating its environment, no less its home (see Figure 1.5). Humans, too, have grown in sophistication to alter the geometry of parts to allow those parts to fit and enable them to interlock into a sound assembly (see Figures 1.2, 1.3, and 1.4). We, uniquely among all species, do this by designing parts in advance, as opposed to on-the-fly (as beavers do), that have the geometric features needed for both ease of assembly and functional performance. We also (albeit, not uniquely) have the ability to "process-in" needed geometric features, particularly for interlocking, after parts have been brought together into position and contact.

Together, these techniques allow integral mechanical attachment to persist as a viable method for joining in the modern as well as in the ancient and even prehistoric world.

1.4 INTEGRAL MECHANICAL ATTACHMENT VS. MECHANICAL FASTENING AND FASTENERS

Integral mechanical attachment and *mechanical fastening* are the two subclassifications of *mechanical joining*. As stated in the preceding section, all mechanical joining brings pieces or parts together to create a device, assembly, or structure strictly by having one part interfere with another and, thereby, prevent unwanted motion between those parts. No atomic-level process (e.g., bonding or diffusion) is involved, and no atomic-level forces are needed. Only mechanical forces are needed.

Once assembled, one part is able to transfer loads or forces to another, if that is what is needed.[4] More importantly, the designer is able to determine in which direction(s) he/she wants parts not to move (and, thus, resist loads or forces) and in which he/she wants them to be able to move.[5] This capability is unique for mechanical joining and can, obviously, not be accomplished by any process that joins parts with atomic-level forces of bonding (e.g., adhesive bonding and welding). By proper design, an assembly can allow relative motion between certain parts in certain directions or absolutely preclude such motion.[6] Examples in which relative motion is wanted abound, with a few recognizable examples being sliding window sashes, x-y traversing machine-tool stages, hinged doors, rotating gears in a gear-train, and many others, including our own bodies through joints in our skeletal system (see Figure 13.27). Only mechanical joining permits this often essential attribute of an assembly or structure. In each example just cited, certain translational and/or rotational motions are permitted, and in many, such motion is either allowed or prevented strictly using geometric features of the mating parts that are integral to those parts, that is, through integral mechanical attachment.

There is however another method to join things mechanically: to cause one part to interfere with and, perhaps, interlock with another. That method is *mechanical fastening.* In *mechanical fastening,* an extra part is employed with the explicit purpose of causing interference (and, possibly, interlocking) between itself and each of at least two other parts. In other words, as opposed to using some geometric feature that is integral to each of the individual parts to be joined to cause interference (and, perhaps, interlocking), a supplemental device, known as a "fastener," is used. The fastener's sole function as a part is to allow mechanical joining, providing interference between itself and each of the parts with which it interacts. Beyond establishing and fixing the position, orientation, and alignment of the parts of a mechanical assembly or structure relative to one another, fasteners often lock the assembly or structure together until it is desirable and/or acceptable for it to come apart. Those fasteners that use threads to hold parts together develop a clamping force in the joint elements. This happens as, say, a nut is turned on the threaded shank of a bolt to cause tightening, and the distance between the top face of the nut and bottom face of the bolt head shortens, applying a squeezing force on the sandwiched joint elements and giving rise to stretching in the bolt shank. This preload is, then, a residual or locked-in stress that serves to hold the joint components together.

---

[4] In fact, there may be instances when it is undesirable to transmit loads or forces between two specific parts and that, too, can be accomplished by appropriate mechanical design and assembly.

[5] Directions in which motion can occur, if there is no restraint to prevent it, are known as "degrees of freedom."

[6] Motion can occur in real, three-dimensional mechanical systems in three orthogonal directions by translation and around three orthogonal axes in rotation, giving six degrees of freedom.

Not every fastener (e.g., not keys in keyways, not pins, not nails, and not many other unthreaded types) develops a residual stress to hold the joint components together, and, in fact, the development of residual stresses in integral attachments is not only unusual but also often regarded negatively. When residual stresses are not developed to hold the parts of an assembly together, what occurs instead is that the components of the joint simply resist being separated when a force is applied to try to separate them. This is most often accomplished using some locking feature of one part with a complementary locking feature of another part (see Section 4.1).

Figure 1.8 schematically illustrates the major differences between a mechanical fastener as a supplemental device whose sole purpose is to allow joining, as opposed to serving first as an essential part for function that also happens to enable joining, and an integral attachment feature that is simply designed and

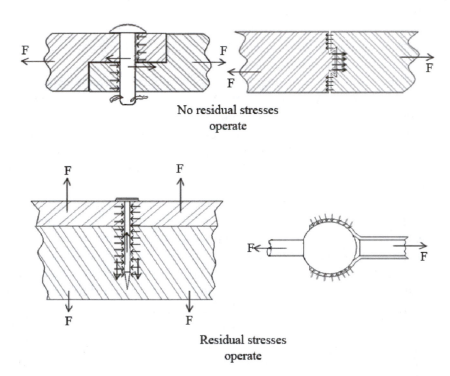

No residual stresses
operate

Residual stresses
operate

FIGURE 1.8 A schematic illustration of some of the differences between the use of integral geometric features of parts in integral mechanical attachment and the use of supplemental devices (i.e., fasteners) in mechanical fastening. Note that not all fasteners and few integral geometric features develop residual stresses to enable and enhance joining. A pin (held in place by a small Cotter pin at its protruding tip) (top left) operates without requiring a residual stress. The applied force F causes a bearing force on the pin at its surface and a shear couple at the shear plane between the joint elements being pinned. Likewise, the dovetail-and-groove joint (top right) develops stresses in the integral features from the applied stress F, but no residual stresses. (In fact, shear stresses also develop in the dovetail and in the groove at their radii.) A nail holds joint elements together (lower left) against an applied force F through the frictional shear force arising from the squeezing force on the nail body by the elastic residual stress in the wood joint elements. The ball in a ball-and-socket elastic snap-fit (lower right) introduces a residual stress from elastic spring back of the socket that helps grip the ball.

processed into parts with a specific function in the proper operation of the structure or assembly of which they are components.

So, there are two methods for accomplishing mechanical joining, with one being far more immediately apparent than the other. *Mechanical fastening* accomplishes joining between parts by, itself, causing strictly mechanical interference (and, perhaps, interlocking) with and between or among those parts. Mechanical fastening employs specialized supplemental parts known as *fasteners* to accomplish the joining. *Integral mechanical attachment,* or simply *integral attachment,* accomplishes joining between parts by using (or, as will be seen later, creating) some geometric feature in mating parts that causes strictly mechanical interference (and, perhaps, interlocking) between or among those parts. Integral mechanical attachment uses specially designed or created (and, occasionally, natural) attachment features that are part of the part. The ability of mating parts to interfere and interlock with one another to maintain position, orientation, and alignment is built into the part beyond its primary function in the assembly.

Without question, mechanical fastening appeared after integral mechanical attachment, probably as an enhancement. Imagine the evolutionary design of an ax when a green stick was split near one end to allow an intentionally sharpened stone to be wedged into place and held by the elastic force of recovery from deflection. Imagine next the frustration when, as the stick dried out or the user became more enthusiastic with his newfound tool, the head flew off of the ax. The now obvious solution, lash the stone to the stick, within the split, using a plant vine or animal sinew (see Figure 1.1d). Mechanical fastening (i.e., the use of a supplemental device—here, the lashing), strictly needed to allow joining, was born!

But, there is another viable solution to our Stone Age ancestor's problem that surely occurred at some point. By forming a hole in the stone, the stick handle could be wedged in and resist even the most vigorous hammering or chopping (see Figure 1.1e). Our prehistoric relative undoubtedly noticed that his/her stone hammer eventually produced a hollow in the flat stone against which he/she ground food. It was only a small step from that observation to purposely making a hole in or through one stone using another stone. In fact, additional tricks like soaking the stick to cause it to swell and lock in place or setting a wedge into the end of the stick where it extends through to the top of the stone head are used to this day. Clearly, there is more than one way to skin a cat or make an ax, and there seems to be no end to the possibilities of integral attachment.

Examples of the use of mechanical fastening are endless and familiar, including: pegging or nailing wood planks; pinning or keying removable metal collars to shafts; bolting steel bridge trusses; riveting aluminum-alloy aircraft airframe structures and skins; stapling sheets of paper; and sewing pieces or plies of cloth, leather, polymeric-matrix composites, and living tissue. Examples of integral mechanical attachment, while they may not come to mind as quickly, also abound and will be familiar in hindsight, including: tongue-and-groove dovetail-and-groove jointing of wood furniture; interlocking cement patio blocks; folded-tab-assembled sheet-metal in old tin-type toys and in modern thin-gauge sheet-metal cabinets; snap-assembled plastic parts of hand-held telephones, lawn-and-yard leaf blowers, and automobile grill trim; and holding our limbs in place with ball-and-socket joints.

TABLE 1.1 Major Methods of Mechanical Fastening (and Associated Fasteners) and for Accomplishing Integral Mechanical Attachment (and Associated Features)

Mechanical Fastening Methods	Associated Fastener
Using threaded fasteners	
– Bolting	– Bolts (without or with nuts)
	– Threaded rods
	– Nuts and lock-nuts
– Screwing	– Machine screws
	– Self-tapping screws
	– Sems
Using unthreaded fasteners	
– Nailing	– Nails, spikes, brads, or tacks
	– Pegs and treenails
– Pinning	– Pins
	– Dowels and biscuits
– Riveting	– Upset (one-piece) rivets
	– Swaged, two-piece rivets
	– Blind or "pop" rivets
	– Self-upsetting rivets
– Retaining or Clipping	– Retaining rings or clips
– Keying	– Keys and keyways
– Other	– Eyelets and grommets
	– Washers and lock-washers
	– Self-clinching fasteners
Using other methods	
– Stitching	– Stitches and sutures
	– Staples
– Lashing or lacing	– Lashes or laces
– Tying	– Knots and splices
– Snap-fit fastening	– Snap-fit fasteners
	– Hook-and-loop fasteners
– Coupling	– Rigid couples
	– Flexible couples
	– Clutches
– Magnetic connections	– Magnets

(Continued)

TABLE 1.1 (*Continued*)

Integral Mechanical Attachment Methods and Associated Features

Using rigid Interlocks

- Tongues-and-grooves
- Dovetails-and-grooves
- Rabbets or dados
- Mortice-and-tenons
- T-slots and Ts
- Shaped rails and ways
- Wedges and Morse tapers
- Shoulders and flanges
- Bosses, lands, and posts
- Tabs and ears with recesses or slots
- Integral keys and splines
- Integral (coarse) threads
- Knurled surfaces
- Hinges, hasps, hooks, and latches
- Turn-buckles
- Collars and sleeves

Using elastic Interlocks

- Integral spring tabs and spring plugs
- Snap slides and clips
- Clamps and clamp fasteners
- Quick-release fasteners
- Integral snap-fits (of all types)
- Interference press fits
- Thermal shrink fits

Using plastic Interlocks

- Setting and staking
- Metal stitching and metal clinching
- Indentation-type and beaded-assembly joints
- Crimping and hemming
- Thermal staking
- Formed tabs

Table 1.1 lists the major methods of mechanical fastening and associated fasteners, as well as the major methods of accomplishing integral mechanical attachment and associated attachment features.

Figures 1.9 through 1.14 show a variety of integrally-attached assemblies and structures, from the height of the Roman Empire to the modern electronic age.

FIGURE 1.9 An artist's rendering of a portion of one of the great aqueducts that brought potable water into ancient Rome in quantities that exceeded those brought into New York City in 1985. Note the use of a specially-shaped keystone at the top center of each arch. Keystones lock into position by their shape and force the rest of the arch stones together to carry downward loads on the arch as thrust loads to the vertical columns. In most Roman structures, hewn stone was embedded in mortar made from volcanic ash mixed with lime and roasted. (Photograph from www.interactiveplayco.com accessible through http://images.search.yahoo.com/ Roman_arches. Despite several attempts, contact could not be made to secure permission to use what might be a copyrighted image.)

FIGURE 1.10 The use of pre-cast concrete block with cast-in interlocking features (left) to construct a wall (right) uses the same principles for joining as found in ancient structures. (Courtesy of the Portland Cement Association, Skokie, IL, with permission.)

FIGURE 1.11 A common method for building log cabins was—and is—to cut notches into the logs near their ends to allow other similarly notched logs to be placed at right angles to these to form strong interlocked joints between front or rear and side walls. (Photograph from www.grandcanyonnorthrim.com following a search for "log cabins" on http://images.search.yahoo.com. Despite several attempts, contact could not be made to secure permission to use what might be a copyrighted image.)

FIGURE 1.12 The use of dovetail-and-groove joints in the assembly of fine wood furniture, here, a drawer front to side pieces. (Courtesy of KraftMaid Cabinetry, Middletown, OH, with permission.)

FIGURE 1.13 The use of formed tabs was common for the assembly of thin-gauge tin-type (actually, thin sheet steel) toys of the post–World War II era. Here, a 1950's vintage "Rocket-man Car" (left) can be seen to have been assembled with folded tabs (right). (Photographs are of an auctioned item on www.eBay.com. Despite several attempts, contact could not be made to secure permission to use what might be copyrighted images.)

1.5   ADVANTAGES OF INTEGRAL MECHANICAL ATTACHMENT

Every material, process, and method has advantages and disadvantages. None is perfect, at least not for every situation. Integral mechanical attachment is no different. It has its advantages, and it has its shortcomings, if not disadvantages. The two pertinent questions are, (1) how does mechanical joining compare to

FIGURE 1.14 Elastic snap-fit features are commonly used for assembling children's toys made from thermoplastics in order to eliminate the potential choke hazard posed by fasteners such as screws. Here, a Rocking Pony is shown that is assembled entirely using molded-in elastic snap-fit features, which cannot be seen because they are located inside the assembly, precisely to prevent young children from tampering with and, possibly, choking on screws. (Courtesy of Fisher-Price, East Aurora, NY, with permission.)

other joining processes and (2) how does integral mechanical attachment compare to mechanical fastening within mechanical joining? Let us look at these one at a time.

Table 1.2 summarizes the general advantages of mechanical joining compared to the processes of welding and adhesive bonding, as well as the specific advantages of integral mechanical attachment within mechanical joining, especially compared to mechanical fastening. Without going into great detail, as there are other excellent references on the subject (Speck, 1997; Messler, 2004), here are the comparisons:

First and foremost, all methods of mechanical joining employ only mechanical forces to accomplish joining. No bonding is necessary or occurs. As such, mechanical joining tends to be simpler to accomplish than other joining processes. While it may be necessary to prepare the joint by drilling holes for pins, bolts, or rivets or to machine keyways or grooves for keys or retaining clips or rings, joints must usually also be prepared (e.g., beveled or grooved) for welding and joint surface must always be scrupulously cleaned mechanically and/or chemically for both welding and adhesive bonding. In addition, many adhesives also require chemical or electrical pre-conditioning of the surfaces to be

TABLE 1.2 Summary of the General Advantages of Mechanical Joining Compared to Welding and Adhesive Bonding and the Relative Advantages or Disadvantages of Integral Mechanical Attachment within Mechanical Joining Compared to Mechanical Fastening

Mechanical Joining	Integral Mechanical Attachment
1. Employs only mechanical forces from interference, with possible interlocking; no bonding	1. Likewise
2. Tends to be simple, but may or may not involve preparation of the joint	2.a. Often involves very simple motions for assembly (push, slide, tilt, twist)
	2.b. Often easy to automate (vs. fastening)
3. Uniquely allows intentional disassembly without damage to parts	3. Likewise
4. Uniquely facilitates maintenance, service, repair, portability, and ultimate disposal (via recycling)	4. Likewise
5. Uniquely allows intentional motion for dynamic devices or structures	5. Some methods allow intentional motion in some direction(s)
6. Causes no changes in material microstructure or composition	6. Likewise
7. Allows the joining of fundamentally different materials to one another	7. Likewise
8. Provides a simple means of imparting structural damage tolerance (beyond material damage tolerance)	8. All methods except formed-in types provide a simple means of imparting structural damage tolerance
9. Is relatively low in cost, requiring only limited operator skill (compared to welding or adhesive bonding)	9. Actual assembly is lower in cost than even mechanical fastening

bonded. Integral attachment is easier to accomplish than most fastening. The motions required to accomplish assembly are extremely simple (see Section 2.4), involving a push, slide, tilt or tip, or twist or spin. As a result, assembly using integral attachment features is almost always much easier to automate than assembly using fasteners.

The fact that mechanical joining involves no atomic-level bonding means that it: (1) uniquely allows intentional disassembly without causing damage to parts; (2) uniquely facilitates maintenance, service, repair, upgrade, portability, and ultimate disposal by recycling; (3) uniquely allows intentional motion in selected degrees of freedom; and (4) causes no changes to the microstructure or composition of the part materials and, thereby, (5) allows the joining of dissimilar materials. Integral attachment, like fastening, allows all of these, but in the case of allowing intentional motion, integral attachments are generally more restrictive than fasteners.

As a final structural advantage, mechanical joining, in general, provides a simple means of imparting damage tolerance to a structure or assembly beyond that of the material, specifically compared to welding, which makes two parts truly one. The fact that there is physical separation (i.e., a break in the elastic continuity) of mechanically-joined parts arrests propagating cracks. A final particular structural advantage of integral attachment is that it adds no weight to the assembly. Fasteners almost always add weight, and as such, are often replaced by welding or adhesive bonding in highly weight-sensitive structures, such as airplanes.[7]

Virtually all of the other advantages of mechanical joining, in general, and integral attachment, in particular, relate to manufacturing. While all mechanical joining tends to be low cost compared to other joining process options, integral attachment offers additional savings. The first source of cost savings with mechanical joining arises from the fact that the skill of the operator needed to perform the joining is low compared to adhesive bonding and, especially, welding. This is even truer of integral attachment assembly than of assembly using mechanical fasteners. After all, it takes no skill to assemble a tongue-and-groove joint, but some skill is required to properly drill and tap holes for bolts, which must be properly torqued. Beyond this, integral attachment reduces the part count in assemblies and structures, which obviously add to assembly labor. No fasteners can mean a tremendous reduction in part count. For example, a modern jumbo-jet contains tens of thousands of parts but contains in excess of a million rivets, each of which takes approximately 2 minutes to fully install (including hole drilling, rivet installation and upsetting, rivet-head shaving, and inspection). The elimination of fasteners with integral

---

[7] As but a simple example, the highly-loaded wing carry-through structure of the Grumman F14 air-superiority fighter would have weighed more than 100 pounds more than its approximately 2,200-pound weight if it had been bolted rather than EB welded.

attachment also dramatically (though not always recognizably) simplifies the logistics of assembly. Assemblers do not need to get, hold, and handle bolts, nuts, and washers, for example. Everything needed to allow joint assembly is integral to the joint parts. Elimination of the need for fasteners eliminates the need for tools to install those fasteners. Finally, and far from incidentally, most methods of integral attachment provide self-keying or self-aligning joints and preclude the need for fixtures and fixturing. Those familiar with manufacturing know that fixturing costs (like all tooling costs) can be high and lead times are always long, often requiring months before production assembly begins.

1.6 POTENTIAL SHORTCOMINGS OF INTEGRAL ATTACHMENT

For every advantage, there is likely a disadvantage of any material, process, or method. Again, integral mechanical attachment is no exception. In fact, not only does integral attachment not have any disadvantages compared to mechanical fastening, but it often does not have some of the disadvantages found in fastening.

Table 1.3 summarizes the general potential disadvantages of mechanical joining compared to the processes of welding and adhesive bonding, as well as the relative advantage or disadvantage of integral mechanical attachment within mechanical joining, especially compared to mechanical fastening. Once again, there are references that discuss the disadvantages of mechanical joining in detail (Messler, 2004; Speck, 1997). Suffice to say here:

TABLE 1.3 Summary of the General Potential Disadvantages of Mechanical Joining Compared to Welding and Adhesive Bonding and the Relative Advantages and Disadvantages of Integral Mechanical Attachment within Mechanical Joining Compared to Mechanical Fastening

Mechanical Joining	Integral Mechanical Attachment
1. With a few exceptions, accidental disassembly can occur without precautions	1. Loosening and/or accidental disassembly does not occur for most methods
2. Stress concentrations are introduced by most methods	2. Some methods reduce stress concentrations vs. fasteners
3. Joints typically allow fluid intrusion or leakage without special precautions	3. Preventing fluid intrusion or leakage can be very difficult
4. Labor intensity associated with fastening can be high	4. Labor intensity is very low
5. Mechanical fasteners can add weight	5. No weight is added
6. Some methods (especially in fastening) may be difficult to automate	6. Automation is generally easy
	7. Up-front costs for molds, dies, etc. may exceed fastening but can easily be amortized by production volume

First, with a few exceptions, accidental loosening or even disassembly can occur in mechanical joints unless precautions are taken to preclude this. It is a virtual certainty in life that if something can be taken apart on purpose, it can come apart accidentally. The loosening of bolts due to vibration, as well as from a force that arises from the rotation of parts in which bolts are used (such as lug nuts in an automobile wheel or bolts in a rotating coupling), is far from unknown or inconsequential. If precautions are not taken to prevent it, bolts can work loose and fall out.[8]

Second, every mechanical fastener gives rise to a stress concentration at the point of force application, as do some types of integral attachments.[9] In some situations (e.g., under cyclic or fluctuating [fatigue] loading), such stress concentrations can lead to premature failure that originates at the fastener or point of attachment. With fasteners, fatigue cracking tends to occur in one of the parts being fastened. With integral attachment features, such as snap-fits (see Chapter 4, Section 4.1), fatigue cracking occurs at the base of the attachment feature, which, of course, is integral to the base part. Therefore, once a crack initiates at the base of an integral attachment feature, it usually propagates to completely separate that feature from the rest of the part.

Many forms of integral attachment operate in such a fashion that they spread the mechanical force for joining, so stress concentration is minimal. A key factor is that integral attachments can be selected to minimize stress concentration, while little (other than the use of washers with bolts or screws) can be done to reduce the stress concentration associated with fasteners. In most cases, it is required holes (or other recesses) that give rise to most of the stress concentration.

Third, all mechanical joints are prone to fluid intrusion or leakage. The fact that there is a physical interface between abutted and joined parts means fluids can pass through, albeit some fluids more easily than others and in some joints more easily than others. When mechanical fasteners, such as bolts (or nuts and bolts), are used, fluid intrusion or leakage can be prevented by using gaskets, seals, or sealants at the joint interface. However, in a lesson learned by the author long ago, any joint that can leak because there is an interface will leak in time! The potential for fluid intrusion or leakage is generally greater for integral attachment methods than for fasteners, as joint clamping forces are much harder to raise to very high levels. As a result, integral mechanical attachment simply does not seem to be applied when fluid intrusion, leakage, or pressure tightness are important.

---

[8] For rotating parts, such as wheels, reverse (i.e., left-handed vs. right-handed) threads are occasionally used. The reason is that if the direction of rotation of the part and the direction needed to tighten the threaded fastener (e.g., nut or bolt) are the same, loosening can occur. If the direction needed to tighten the fasteners is opposite the direction of rotation of the part, the fasteners actually have a resultant force trying to tighten them further.

[9] Among integral attachments, snap-fit features (discussed in Chapter 4) are most apt to cause some stress concentration, as can some forms of deformed-in attachments (discussed in Chapter 5).

A fourth disadvantage of fasteners that is almost always a non-issue with integral attachments is added weight for joining. Being integral to the part, attachment features add nothing or virtually nothing to the part weight.

Two other (i.e., fifth and sixth) disadvantages of mechanical fastening that are much less problematic with integral mechanical attachment relate to manufacturing/production assembly. The first is that the labor intensity required to install fasteners can be very high. The proper installation of bolts, for example, requires drilling (and perhaps reaming and/or tapping) of holes, insertion of bolts (often with washers), application and controlled tightening of nuts (or simply the bolt, if nuts are not needed because the part is internally threaded), torque-level verification, and, possibly, application of sealants, locking adhesives, corrosion-protective coatings, etc. All this takes time—and costs money. Similarly high labor intensity is associated with the proper installation of upset rivets, where hole preparation, rivet insertion, rivet upsetting, rivet-head shaving (for aerospace applications), and inspection typically take 2 minutes total. For a jumbo-jet airliner, in which more than 1 million rivets are used, this amounts to 2 million minutes or more than 30,000 labor hours! Assembly time for most integral attachment methods is extremely short, as no holes are needed, no supplemental pieces are needed, no fixturing is needed, no tools are needed, and as will be seen in Chapter 2 (Section 2.4), assembly motions are very simple.

The second manufacturing-related disadvantage of many of the higher performance fasteners, such as bolts, 1-piece upset, and 2-piece high-shear-strength and blind rivets, is that automation of the fastening process can be difficult to impossible. Required dexterity is the greatest problem, followed by often difficult accessibility. For integral mechanical attachment, the fact that the attachment features are part of the parts to be joined, means accessibility is rarely a problem. The parts are simply designed to form a joint without excessive access. More importantly, the motions needed to cause most integral attachments to engage (and possibly lock) are extremely simple for both humans and robots, for example. These motions are "push," "slide," "tilt" (or "tip"), and "twist" (or "spin"). Besides being simple, the forces needed to cause proper engagement (and possibly locking) are low in comparison to the forces needed to cause separation. In fact, one of the major driving forces for the introduction of integral snap-fit features in plastic parts was to reduce part count and facilitate assembly, particularly using simple automation. Integral mechanical attachment was the single greatest enabler for design for assembly (Boothroyd, 1980, 1992).

The sole seeming shortcoming of integral mechanical attachment is the added cost for creating the added geometric attachment features on parts. However, it turns out that for even the most geometrically-complex features (such as various designs of snap-fits; Chapter 4, Section 4.2) widely used in plastic-part assemblies, the actual added cost to molds and dies is small, and even that cost is easily and effectively amortized with typical production volumes (Bonenberger, 2000).

So, integral mechanical attachment has all of the advantages of mechanical fastening, and more. It also has fewer of the shortcomings of fastening, and none of significance of its own. No wonder this oldest of all methods for joining has not only persisted over millennia but has experienced multiple resurgences, as it is again today.

1.7 SUMMARY

Integral mechanical attachment is one of the two sub-classifications of mechanical joining, the other being mechanical fastening. In all mechanical joining, strictly mechanical forces are used to hold parts together in an assembly or structure. The origin of these forces is the physical interference between abutting or mating parts, so that those parts are held in the proper locations, in the proper orientations, and with the proper alignment. To resist forces that attempt to pull the assembly apart, the fasteners or integral attachments may also need to cause interlocking between the parts or between the parts and the fastener. In fastening, a specialized part, whose sole function is to cause this needed interference and/or interlocking between parts to be joined, is required. Such supplemental specialized parts are called *fasteners*. In integral attachment, the actual shapes of the parts to be joined are designed and/or altered to create a geometric feature on each part that causes interference and/or interlocking. Such features are called *integral attachment features* or, in some cases, simply *attachment features*.

Integral mechanical attachment offers all of the advantages afforded by fasteners, as well as some of its own. Examples of major advantages of both include: the unique ability to allow intentional disassembly; the consequential capability to facilitate maintenance, repair, modification, or ultimate disposal; the ability to join any material to itself or to another material and not change the microstructure or composition in the process; and simplicity and low cost compared to welding and adhesive bonding. Integral mechanical attachment has no shortcomings not found with mechanical fastening, and usually, any such shortcoming is less severe. Examples of some general shortcomings are the development of stress concentrations at discrete points of joining; the potential for fluid intrusion or leakage, unless special precautions are taken; the addition of weight to the assembly or structure because of any fasteners; and varying ease of automation, depending on the specific technique.

Because of these advantages and freedom from any serious shortcoming, integral mechanical attachment has not only persisted from prehistoric times but has enjoyed resurgences at various times in history, including what may be a near-revolution today.

REFERENCES

Bonenberger, P.R., *The First Snap-fit Handbook,* Hanser Gardner Publications, Cincinnati, OH, 2000.

Boothroyd, G., *Design for Assembly—A Designer's Handbook,* University of Massachusetts, Amherst, MA, 1980.

Boothroyd, G., *Assembly Automation and Product Design,* Marcel Dekker, New York, NY, 1992.

Brandon, D., and Kaplan, W.D., *Joining Processes: An Introduction,* John Wiley & Sons, Ltd., Chichester, England, 1997.

Messler, R.W., Jr., *Joining Materials and Structures: From Pragmatic Process to Enabling Technology,* Elsevier/Butterworth-Heinemann, Boston, MA, 2004.

Speck, J.A., *Mechanical Fastening, Joining, and Assembly,* Marcel Dekker, New York, NY, 1997.

2 CLASSIFICATION AND CHARACTERIZATION OF INTEGRAL MECHANICAL ATTACHMENTS

2.1 WHY CLASSIFY THINGS AT ALL?

Human beings facilitate learning and memory by classifying what they experience and what they need to know. It is deeply embedded in the cognitive process. Somehow, toddlers who see their first dog and hear from their mother or father that "it's a doggie!" immediately store that name with the vision and, thereafter, seem to recognize other dogs, despite differences that can be significant. Unfortunately, the initial learning isn't perfect. The toddler tends to call every furry four-legged animal with ears and a tail a "doggie." They have clearly classified the first dog they saw by certain obvious features, attributes, or characteristics. The "doggie" has four legs, so all four-legged animals might be identified as "doggies." But, the "doggie" is also furry, so, eventually, only furry four-legged animals are identified as "doggies." And so the process of classification for learning and memory goes. It evolves to allow greater and greater differentiation based on more and more subtle similarities and/or differences.

Materials, processes, and methods are also classified to facilitate learning and recall. But, beyond learning and recall, we classify things based on and in order to see familial relationships. We look for and use obvious similarities and obvious differences to classify. From these, we evolve a hierarchy or, more properly, a taxonomy; from the Greek *taxis,* meaning "arrangement."

Something seemingly derived directly from another in a taxonomy has a "daughter–mother" relationship. Something derived from the combination of two things at the next-higher level in a taxonomy has a "child–parents" relationship. Two or more things with the same origin at the next-higher level of a taxonomy have a "sibling" relationship. And, not surprisingly, there are other, more distant relationships, such as "grandparents," "aunts," and "cousins."

So, in this chapter, integral mechanical attachments is classified, within an overall taxonomy of joining methods and within themselves as a related group. The purpose of this classification is for the identification, recognition, and use of key attributes or characteristics to help understand the principles of operation, the expectations from performance, and the potentials for application.

2.2 INTEGRAL ATTACHMENTS' PLACE WITHIN THE TAXONOMY OF JOINING PROCESSES AND METHODS

At the most fundamental level (which appears as the highest level in a taxonomy, because of the familial relationship), joining can be accomplished using one or more of three fundamental forces: (1) mechanical forces, (2) chemical forces, or (3) physical forces (Brandon and Kaplan, 1997; Messler, 1993). These major categories are shown in Figure 2.1, as a taxonomy of major joining processes, including hybrid and variant processes.

As described in Chapter 1, mechanical forces arise in joining from the physical interference between two or more solid bodies; at which point, one body prevents the movement of the other(s) in the direction in which the combination was brought into contact. If separation of these bodies is made impossible by the development of a mechanical retaining force from a physical restraint, the parts are said to be interlocked, and the assembly stays together until separating forces are applied that exceed the retaining or retention force. Besides just having to overcome blocking forces against movement, it may also be necessary to overcome actual locked-in or residual stresses, depending on the type of mechanical joining. These forces of interference and, possibly, interlocking, can be obtained solely from the shapes of the parts being joined or by the shape of a specially-designed part whose sole function is to cause interference and, possibly, interlocking between itself and at least two other parts. The former method of accomplishing *mechanical joining* is known as *integral mechanical attachment* (see Chapter 1, Section 1.3). The latter method of accomplishing mechanical joining is known as *mechanical fastening* (see Chapter 1, Section 1.4). These two sub-divisions of mechanical joining are shown in Figure 2.1.

Chemical forces are what enable adhesive bonding. In *adhesive bonding,* a special substance is used as the agent between two materials (or parts that those materials comprise) to cause each part to adhere to the special substance and, thereby, to each other, indirectly. The active substance serves as a compatible

Joining Processes									
Using Mechanical Forces						Using Chemical Forces		Using Physical Forces*	
Primary Processes									
Integral Mechanical Attachment			Mechanical Fastening			Adhesive Joining		Welding	
Plastic Interlocks	Elastic Interlocks	Rigid Interlocks	Unthreaded Fasteners	Threaded Fasteners	Other Fasteners	Organic Adhesives	Inorganic Adhesives	Non-fusion Welding	Fusion Welding
Secondary Processes									
						Solvent Cementing	Cementing/ Mortaring		Brazing/ Soldering
Hybrid Processes									
			Rivet-Bonding				Weld-Bonding		Weld-Brazing
Variant Processes									
			Thermal Spraying						

* In the context of joining, "physical forces" refer to those forces arising from electromagnetic sources in atoms.

FIGURE 2.1 A general taxonomy of all joining processes organized by the basic type of force involved, including hybrids and variants.

intermediary between two materials that, themselves, may not and need not be compatible with each other. This sticking together, or adhesion, is chemical in nature and has its origin in what is known as *surface adsorption*. In *surface adsorption*, some of the atoms or molecules of the intermediate substance form actual atomic-level bonds with some of the atoms or molecules of the materials being joined.[1] The active substance or agent is known as the *adhesive*, while the materials being joined are known as the *adherends*.

The chemical bonding that occurs in adhesive bonding is usually of an inherently weaker secondary type, as opposed to a stronger primary type. Most often, the bonding that occurs in adsorption is van der Waals' bonding or hydrogen bonding but can be of other similar types. In all secondary types of bonding, the bonding occurs between molecules and not atoms. In *van der Waals' bonding,* the force of attraction that holds the molecules together is always the result of the dipole character of the molecules involved in the bond. Such a dipole can be inherent in the molecule because of its atomic makeup and shape. It is the result of a permanent asymmetry of negative-electron and

---

[1] What is meant by "some of the atoms or molecules" is really all of the atoms or molecules of certain, but not necessarily all, ingredients, as opposed to some, but not all, of the atoms or molecules of any particular ingredient. If chemical bonding occurs between two atomic or molecular species, it will tend to occur completely. If such bonding does not tend to occur between two atomic or molecular species, it will not occur at all. Usually, only some component of the intermediate substance and some component of the materials being joined actually form chemical bonds.

positive-nuclear charge distribution. Alternatively, a dipole character can be induced into a molecule in which the negative-electron and positive-nuclear charge distribution is normally symmetrical, but is perturbed to become asymmetrical when one molecule approaches another. The force of attraction between dipoles is purely electrostatic or coulombic. In short, van der Waals' bonding arises from either permanent or induced dipoles in molecules. It is also possible that a hydrogen atom in a molecule serves to bridge between and link two substances. The force of attraction arises from the positively-charged hydrogen ion, which is bound to or part of the molecules of one material and the negative outer electrons of an atom that is part of the molecules of the other material. When this occurs, the resulting bonding is known as *hydrogen bonding*. Again, the force of attraction is electrostatic and is stronger than for van der Waals' bonding, but weaker than primary bond types (e.g., ionic, covalent, and metallic).

The formation of primary chemical bonds between an adhesive and an adherend is possible and does occur for some adhesive–adherend systems. But, this primary bond formation requires an actual chemical reaction and compound formation (as a bonding reaction layer) to occur between the adhesive and the adherend, and this is usually rare. The most common examples in which it does occur are less recognizable as adhesive bonding and involve the use of inorganic ceramic adhesives to react with and join ceramic or metallic adherends.

What is common to all adhesive bonding is that the adsorption occurs at the surfaces of, or interface between, the adhesive and each adherend (Pocius, 1997). As such, the forces per unit area leading to adhesion, though small compared to the mechanical forces per unit area that operate in mechanical joining, act over much larger areas. This spreads loading and generally reduces stress concentration compared to mechanical joining methods. Load-carrying capability can be made high with adhesive bonding by causing bonding to occur over a large area.

The process of welding involves the physical force of attraction between atoms, which is electromagnetic in its origin. All atoms, except those of the inert gases, seek to achieve a stable electron configuration by filling their outermost electron levels or shells in accordance with rules of quantum physics (Callister, 2005). Some atoms do it by giving up or taking on electrons to form positive or negative ions, respectively, and then bonding to one another due to the resulting electrostatic (coulombic) attraction. This primary bond type is *ionic bonding*. Alternatively, some atoms achieve stable electron configurations by sharing electrons with other atoms. When they do so intimately with one or a few other atoms of the same or different elemental species, the primary bond type is *covalent bonding*. When they do so in a more extended manner with as many atoms of the same or different elemental species as are able to surround them, the primary bond type is *metallic bonding*. In all three cases, the resulting bonding is much stronger than it is for secondary bond types, so the resulting joining force is much greater. Regardless, the force of attraction is purely electromagnetic in origin, and is, thus, physical in origin.[2]

Welding can lead to joining based on this physical source of force in ceramic materials using ionic or covalent or mixed ionic-covalent bonding, in metals using metallic bonding, and in glasses using covalent bonding. Welding can also lead to joining in certain polymers (most notably, thermoplastic types), but the atomic- or molecular-level origin is more complicated, and involves substantial degrees of physical entangling of the long-chain molecules that make up polymers beyond just bond formation.

Another example of the use of pure electromagnetic forces in joining occurs for the special case of *magnetic joining*. While not generally thought of as a joining process, it certainly is one. The use of magnets to clamp parts together during machining or in tooling is not unusual (e.g., magnetic chucking). Magnets are also used to hold cabinet doors closed against accidental or casual opening and are used to pull refrigerator doors tight against a rubber seal. Finally, the use of decorative magnets to hold our children's first drawings, good report cards, postcards, or reminder messages on the refrigerator is commonplace in most homes.

As described in Chapter 1, Section 1.4, and as shown in Figure 2.1, mechanical joining is sub-divided or sub-classified into mechanical attachment (using integral features of parts to accomplish joining) and mechanical fastening (using supplemental devices, known as *fasteners*, to accomplish joining). Adhesive bonding is usually not sub-divided, although some classification schemes specific to adhesive bonding sub-divide adhesives into synthetic organic (polymeric) and inorganic (ceramic) types. A particularly noteworthy and appropriate sub-division of adhesive bonding is solvent cementing. *Solvent cementing* involves the partial dissolution of two like polymeric adherends by a suitable solvent and the intimate intermixing of material from each with the other. Once the solvent evaporates, the resulting bond is virtually identical to the surrounding base materials, and the resulting joint is more like that produced by welding than by an adhesive. As such, solvent cementing, though a sub-division of adhesive bonding, is properly positioned between welding and adhesive bonding, not as a hybrid, but as a "child" of the two.

Two sub-classifications of welding are fusion and non-fusion welding. In *fusion welding,* two materials are caused to form primary chemical bonds by bringing them into their molten state, causing them to mix, and allowing them to solidify and form bonds. Once solidified, the two materials become as

---

[2] In actuality, it is this same physical force (i.e., of atoms attracting one another to form bonds with other atoms so that each achieves a stable electron configuration) that leads to chemical reactions and, thus, is the basis for some adhesive bonding. When this occurs in adhesive bonding, the difference between it and welding breaks down. A particularly good example is solvent cementing, described later in Section 2.2.

one across the weld. In *non-fusion welding,* two materials form primary bonds simply by forcing atoms of each into intimate contact with one another, without melting or solidification having to occur. Once atoms are brought close enough together, they will bond to achieve stable electron configurations. There may also be fairly significant inter-diffusion of atoms (and associated mass transport) from one base material into the other or from the filler into the base materials and vice versa.

Brazing is a sub-classification of welding in which a filler (or intermediate material), different than the base material(s) to be joined, is used to join that (those) base material(s) without causing any melting of the base material(s). The filler is melted, however. Primary bonding occurs as the result of inter-diffusion of some elemental component(s) of the molten filler and some elemental component(s) of the solid base material(s), and, occasionally, the formation of some actual chemical reaction layer. *Soldering* is a further sub-division of brazing and is a rather artificial one at that. The only difference between brazing and soldering is the temperature at which the fillers required melt. In brazing, by general consensus (as opposed to anything real), the fillers melt above 450°C (840°F), while in soldering, fillers melt below this temperature. In both cases, the distribution of the filler through the joint, once melting occurs, is supposed to occur by capillary action. Also, for both processes, it is essential that the base materials not melt at all (i.e., that the liquidus temperature of the filler not exceed the solidus temperature of any base material). These sub-divisions of welding are shown in Figure 2.1.

Farther down the taxonomy of joining processes, various *hybrid joining processes* arise from the combination of the two next-higher processes (Messler, 2004). In other words, hybrid joining processes are "daughters" of two "parents" processes. Examples are *rivet-bonding* (between mechanical fastening and adhesive bonding) and *weld-bonding* (between welding and adhesive bonding). *Weld-brazing* is another hybrid between welding and brazing, which are not actually at the same level in the taxonomy.

Finally, there are a couple of *variant processes* (Messler, 2004). These are related to higher level joining processes but have some unique qualities or attributes of their own and can be difficult to precisely classify. The two examples are braze welding and thermal spraying. *Braze welding* uses a filler that melts to join base materials that do not melt, but the filler does not distribute within the joint by capillary flow. Rather, the filler is melted within a pre-prepared groove between the base materials. Another process variant is thermal spraying. *Thermal spraying* appears, in different embodiments and in different applications, to be related to welding or adhesive bonding or even mechanical joining, in various combinations or two or all three (American Welding Society [AWS], 1985). These hybrids and variants are shown in Figure 2.1 as lying between the higher level processes or sub-processes.

Within mechanical fastening, a final sub-classification (at the lowest level shown in Figure 2.1) is normally-threaded versus unthreaded fasteners

(Messler, 1993, 2004; Parmley, 1987). Within mechanical attachment, as will be seen in Sections 2.3 and 2.4, attachment features can be further sub-classified.

The pertinent question for integral mechanical attachment methods is: what is the taxonomy within that general joining approach?

2.3 A CLASSIFICATION SCHEME FOR INTEGRAL ATTACHMENTS BASED ON FEATURE CHARACTER OR OPERATION

The most common basis for classifying things is by some key characteristic or characteristics. This often allows meaningful comparison of similarities and differences but may not, depending on whether the chosen characteristic relates to something fundamental about the item being classified. As an example, one might decide to classify animals based on some obvious characteristic, such as the number of legs each has. While this might seem logical, the resulting classification would really reveal very little that is truly meaningful about the animals. It might or might not even reveal familial relationships over long periods (e.g., by evolution). A bird and a human each have two legs, yet there, the similarity largely ends. In fact, humans, being mammals, have more that is fundamentally common with many four-legged animals than they do with two-legged birds. Hence, when things are classified, the basis for classification should be on some essential feature(s) or characteristic(s).

For integral mechanical attachments, all of which use geometric features of parts to be joined to cause physical interference against movement in certain translational and/or rotational directions and, possibly, interlocking against separation, it becomes apparent that there are three fundamental classes of such attachments: (1) rigid interlocks, (2) elastic interlocks, and (3) plastic interlocks (Messler and Genc, 1998).[3] With just this knowledge, one is able to understand how a particular feature operates to cause joining, in what materials such a feature can likely be produced, how such a feature can be produced in a part, and how such a feature can be expected to perform in terms of load-carrying capability and reliability from the standpoint of permanence. This understanding is crucial for design and the designer, for manufacturing and the process engineer, and for product function, service, and durability or robustness.

Each class of interlocks, rigid versus elastic and plastic, is described in Sections 2.5 and 2.7, and is treated in depth in Chapters 3, 4, and 5, respectively. But, very briefly, here's the basis for this scheme of classification.

---

[3] In this particular context (i.e., for classification), integral mechanical attachments are referred to as "interlocks." The reason is that in most, though not all modern applications, a key function required of an integral attachment is that it provides resistance to the accidental or unintentional separation of parts of the assembled entity. As will be seen in Section 2.5, the motions to cause parts with integral attachment features to engage and lock with one another involve a simple push, slide, tilt, or twist, and the force needed is normally very low, particularly compared to the force needed to cause subsequent separation.

If one part or a portion (i.e., a specific geometric feature) of one part interferes with another part or portion of another part such that applied loads tending to cause movement in an unwanted direction or directions (including separation) are resisted by those parts or features remaining rigid, the attachment is said to be a *rigid interlock*. By "rigid" is meant that the actual features preventing further movement and/or causing interlocking against part separation do not deflect elastically and do not deform plastically under reasonable forces. In other words, resistance to movement, including separation, is enabled by the interlocking features remaining rigid against forces that could cause movement or separation. Obviously, for any material of construction, a force can be imposed that is high enough to cause interlocked parts to separate by breaking the parts themselves. Thus, what is really meant by rigid interlocks is that they resist separation under applied separating forces without deflecting elastically or deforming plastically before the interlocking features or parts themselves fail by fracture (Figure 2.2a and b). Not surprisingly, rigid interlocks, when used correctly in the right material(s), are capable of resisting the greatest stresses of all integral interlocks or attachments.

If one part or specific geometric feature of one part interferes with another part or specific feature of another part such that applied loads tending to cause movement in an unwanted direction or directions (including separation) are resisted by those parts or features acting elastically, the attachment is said to be an *elastic interlock*. By "elastic" is meant that at least one of the actual features in a mating and interlocking combination (see Chapter 4, Section 4.1) must be caused to elastically defect during insertion of one part/feature with its mate. Once the two parts in the mating pair are engaged, an elastic interlock may relax by recovering from the deflected state or may remain deflected and under a residual elastic stress. Elastic interlocks normally operate by deflecting elastically during assembly, but almost always resist separation under applied separating forces by failing either by fracture as the feature literally breaks or by severe plastic deformation as the feature severely deforms (Figure 2.2c and d). The load-carrying capability of elastic interlocks, though generally not as great as that for rigid interlocks, can be high, provided failure of the interlocks occurs by fracture versus elastic deflection in the direction opposite that causing part engagement for assembly. As will be seen in Chapter 4, it is always best design practice to use elastic interlocks in such a way (i.e., location and orientation) that they are not loaded in the direction opposite that used for their engagement (Genc et al, 1998).

If one part interferes with another part to resist applied loads tending to cause movement in an unwanted direction or directions (including separation) by having portions of the two mating parts plastically deform into or around each other, the attachment is said to be a *plastic interlock*. Notice that these interlocks are distinctly different than their rigid and elastic cousins in a taxonomy of integral mechanical attachments. Rather than being designed in and

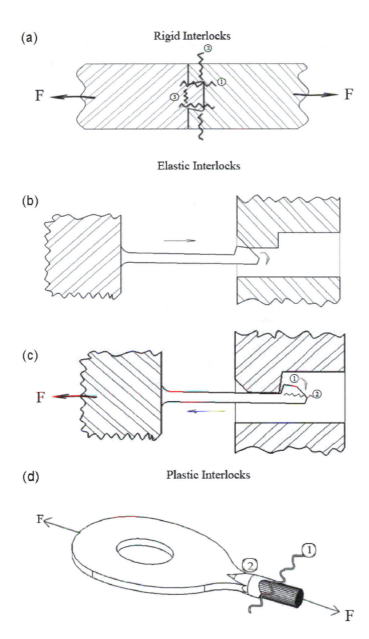

FIGURE 2.2 Operation of a typical rigid interlock and examples of brittle fractures at Points 1, 2, or 3 (a); operation of a typical elastic interlock and examples of severe plastic deformation and "camming over" at Point 1 and fracture at Point 2 (b and c); and operation of a typical plastic interlock and examples of failure by either part fracture at Point 1 or either loss of grip or severe plastic deformation at Point 2 (d).

pre-fabricated into parts to be joined, they are created in the parts once they are properly positioned, oriented, and aligned. These interlocks are said to be "plastic" because they are created by plastic deformation, actually co-deformation, of at least one of the mating parts into or around the other (see Chapter 5).

Plastic interlocks develop a holding force or stress in two ways. First, there is always some level of locked-in or residual elastic stress. After the locking feature is created by a suitable deformation process, there is some inevitable "spring-back" from the elastic portion of the stress causing deformation. This stress can lead to a frictional force that helps hold the parts in position if the stress acts in the proper direction. Second, there is the additional possibility (depending on the particular plastic interlocking method) that material from one part surrounds the other mating part and literally locks the two together by the physical interference that results. Plastic interlocks can fail only when the applied force leads to further plastic deformation in a direction that causes unwanted movement, possibly including separation, or fracture (Figure 2.2c and d). The load-carrying capability of plastic interlocks can be as great as the yield strength of the weaker material involved in the mating pair of parts, at which point the deformed-in locking feature fails by deformation.

Figure 2.3 gives a simple taxonomy for integral mechanical attachments based on this rigid, elastic, and plastic characterization, including general sub-types.

2.4 AN ALTERNATIVE CLASSIFICATION SCHEME BASED ON METHOD OF FEATURE OR JOINT CREATION

Rarely is there only one way—or scheme—by which things can be classified. Ignoring schemes that have no particularly sound basis,[4] there are likely different schemes for different purposes. One of the best, and particularly relevant, examples is the difference by which design engineers as opposed to manufacturing engineers might prefer to classify parts.

A design engineer would tend to classify what is commonly called a "waffle plate"[5] as a thin plate having longitudinal and transverse integral stiffeners to provide resistance to buckling at minimum weight. He or she does so because structural performance or function is most important to him or her. A manufacturing engineer would tend to classify the same waffle plate as a thick plate having an orthogonal array of pockets that do not fully penetrate the plate. He or she prefers this classification because the method and ease of manufacturing is most important to him or her. The former thinks about what material and/or shape is needed to obtain the needed function. The lat-

---

[4] A "sound basis" for classification is one that uses some fundamental and inherent property(ies), attribute(s), or characteristic(s). But, not surprisingly, what is fundamental in one situation may not be fundamental in another. While whether an animal has a backbone (i.e., is a vertebrate) or not or whether it bears live young or lays eggs may be fundamental to a zoologist, size would be much more fundamental to an animal transporter.
[5] Like a breakfast waffle, a "waffle plate" consists of a horizontal and vertical array of pockets, with thin bottoms, separated by thick horizontal and vertical ribs.

Integral Mechanical Attachments

Rigid Interlocks	Elastic Interlocks	Plastic Interlocks
Operate by remaining rigid; rarely employ residual stresses	Operate by deflecting elastically; sometimes employ residual stresses	Operate by being deformed plastically; rarely require residual stresses
Tongues-and-grooves	Integral spring tabs	Setting
Dovetails-and-grooves	Spring plugs	Staking
Rabbets or dados	Snap slides	Metal stitching
Mortise-and-tenons	Snap clips	Metal clinching
T-slots and Ts	Clamp fasteners	Indentation-type joints
Shaped rails and ways	Clamps	Beaded-assembly joints
Wedges and Morse tapers	Quick-release fasteners	Crimping
Shoulders and flanges	Integral snap-fit features	Hemming
Bosses, lands, and posts	Interference press fits	Thermal staking
Tabs and ears	Thermal shrink fits	Formed tabs
Integral keys and splines		
Integral threads		
Knurled surfaces		
Hinges		
Hasps		
Latches		
Hooks		
Turn-buckles		
Collars and sleeves		

FIGURE 2.3 Simple taxonomy for integral mechanical attachment schemes and attachments organized based on methods of operation, including general sub-types.

ter thinks about what material has to be removed or added or otherwise altered to produce the needed item. Both schemes have meaning, and both are based on an important attribute or characteristic or feature of the waffle plate. However, the design engineer would talk about the "ribs" as the key feature, while the manufacturing engineer would talk about the pockets as the key feature.

In integral mechanical attachment, the design engineer would likely prefer the classification scheme based on feature character or operation, just as described in Section 2.3. The manufacturing engineer, on the other hand, would likely prefer a classification scheme based on the method by which attachment features might be created. Hence, an extremely useful alternative scheme for classifying integral mechanical attachments is based on whether the integral features are designed in and processed into detail parts prior to their assembly or are processed in (i.e., created) after part positioning during the assembly process.

Figure 2.4 shows an alternative taxonomy for integral mechanical attachments based on how features or actual joints are created by interlocking features that occur naturally, are designed and fabricated into detail parts, or are processed into the parts after assembly, including some major sub-divisions.

Integral Mechanical Attachments

Naturally-occurring Features or Joints	Designed-in Features or Joints	Post-assembly Processed-in Features or Joints
Rigid Interlocks	Rigid Interlocks	Rigid Interlocks
Mating shapes (stone) Entangling shapes (wood) Gripping surfaces (stone)	Protrusions/recesses Cast (metal/ceramic) Molded (polymer/glass) Forged (metal) Extruded (metal/polymer) Machined (metal) Press fit (metal/polymer*) Shrink fit (metal) Cast or molded-in inserts (metal/ceramic/polymer) Welded-on features (metal/polymer/glass)	None
Elastic Interlocks	Elastic Interlocks	Elastic Interlocks
Entangled shapes (wood) Hook-and-loop (burrs)	Snap-fit features Molded (polymer) Machined (metal) Extruded (metal/polymer) Cast or molded-in inserts (polymer/metal/wood)	None
Plastic Interlocks	Plastic Interlocks	Plastic Interlocks**
?	None	Formed cold Conformed features (metal) Co-formed features (metal) Formed hot Conformed features (metal/polymer/glass) Co-formed features (metal/polymer)

* Includes wood
** All operate while remaining rigid

FIGURE 2.4 An alternative taxonomy for integral mechanical attachment organized based on how interlocking features and/or joints are created.

2.5 RIGID VERSUS ELASTIC INTEGRAL ATTACHMENT METHODS

Naturally-occurring objects (e.g., stones, sticks, and bones, teeth, and tusks of dead animals) allowed our prehistoric ancestors to create tools, weapons, shelters, and other devices and structures by taking advantage of either of two properties of these objects. Because of the material of which they were composed and/or the size and shape they had, these naturally-occurring objects behaved in either a rigid manner or an elastic manner. When they were rigid (e.g., as in the case of stones), they created rigid interlocks. When they were elastic (e.g., as in the case of forked green sticks), they may have created elastic interlocks (or, for the case of green sticks, which behave rigidly to a point, they may have created rigid interlocks). This basic division of integral mechanical attachment features, interlocks, or methods has persisted through the eons until today. There are still rigid interlocks and elastic interlocks. The difference is that today (and for a long time before today) we are not restricted to the use of objects with natural shapes to accomplish joining by integral attachment. Rather, we can design in and create the shapes needed to allow and enable joining.[6]

Rigid interlocks or *rigid integral mechanical attachments* are those that have a naturally-occurring geometry or, more often, have a designed-in and pre-fabricated geometry that allows integral mechanical attachment. All rigid interlocks consist of two opposing complementary shapes that can be caused to engage with each other with some simple motion in some direction. Once fully engaged, the two objects physically interfere with each other to successfully carry a load or loads in one or more directions different from the direction opposite the one in which they were caused to engage. The concept is shown in Figure 2.5, in which one part with a protrusion is slid into a receiving groove or recess in the mating part. Once fully engaged,[7] one part with the other, the joined assembly (i.e., individual and combined parts) should be able to sustain and transmit loads in all directions except opposite the direction that caused engagement. Depending on the design of the actual joint created, it may or may not be possible to continue to cause motion in the same direction in which engagement was accomplished. If desired, assembled rigid parts can subsequently be locked together to prevent motion (i.e., separation) in the direction opposite that used to cause engagement by adding some additional part to serve as a locking member. Figure 2.5 shows how a peg could be added to secure rigid interlocking features against disassembly.

In being rigid, interlocking parts or features are expected to operate well within the elastic limit of the weaker part or feature. Exceeding the elastic limit of the weaker part or feature would cause the interlock to fail by causing the material(s) to fail in either a ductile or a brittle fashion. Ductile failure would occur if, when loading, caused the stress to exceed the elastic limit, the material deformed plastically, whereas brittle failure would occur if the material fractured. This characteristic of rigid interlocks suggests that they are best used

---

[6] In fact, we are increasingly able to also design and synthesize materials to have precisely the properties we want (i.e., we can create materials with functionally-specific properties).

[7] In the context of this treatment, "engaged" or "fully engaged" means that one part has been caused to interact with another part until those two parts have come into the proper/desired position, orientation, and alignment so they are, or could be, locked together if desired.

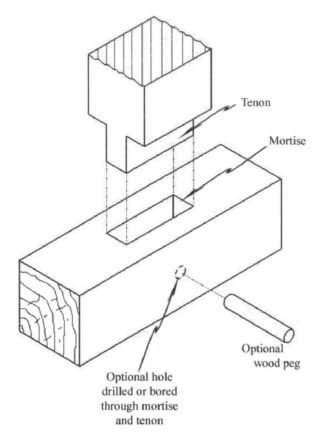

FIGURE 2.5 A schematic of a mortise-and-tenon joint. Note how a wooden peg could be inserted into a hole drilled or bored all the way through the mortise and the tenon (as shown by the dashed circle on the mortise only) to secure the joint against disassembly.

in materials that do not exhibit much plasticity (if any) and have high elastic limits, and in parts that should not ever deform if they are to function properly. Candidate materials include ceramics (including stone, clay units or bricks, cement and concrete, and engineered or advanced ceramics, such as alumina, magnesia, silicon carbide, etc.), wood, brittle or simply very high strength metals (such as cast irons or cast aluminum, magnesium, or bronze alloys, or quenched and tempered high-strength low-alloy or tool steels), and rigid polymers (typified by many thermosetting types and some thermoplastic types). Use in reinforced polymer-matrix composites, as well as in metal-matrix, ceramic-matrix, and carbon-matrix composites, is also feasible but not yet seen very often.

Examples of rigid interlocks abound and are described in detail in Chapter 3.

Rigid interlocks rarely rely on locked-in or residual stresses for their proper operation, although some might. Rather, they resist loads that try to cause part movement by developing a reaction force.

Figures 2.6 and 2.7 show two modern examples of rigid interlocks: one in pre-cast cement or concrete units and one in pre-shaped wood joints in architecture.

Elastic interlocks are those that occasionally have a naturally-occurring geometry or, more often, have a designed-in and pre-fabricated geometry that allows integral mechanical attachment. In either case, the part with the elastic attachment feature is engaged with a rigid (or, at least, relatively more rigid) feature in an opposing and mating part to cause physical interference and, usually, interlocking of the two parts. The critical characteristic of these

FIGURE 2.6 Rigid interlocking features can be incorporated into pre-cast concrete units to facilitate the construction of structures, such as retaining walls, without a need for mortar. Two examples of such retaining walls are shown here; one (right) more decorative than the other (left). (Photographs courtesy of the Portland Cement Association, Skokie, IL; used with permission.)

FIGURE 2.7 Notched logs in many log structures form rigid interlocks where one intersects or crosses the other at an angle, as shown in this gateway at the entrance to a summer home, itself constructed from notched logs, in Ocean Point, near East Boothbay, on Linekin Bay in Maine. (Photograph by the author, Robert W. Messler, Jr.; used with permission.)

engagement features is that at least one (and usually only one[8]) is explicitly designed to be capable of deflecting elastically. Once the elastic feature on one part fully engages with the more rigid feature on the mating part in the assembly, the deflected feature elastically recovers, at least partially, to cause interference, interlocking, and joining. Figure 2.8 shows, generically, how an elastic interlock works.

In some cases, elastic interlocks rely on locked-in or residual stresses to resist unwanted movement, including separation, but not always (see Chapter 4, Section 4.1). That same elastic feature must be caused to elastically deflect in the opposite direction to permit disassembly. Any direction of motion other than that opposite the insertion or engagement direction is prevented by the rest of the geometry of the mating parts. Further, motion in the direction opposite the insertion direction, known as the *retention direction,* is usually precluded unless and until the elastic feature is reverse-deflected or damaged.

Being that they are intentionally designed to allow elastic deflection in some direction but not in all directions, elastic attachment features or interlocks provide the most effective joining when they are caused to resist applied service loads or internally generated loads in directions orthogonal to the direction in

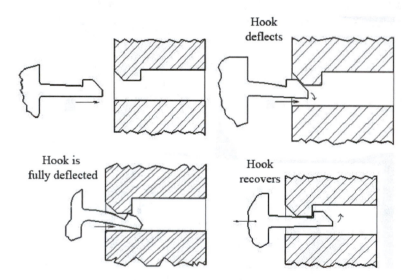

FIGURE 2.8 A schematic showing how a cantilever hook operates with a rigid catch. At the upper left, the hook is moved toward the catch to have its slanted face slide along the inclined ramp on the catch and deflect the beam elastically at the upper right. At the lower left, the hook is pushed further into the catch until, as shown at the lower right, the hook clears the back edge of the catch, elastically recovers to its undeflected position, and locks the pair.

---

[8] The case in which both parts or features in a mating pair deflect elastically is very complicated to analyze but is treated in a some design manuals (e.g., Bayer Polymer Division, 1992).

which the part was inserted into its mate. Resistance to release in the so-called "retention direction," opposite the insertion direction, is characteristically much greater than the force needed to accomplish insertion, full engagement, and interlocking. In fact, loads that act along the same Cartesian axis, but with the opposite sense, as the insertion direction, can usually cause unintended and unwanted disengagement of the elastic feature from the more rigid feature in the locking pair, unless some special precautions are taken to prevent this (see Chapter 4, Section 4.6). Resistance to movement in directions orthogonal to the retention direction can be made as great as desired by proper design of the rest of the mating part shapes.

The essential characteristic of elastic interlocks to be able to defect to engage suggests that they are best used in materials that exhibit elasticity and in shapes that exhibit flexibility, but not unacceptably low moduli or excessive flexibility, as then parts would deflect too easily. Candidate materials include most metals, the more rigid (as opposed to highly elastomeric) plastics or polymers, wood (although drying over time can cause the elasticity of wood to be lost), and reinforced polymer-matrix and metal-matrix composites. In metals, shape is often a key to deflection through moment of inertia, so such interlocks are found in sheet-gauge parts or thin sections or protrusions on castings, extrusions, and machined parts.

Examples of elastic interlocks abound and are described in detail in Chapter 4.

Figures 2.9 and 2.10 show several modern examples of elastic interlocks. So-called *snap-fit attachment* is widely used to assemble toys for children, especially toddlers. In modern office furniture, a variety of different elastic interlocks are employed in both plastic and metal parts to allow easy snap assembly.

2.6 FORCES AND MOTIONS FOR ASSEMBLY OF RIGID AND ELASTIC INTERLOCKS

Two characteristics of rigid and elastic interlocks make them particularly attractive as integral mechanical attachment methods. The first is that the force needed to cause part-to-part (which really means feature-to-feature) engagement and interlocking (when it occurs) is generally low. For loose-fitting rigid interlocks, this force can be extremely low, as not even part-against-part friction may be significant. For tight-fitting rigid interlocks wherein one part feature fits tightly with the other, the force of friction between surfaces must be overcome. This is usually not very great, although proper operation of the interlock once parts are fully engaged may not allow the use of lubricant during assembly. In other applications, however, lubricant is used not only to facilitate assembly but also to allow continued proper operation of the joint. A prime example is in the use of rigid interlocks in the guides or ways of machine tools, where a working table, for example, must freely move relative to the machine's base when the

FIGURE 2.9 The use of elastic interlocking "snap-fit" features in children's toys molded from thermoplastic allows easy assembly by parents, difficult disassembly by children, and safety from any potential choke hazard from loosened fasteners. Here, an assembled Little People Farm™ is shown. (Photograph courtesy of Fisher-Price, East Aurora, NY; used with permission.)

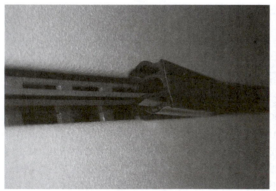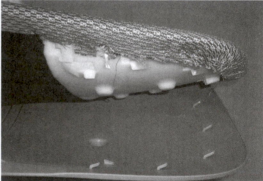

FIGURE 2.10 The use of elastic interlocking features in the assembly of modern office furniture is shown by: (1) flexible plastic extrusion that locks together Privacy Wall™ panels, on-site, while also serving as the trim (top left); (2) the connection of an upholstery sub-assembly to the outer shell for a Leap™ chair as molded-in plastic tabs on the outer shell mate with molded-in "dog houses" on the upholstered sub-assembly (top right);

FIGURE 2.10—cont'd and (3) the use of formed-in snaps in sheet metal to allow the dark piece of a lateral file drawer to snap into place with the side flanges once it is rotated downward (bottom). (Photographs courtesy of Steelcase, Inc., Grand Rapids, MI; used with permission.)

table is driven by an x-y traversing system. In extreme cases, usually for the highest load-bearing or most highly stressed applications, rigid interlocks may have to be forced together, as in press-fits. Here, one part may actually have to elastically or even plastically deform the other, albeit usually only slightly.

For elastic interlocks, design of the engaging features often intentionally limits the force needed to cause engagement. However, it should come as no surprise that in order to have certain types of elastic interlocks work properly, they must operate as springs to apply a sustained clamping force. In such cases, the insertion force can be high, in order to cause the elastic feature to deflect enough to develop the needed clamping force. In spring-type elastic interlocking, locked-in or residual stresses are used to maintain assembly. In most other types, such residual stresses are not used and are not desirable.

A major advantageous attribute of elastic interlocks, in particular, is that they are capable of resisting part separation with from 2 to 20 times (or more) the force needed to cause engagement during assembly. In other words, the retention force provided by elastic interlocks is always higher and often much higher than the insertion force.

The second attractive characteristic or attribute of both rigid and elastic interlocks is that they require only very simple motions to cause proper engagement and interlocking. Four fundamental simple motions used are: (1) push, (2) slide, (3) tip or tilt, and (4) spin or twist.

A "push" involves motion only in the direction needed to cause part-to-part engagement and locking. When a push motion is needed/used, it is not necessary for the opposing parts of the ultimate assembly to be in contact throughout the process of assembly. However, toward the end of the assembly process, the two parts must come into contact, and the push motion must usually continue until the parts are fully engaged and, possibly, are interlocked by some type of catch mechanism (see Chapter 4, Section 4.3). Application of force to cause insertion during assembly usually rises at the point that part-to-part contact occurs.

A "slide" involves motion only in the direction needed to cause part-to-part engagement and locking. However, when a slide motion is needed/used, it is necessary for the opposing parts of the ultimate assembly to be in contact throughout the process of assembly. The part being inserted by a sliding motion is placed in contact with the opposing part, which is usually stationary or fixed. (Such a part is typically referred to as a "base part" in assembly language.) Once the two parts are in contact, motion in the directions orthogonal to that needed for engagement are usually prevented by the geometry of the base part. The actual motion is otherwise identical to a "push."

A "tip" or "tilt" involves a rotation about a hinge that is usually more virtual than real. The part to be inserted into the base part is moved into position—and proper orientation—so that one edge contacts the base part along a line, with the rest of the part tipped upward and away from the base part. The part being inserted is then tipped toward the base part by a rotation about this line of contact until it is properly seated within the base part. Normally, a tip motion is used because the part being inserted will nest within some geometric features of the base part commonly known as "locating" or "locator features." Once seated, or fully engaged, the newly inserted part is restrained from moving in the two-dimensional plane separating the two parts. A tip motion is usually used either because of some geometric feature(s) of the base part that limits access of the part being inserted for simple push or slide motion *or* because bringing the two parts into contact along a line stabilizes the part being inserted.

Finally, a "spin" or "twist" involves rotation of the part being caused to engage with a base part around some axis normal to the eventual plane of contact between the two parts.

Figure 2.11 schematically illustrates the four fundamental simple motions used for the assembly of parts using rigid or elastic interlocks.

Some designers presume a fifth simple motion, namely a "pivot." In fact, a pivot motion is the combination of a "push" followed by a "tip." Hence, while it is, indeed, simple and widely used, it is not a fundamentally different motion from the other four.

As will be seen in Chapter 4, Section 4.5, simple motions are often combined to enhance the security or attachment, thereby preventing accidental disassembly.

Recognition and designation of the aforementioned simple motions, as well as the "pivot" motion, is properly attributed to P.R. Bonenberger from work done at General Motors in the late 1980's and early 1990's (Bonenberger, 2000). The extension of these to rigid interlocks, as well as the classification of integral mechanical attachments as rigid, elastic, or plastic types, is properly attributed to Messler and Genc (1998).

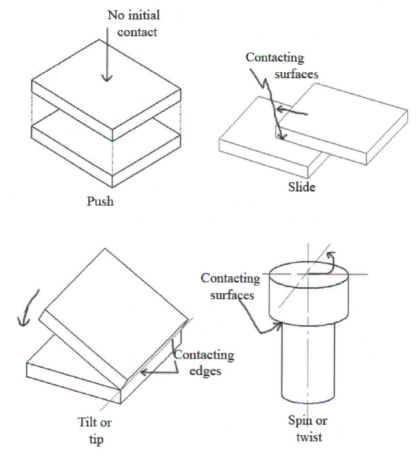

FIGURE 2.11 Schematic illustrations of the four fundamental simple motions used in the assembly of rigid or elastic interlocks.

The combination of simple motions and low forces to accomplish the assembly of parts using rigid or elastic integral mechanical attachments is the major driving force, and enabler, for the wide use of this method of joining with automation, particularly robots. Robots are most adept at performing very simple motions and are best suited to applying only low forces, albeit doing both tirelessly!

2.7 PLASTIC ATTACHMENT METHODS

It is possible to join one part to another mechanically not by using supplemental parts known as *fasteners,* but by using integral mechanical attachments. However, it may, at the same time, not be possible or practical to design and process in the features needed to accomplish such mechanical joining beforehand (i.e., in the detail parts before their actual assembly). This is the case for what is known as "plastic interlocks."

Without question, when it is necessary or desirable to join parts of an assembly or structure using integral mechanical attachments (or any other process, for that matter), the designer must always make some provisions for this. For joints that are to be mechanically fastened, welded, or adhesively bonded, design details are needed on the locations of joints, configuration of joints (including all critical dimensions, surface finishes, and essential preparations, as callouts), and methods of joint production (including details on the finished joint). For welding, this means the minimum and maximum size of weld crowns and roots for full-penetration welds and of fillet size and minimum depth of penetration for fillet welds. For adhesively bonded joints, it means the final thickness of adhesive, known as "bond-line thickness." And, for certain fasteners, like countersunk rivets in aircraft air-passage surfaces, it means limiting the amount of stick-out or recess above or below the skin surface of aerodynamics.

For rigid or elastic integral mechanical attachments or interlocks, these are designed in, with locations, types, and detailed geometry and dimensions all called out in the design. For plastic interlocks, the designer must stipulate the method of interlocking feature production, along with locations and, perhaps, sizes. The reason is that it is necessary or simply more convenient to create this type of attachment after parts are brought into contact and full engagement. Actual interlocking is accomplished by causing some plastic deformation in the mated parts.

Plastic interlocks or *plastic integral mechanical attachments* involve the actual plastic deformation of one part of an assembly into or around another part to lock the two together with purely mechanical forces arising from the resulting physical interference created by the deformation. As such, plastic interlocks are those that are processed into parts once they are brought into contact in the positions and orientations in which they are to remain in order to function properly. Interlocking features are, in all cases, created at the interface(s) between abutting parts by plastically deforming one or both part materials. What happens is that precisely interlocking protuberances and recesses are formed simultaneously (i.e., are co-formed) or, in some methods, protuberances or recesses are created down into existing recesses or over existing protuberances (i.e., the parts are conformed), respectively. Once formed, such interlocks tend to be quite strong, not simply because it would be necessary to undo the interlocks with a counter-force of the same magnitude as needed to produce the interlock, but because in metals, at least, there has been some strain hardening of the material during its deformation as well (see Figure 7.3).

What holds parts joined by plastic interlocks is not only the physical interference and interlocking of co-formed or conforming features, but also the added squeezing force or stress from the elastic component of the force or stress used to create the interlock plastically. This force is really due to friction between the intimately contacting surfaces of the two co-formed or conforming

Conformed Plastic Interlocks

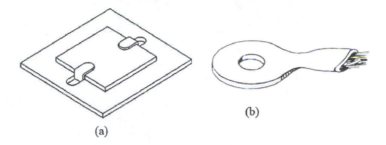

(a)

(b)

Co-formed Plastic Interlocks

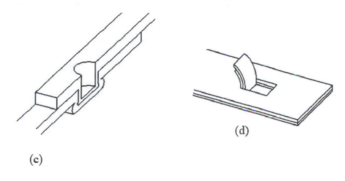

(c)

(d)

FIGURE 2.12 Schematic illustrations showing examples of conforming and co-formed methods for creating plastic interlocks; conforming locking tabs (a), conforming crimped connector body on wires (b), co-formed Tog-L-Loc™ metal clinch (c), and co-formed punched tabs (d).

parts arising from locked-in or residual stresses. For this reason, care must be taken to not lose these residual stresses during stress-relief or tempering heat treatments on the assembly.

Figure 2.12 schematically illustrates how plastic interlocks can be created between mating parts by co-forming the actual interlocking features or by conforming the parts to create interlocking features.

Since plastic deformation is essential for this type of integral mechanical attachment, this method is restricted to materials that can be easily plastically deformed *and* retain their newly created shape(s). The best example is metals, especially soft ductile (or malleable) ones. However, polymers (especially, but not only, thermoplastics) can also be joined this way, as can glasses.

Examples of plastic interlocks abound and are described in detail in Chapter 5.

Figure 2.13 shows how plastic interlocks, known as "hems," can be produced in the body panels of modern automobiles to lock decorative finished outer skins to structural inner skins or stiffeners.

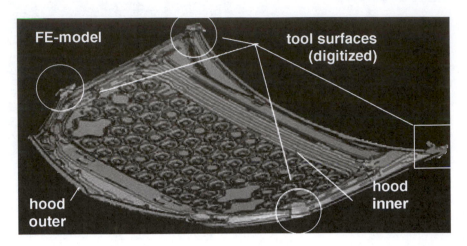

FIGURE 2.13 An artist's rendering showing a plastic interlock, known as a "hem," produced in metal body panels of an automobile. Here, the use of hems, or the process of "hemming," is shown for the hood of an automobile (viewed from the underside), with the finished outer panel being interlocked by plastically deforming it around the structural inner panel around the latter's perimeter. (Courtesy of General Motors Corporation, Detroit, MI; used with permission.)

2.8 ATTACHMENT METHODS VERSUS MATERIALS: IS IT THAT SIMPLE?

In Sections 2.5 and 2.7, rigid and elastic interlocks and plastic interlocks were introduced, respectively. In each case, the point was made that whether such an interlock was to obtain its locking by remaining rigid throughout assembly and service, by deflecting elastically to allow assembly and, often, partially recovering for sustained service, or by being created once parts were in proper position and orientation by plastically co-forming the parts at their interface or conforming one part over or into the other, influenced greatly the types of metals or other materials for which each could be considered. While the inherent ability of a metal used to make a part remain rigid, behave elastically, or form plastically following assembly is clearly important, the relationship between integral mechanical attachment methods and materials is not that simple.

In all design, there are multiple goals, which include (1) attainment of functionality, (2) provision of ease of manufacturing, (3) minimization of life-cycle cost,[9] and (4) provision of aesthetics, when that is required. Without question, however, first among these must be the attainment of functionality. What good are the others if the finished object doesn't function as required? So, in integral mechanical attachment (and all joining), attainment of required functionality is

---

[9] Total life-cycle costs include the costs of all raw materials used (including scrap), the cost of fabricating those materials into needed product forms (rolled plates, castings, forgings, extrusions, etc.), the cost of all processing required to produce the finished article (including the cost of the energy needed), the cost to maintain and otherwise support the product through its warranted service life, and amortized costs of research and development (R&D), design, marketing, sales, administration, etc.

paramount. That said, there are usually several ways of obtaining needed functionality. When that is the case, the other goals of good design become important, with ease of manufacturing often being next.

The fundamental inextricable interrelationship among a material's structure, properties, and processing quickly leads to the obvious conclusion that some methods of integral mechanical attachment are more suited to certain types of materials. As a corollary, of course, certain materials are obviously better suited to being joined by certain methods of integral mechanical attachment (or other processes, for that matter!).

Because the second half of this book deals with the integral mechanical attachment of various materials, type by type, suffice to say now that there are strong associations between the attachment method and the material.

Figure 2.14 gives the general relationships or associations between materials and integral mechanical attachment methods schematically. Virtually all ceramics, many metals, and the harder polymers, as well as some composites of these materials, are suited to rigid interlocks. Most metals, very few ceramics, and all but the softest polymers (e.g., elastomers), as well as some composites, are suited to elastic interlocks. The softer metals, all thermoplastic polymers, and few if any ceramics, as well as some polymer-matrix and metal-matrix composites, are suited to plastic interlocks.

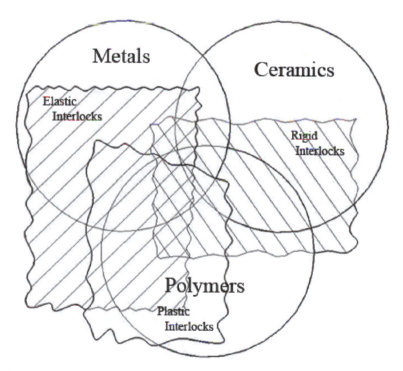

FIGURE 2.14 A schematic representation, in the form of a Venn diagram, of the relationship or association between materials and various integral mechanical attachment methods.

2.9 SUMMARY

An essential need exists in humans to classify things around us. Doing so facilitates our learning and our recall through associations and aids our extension of knowledge to the solution of problems by using recognizable familial relationships that arise from, besides being the basis for, a taxonomy.

Integral mechanical attachment is one of the two major sub-divisions of *mechanical joining,* the other being *mechanical fastening.* Both sub-processes use only mechanical forces (as opposed to atomic- or molecular-level bonding) to hold parts together. In integral mechanical attachment, geometric features that are part of the parts being joined (not separate specialized parts intended solely to accomplish joining, and nothing else, known as *fasteners*) cause physical interference upon part-to-part engagement and often result in interlocking to prevent part separation. Through interference and interlocking, parts are able to sustain or transfer loads without moving in unwanted directions, while, possibly, allowing relative motion in one or more other selected direction(s).

Integral mechanical attachments can be logically classified as *rigid, elastic,* or *plastic interlocks* based on how the feature actually responsible for attachment operates, performs, or, for the latter type, is created. Rigid and elastic interlocks are *designed-in* and fabricated into the detail parts well before their assembly, while plastic interlocks are *processed-in* features that cause interlocking of parts after their assembly. Such features may be *co-formed* at the interface between mating parts in the assembly or can be created as one part is made to *conform* over or into the other. Examples of all three abound in both the ancient and the modern world.

Finally, there is a logical, but not strict, relationship between the various types of interlocks and the major classes of materials in which they best operate, and vice versa.

REFERENCES

American Welding Society, *Thermal Spraying: Theory, Practice, and Applications,* American Welding Society, Miami, FL, 1985.

Bayer Polymers Division, *Plastic Snap-fit Joints: A Design Guide,* Bayer Polymers Division (formerly, Miles Mobay), Pittsburgh, PA, 1992.

Brandon, D., and Kaplan, W.D., *Joining Processes: An Introduction*, John Wiley & Sons, Ltd., Chichester, England, 1997.

Callister, W.D., Jr., *Fundamentals of Materials Science and Engineering: An Engineering Approach,* 2nd ed., John Wiley & Sons, Inc., New York, NY, 2005.

Genc, S., Messler, R.W., Jr., and Gabriele, G.A., "Selection issues for injection molded integral snap-fit locking features," *Journal of Injection Molding Technology,* 1(4), 1998, 217–223.

Messler, R.W., Jr., *Joining of Advanced Materials,* Butterworth-Heinemann, Stoneham, MA, 1993.

Messler, R.W., Jr., *Joining Materials and Structures: From Pragmatic Process to Enabling Technology,* Elsevier/Butterworth-Heinemann, Burlington, MA, 2004.

Messler, R.W., Jr., and Genc, S., "Integral micro-mechanical interlock joints for composite structures," *Journal of Thermoplastic Composite Materials,* 11(5), 1998, 200–215.

Parmley, R.O., *Standard Handbook of Fastening & Joining,* 2nd ed., McGraw-Hill, Inc., New York, NY, 1987.

Pocius, A.V., *Adhesion and Adhesives Technology,* Hanser/Gardner Publications, Inc., Cincinnati, OH, 1997.

3   RIGID INTEGRAL MECHANICAL ATTACHMENTS OR INTERLOCKS

3.1   HOW RIGID INTERLOCKS WORK

Integral mechanical attachments or integral interlocks are geometric features of two or more parts that when caused to come into contact and fully engage, prevent relative motion in certain directions of translation and/or rotation, including, when features physically interlock, unwanted separation in any or certain directions (see Chapter 2, Section 2.2). The interlocking features can pre-exist in the parts being brought together, having been created in the detail parts during their fabrication, or can be created once the parts are properly positioned in the intended assembly. The former situation occurs for both rigid and elastic interlocks, while the latter occurs for plastic interlocks (see Chapter 2, Section 2.3). *Rigid integral mechanical attachments* or *rigid interlocks* accomplish and retain assembly of parts by retaining their individual overall shapes *and* the shapes of the actual attachment or interlocking features throughout the process of part engagement, feature interlocking, and subsequent service as the assembly. Neither the parts nor the actual geometric features that allow and enable engagement and interlocking are intended to temporarily deflect as they do in elastic interlocks or permanently deform as they do in plastic interlocks.

To resist both temporary noticeable elastic deflection and permanent plastic deformation, rigid interlocks must be designed to remain rigid through the proper combination of their geometry (i.e., mass and shape-induced moment of inertia) and material (i.e., yield or flow strength for inherently ductile materials, fracture strength for inherently brittle materials, and elastic and/or shear modulus for all materials[1]). From the material standpoint, this demands that the material used in parts containing rigid integral attachment or interlocking features must have a high elastic limit, should preferably have a high elastic modulus in tension and shear,[2] and should be easy to process into shapes to allow the needed, sometimes complex, geometry of parts and attachment features to be produced.

While quite different fabrication processes and techniques must be used, most metals, virtually all ceramics (including cement, concrete, and natural and engineered stone[3]), most glasses, and the more inherently strong and hard (i.e., rigid) polymers can be used. Not surprisingly, metal-matrix, ceramic-matrix, carbon-matrix, and most polymer-matrix reinforced composites have the desired high elastic strength and modulus but often lack the needed fabricability. Also, such materials, as well as wood (a natural composite), can suffer from weakness in directions transverse (orthogonal) to reinforcing phases or constituents that are unidirectionally aligned.

The way rigid attachment or interlocking features operate is by having one part or part feature bear against another or others. The actual feature allowing attachment and/or interlocking may carry loads in tension or in shear (as different fasteners do), but there is also always a compressive bearing load or stress at the actual part-to-part or feature-to-feature interface. Thus, the bearing strength of the materials used in interlocking parts is also important.[4]

Figure 3.1 schematically illustrates how a rigid integral mechanical attachment or interlock works by developing and resisting a bearing load as it carries a tensile or shear load, or both.

Another normally unique characteristic of how rigid interlocks operate is that the only time they are under loading is when an externally-applied or internally-generated[5] load tries to cause movement between the engaged and interlocked parts. There normally is no locked-in or residual stress operating in properly designed and fabricated rigid interlocks. This tends not to be the case for elastic interlocks and is not the case for plastic interlocks (see Chapter 2, Section 2.3).

---

[1] The elastic modulus (E) and shear modulus (G) are related through Poisson's ratio (ν) for the material; as $G = E/2(1+\nu)$. Poisson's ratio generally has values between 0.3 and 0.5.

[2] For polymeric materials, the added requirement of high modulus is critical, as a high elastic limit ensures elastic versus plastic deformation but usually is accompanied by excessive flexibility as well.

[3] Natural stone, such as granite, is found in nature. Engineered stone is natural stone that has been crushed, mixed (possibly with other types of stone, for other properties), and reconstituted into a shape by employing a polymeric (usually thermosetting) binder. Examples are DuPont Zodiaq (Wilmington, DE) and Silestone (Stafford, TX) used for kitchen countertops.

[4] The bearing strength of a material is a complex, albeit intrinsic, property. It is affected by the yield strength of a metal or the flow strength of a polymer or glass, and the fracture strength, the hardness, and the modulus of the material. It is also very much affected by the tendency of the material to continue to deform under a constant sustained stress, which metals and ceramics only do at temperatures above about half their absolute melting temperature, but polymers do at much lower temperatures.

[5] Internally-generated loads can arise from self-induced inertial loads in a structure or assembly, from prime movers operating within or on a structure or assembly, or from non-uniform thermal expansion/contraction due to temperature gradients or from differential coefficients of thermal expansion (C.T.E.s).

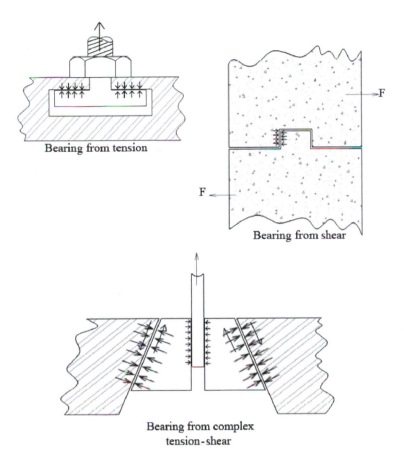

FIGURE 3.1 Schematic illustrations of how rigid integral mechanical attachments develop a bearing load as they carry a tensile load (upper left), a shear load (upper right), or both (bottom).

3.2 SUB-CLASSIFICATION SCHEMES FOR RIGID INTERLOCKS

Just as there are two different schemes for classifying integral mechanical attachments or interlocks in general (see Chapter 2, Sections 2.3 and 2.4), there are two parallel schemes for sub-classifying rigid integral mechanical attachments or rigid interlocks within themselves. One is based on the way in which the features of the interlocks are intended to achieve their joining function and the other is based on the way in which the features themselves are created in, on, or around the parts. Let's look at each sub-classification scheme, beginning with that based on intended means for achieving function.

Figure 3.2 shows the taxonomy for sub-classification of rigid integral mechanical attachments or rigid interlocks based on the way in which the actual features responsible for allowing and enabling part engagement and interlocking are intended to operate. The scheme is divided into five major categories or

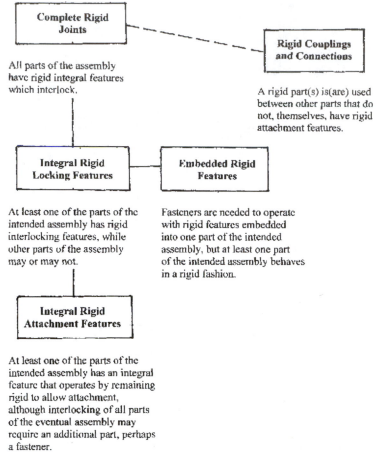

Complete Rigid Joints

All parts of the assembly have rigid integral features which interlock.

Rigid Couplings and Connections

A rigid part(s) is(are) used between other parts that do not, themselves, have rigid attachment features.

Integral Rigid Locking Features

Embedded Rigid Features

At least one of the parts of the intended assembly has rigid interlocking features, while other parts of the assembly may or may not.

Fasteners are needed to operate with rigid features embedded into one part of the intended assembly, but at least one part of the intended assembly behaves in a rigid fashion.

Integral Rigid Attachment Features

At least one of the parts of the intended assembly has an integral feature that operates by remaining rigid to allow attachment, although interlocking of all parts of the eventual assembly may require an additional part, perhaps a fastener.

FIGURE 3.2 A hierarchical classification scheme for rigid interlocks based on intended feature operation.

classes: (1) completely rigid interlocking joint elements or complete rigid joints; (2) integral rigid locking features; (3) integral rigid attachment features; (4) embedded rigid fastening features; and (5) rigid couplings and connections. These classes are arranged in Figure 3.2 to show some hierarchical, if not truly familial, relationships. What is meant by this is that the higher-level classes constitute a more complete joint, all of the components of which operate as rigid members. The further down in the hierarchy a class is located, either the less the rigid features in that class operate by themselves to allow and enable full attachment and any true locking *or* the more they depend on the use of another part or other parts to complete the assembly and allow full achievement of function.

The various classes listed in the preceding paragraph, and shown in Figure 3.2, are described individually in Sections 3.3 through 3.7. Each class is char-

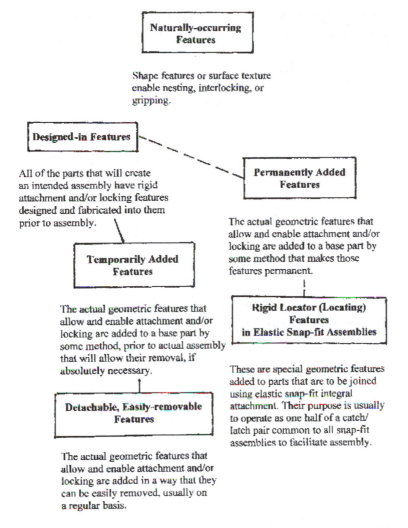

Naturally-occurring Features

Shape features or surface texture enable nesting, interlocking, or gripping.

Designed-in Features

All of the parts that will create an intended assembly have rigid attachment and/or locking features designed and fabricated into them prior to assembly.

Permanently Added Features

The actual geometric features that allow and enable attachment and/or locking are added to a base part by some method that makes those features permanent.

Temporarily Added Features

The actual geometric features that allow and enable attachment and/or locking are added to a base part by some method, prior to actual assembly that will allow their removal, if absolutely necessary.

Rigid Locator (Locating) Features in Elastic Snap-fit Assemblies

These are special geometric features added to parts that are to be joined using elastic snap-fit integral attachment. Their purpose is usually to operate as one half of a catch/latch pair common to all snap-fit assemblies to facilitate assembly.

Detachable, Easily-removable Features

The actual geometric features that allow and enable attachment and/or locking are added in a way that they can be easily removed, usually on a regular basis.

FIGURE 3.3 A hierarchical classification scheme for rigid interlocks based on method of feature creation.

acterized and the major embodiments or types are given and described by intended function and application.

Figure 3.3 shows the taxonomy for sub-classification of rigid integral mechanical attachments or rigid interlocks based on the way in which the actual features responsible for allowing and enabling part engagement and interlocking are created during either detail part or assembly (i.e., joint) creation. The scheme is divided into six major classes: (1) naturally-occurring features; (2) designed-in features; (3) permanently-added features; (4) temporarily-added features; (5) easily-detachable or easily-removable features; and (6) rigid locating features in elastic snap-fit attachment. As for the previous classification scheme based on intended feature function, these classes are arranged to show some hierarchical, if not truly familial, relationship. In this case, the further down in the hierarchy a class is located, the less permanent and more temporary the joints, if not the constituent parts and features, can be expected to be. At the lowest level, a key characteristic or attribute sought after in the joined assembly is the ability to make or break the engagement and locking at will for both planned and unplanned purposes. The class containing rigid locating features (also found within integral rigid attachment features in Figure 3.2) for use in elastic snap-fit assembly represents a special case in which the need for parts to remain together in an assembly until and unless they are intentionally disassembled varies.

The various classes listed in the preceding paragraph, and shown in Figure 3.3, are described in Section 3.9. Each class is characterized by its preferred areas of application. The various embodiments or types of rigid interlocks in Figure 3.3 are given in Sections 3.3 through 3.7. The special class of rigid locating features used with elastic snap-fits is described in detail in Section 3.8.

3.3 COMPLETELY RIGID INTERLOCKING JOINT ELEMENTS OR COMPLETELY RIGID JOINTS

At the highest level of the hierarchy of sub-classifications within rigid interlocks, based on how the parts and actual integral attachment features are intended to produce a joint, is the class of *completely rigid interlocking joint elements* or *completely rigid joints* (see Figure 3.2). This class is characterized by all of the parts comprising the joint having designed-in features that allow and enable part-to-part engagement *and* automatically enable interlocking in selected directions. Nothing beyond the mating parts of the joint to be created is needed to create the joint. The parts comprising the completely rigid interlocking joints tend to be robust and the joints tend to be strong and reliable.

The major embodiments or types of completely rigid interlocking joint elements are:

- Tongue-and-groove joints
- Dovetail-and-groove joints
- Rabbet (or dado) joints
- Dovetail-finger joints
- Mortise-and-tenon joints
- T-slot joints
- Shaped (machine) rails and ways
- Integrally-threaded parts
- Tabbed fittings and bayonet fittings
- Key-and-slot joints

Each type is described in the following subsections.

3.3.1 Tongue-and-Groove Joints

Tongue-and-groove joints consist of mating joint elements with a long, shaped- or profiled-protrusion, known as a "tongue," along one or more edges of one element, and a similarly long, shaped- or profiled-groove along one or more edges of the mating element. Tongues and grooves come in a variety of profiles or shapes, including square grooves, U-grooves, V-grooves, trapezoidal-grooves, and Z-grooves or stepped joints, singly and as multiples, the latter for supporting higher loads. In all cases, mating pairs of joint elements are assembled by a simple push motion (see Chapter 2, Section 2.6) applied perpendicular to the length of the tongues and grooves and in the plane of the normally flat joint elements. Because of their design and method of assembly, tongue-and-groove joints resist out-of-plane shear but do not resist shear parallel to the joint or tension in the direction opposite that which caused engagement. These joints can be enhanced in strength and resistance to movement in directions not constrained by the joint's geometry using adhesives or various mechanical fasteners.

The U-groove and Z-groove are most commonly used in wood, although the other profiles could be produced and used. The V-groove, trapezoidal-groove, and Z-groove (or stepped-joint) are all used in cement, concrete, natural and engineered stone, clay units (such as bricks and ceramic tiles), and in engineered or advanced ceramics (such as refractory fire brick). All three can also be found used in metal. A unique form of tongue-and-groove used in cement and concrete is the scored joint. The scored joint consists of multiple, parallel grooves or serrations, normally having a V-shape. A square-profiled scored joint is sometimes used with metal and in wood.

3.3.2 Dovetail-and-Groove Joints

Dovetail-and-groove joints, like their close cousins, tongue-and-groove joints, employ complementary-shaped or profiled joint elements. One element consists of a profiled-protrusion along one entire edge. The other element consists of a similarly-profiled groove along one entire edge along which mating is to occur. Because the dovetail (named for the shape of a dove's tail when splayed) has re-entrant angles, as does the mating groove, this type of joint can only be assembled by using a sliding motion in a direction parallel to the joint interface. Once assembled, the dovetail-and-groove joint resists both out-of-plane shear and in-plane tension normal to the length of the joint. The dovetail profile generally provides greater strength than the simpler-shaped tongue-and-groove profiles but is more difficult to produce. Because of the greater complexity of this geometry compared to those of tongue-and-groove joints, dovetails are rarely used on more than one edge, although it would not be impossible.

Dovetail-and-groove joints are used in wood for both building and furniture construction, and in metal in machine rails and ways (described later in Section

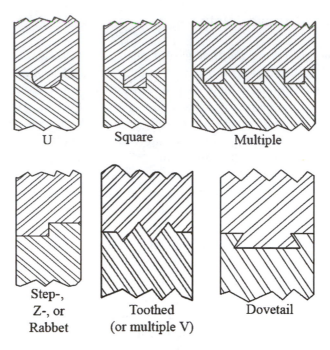

FIGURE 3.4 Schematic illustrations of different tongue-and-groove and dovetail-and-groove linear joints (i.e., for end-to-end or edge-to-edge joining).

3.3.7), and can be used in inherently rigid polymers. Because of the sharp radii at the root of the tail and the groove, this type of joint is not used with inherently brittle ceramics.

Tongue-and-groove and dovetail-and-groove joints are used to make end-to-end, linear splices and side-to-side, lateral assemblies. Related, but slightly different or altered, rigid feature shape combinations are needed to make corner (L-) joints, T-joints, or cross-joints.

Figure 3.4 schematically illustrates several different tongue-and-groove joints, as well as a dovetail-and-groove joint.

Figures 3.5 and 3.6 show examples of tongue-and-groove and dovetail-and-groove joints used in assembly of hardwood floors and fine wood furniture, respectively.

Three types of completely rigid interlocking joints are used at corners (i.e., L-joints) or T-intersections (i.e., T-joints); rabbet joints, dovetail-and-groove or dovetail-finger joints, and mortise-and-tenon joints.

3.3.3 Rabbet Joints (or Dado Joints)

Rabbet joints, also known as *dado joints,* consist of a long Z-shaped protrusion that normally is half the thickness of the joint element and fits into a recess of the same shape and dimensions as the protrusion's profile. When assembled,

FIGURE 3.5 Photographs showing how hardwood flooring (here oak) employs milled tongues and mating grooves along its edges and at its ends (left) to achieve interlocking (right). To prevent unwanted up and down movement of the flooring, the tongues are blind toe-nailed to the under-flooring using thin brads. (Photographs taken by Sam Chiappone for the author, Robert W. Messler, Jr.; used with permission.)

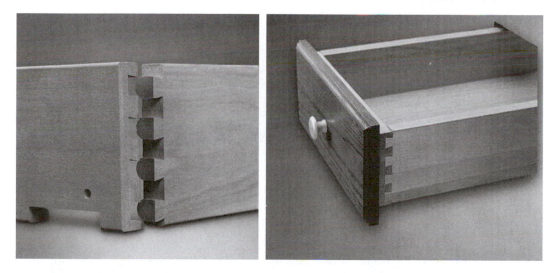

FIGURE 3.6 Dovetail-and-groove joints are milled into wood details (left) to allow simple and secure joining at corners in fine wood furniture (right). (Photographs courtesy of KraftMaid Cabinetry, Middlefield, OH; used with permission.)

using a simple push motion, the back-set vertical portion of the rabbet's profile abuts the face of the mating joint element containing the recess, creating either an L-(corner) or T-joint. These joints resist out-of-plane shear loads for the member containing the protrusion and in-plane shear loads for the member containing the slot. Rabbet or dado joints are widely used in wood and are occasionally used in metal. They are also found in pre-cast concrete units, particularly steel-reinforced concrete units such as those used on the vertical stanchions of concrete bridges.

3.3.4 Dovetail- or Dovetail-Finger Joints

When dovetail-shaped protrusions are made on one joint element, a corner joint can be created by appropriately reorienting the mating dovetail-shaped grooves in the other element. When this is done, the resulting joint, often called a *dovetail-finger joint,* is very strong and able to resist shear stresses perpendicular to and out of the plane of the joint, as well as tensile stresses in the direction of the element with the dovetail protrusions. It is also possible to use the dovetail-and-groove to produce T-joints. These joints are very common in wood furniture and, to a lesser extent, in architecture. This joint configuration is not found in metals.

3.3.5 Mortise-and-Tenon Joints

Mortise-and-tenon joints consist of a rectangular tang (called a "mortise") located along the centerline of one end of a joint element. This is fitted into a suitably sized and rectangular-shaped recess in the other element of the joint (called a "tenon") oriented at right angles to the first element. This joint type allows the construction of a T-joint or L-corner joint or even a cross-joint. Both through recesses (for *through mortise-and-tenons*) and blind recesses (for *blind mortise-and-tenons*) are made. Assembly, in all cases, is by a simple push. The resulting joints resist shear in both directions orthogonal to the direction of mortise insertion into the tenon. Mortise-and-tenon joints are the strongest of the corner joints and splice joints. Separating forces can be resisted by pegging (i.e., fastening) the mortise through the tenon. These joints are used exclusively in wood, especially for heavy construction.

Figure 3.7 schematically illustrates rabbet or dado joints, dovetail-finger joints, and mortise-and-tenon joints, which can actually be used for making linear splices, corners or tees. Figure 3.8 shows the use of a mortise-and-tenon joint in the construction of a wood building.

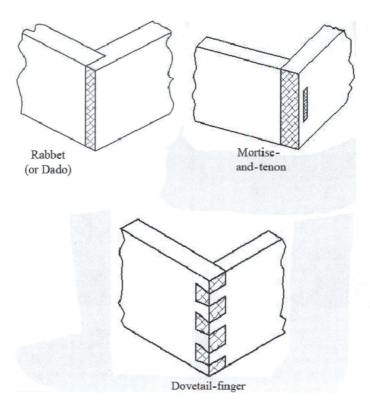

Rabbet
(or Dado)

Mortise-
and-tenon

Dovetail-finger

FIGURE 3.7 Schematic illustrations of rabbet-, dovetail-finger-, and mortise-and-tenon joints.

3.3.6 T-slot Joints

T-slot joints consist of a long, normally-fixed joint element containing a long inverted T-shaped recess or slot, into which is inserted a T-shaped pin or rail. More often than not, a simple T-shaped pin is used, in which case joint assembly is accomplished by rotating the head of the T-shaped pin to align with the narrow opening at the top of the slot (which is at the bottom of the vertical leg of the inverted T-shaped recess), inserting the pin down into the recess until its head can be rotated 90 degrees to prevent pull-out from the overhang of the inverted T-slot. Assembly motion thus involves a simple push followed by a simple twist. Alternatively, it is possible to have the T-shaped protrusion be in the form of a long rail, but this is unusual except in machine ways (see Section 3.3.7).

Once assembled, this joint configuration resists shear perpendicular to the slot within the plane of the normally fixed joint element, as well as tension attempting to pull the T-pin or rail out of the T-slot or recess. T-slots are used so that the pin's vertical leg points either into or out from the T-shaped recess.

T-slots are used in metal but are also commonly used in cement or concrete. When used in cement or concrete, such as in the floor of a machine shop, the

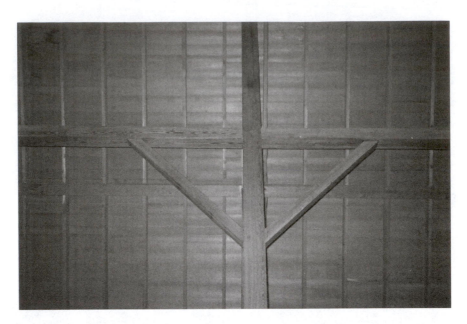

FIGURE 3.8 Mortise-and-tenon joints were—and are—commonly used in the erection of heavy wood structures. The structural framework of Our Lady of the Sacred Heart Catholic Church in Boothbay Harbor, ME, employ mortise-and-tenon joints, as well as other wood joinery techniques used by wood ship builders that constructed the church. While difficult to see, the Y-shaped members join the horizontal and vertical members with mortise-and-tenon joints augmented with wood pegs. (Photograph by the author, Robert W. Messler, Jr.; used with permission.)

inverted-T–shaped recess is cast in and a formed sheet-metal liner or sleeve is used inside the wall of the T-slot. This allows the bearing force of the T-pin to be better tolerated. A very common application is for mounting heavy machines to the floor of a shop without having to drill or otherwise excavate the cement or concrete. The machine is positioned over pre-existing T-slots in the floor, and T-pins with threaded ends are inserted into the slot so that they pass through holes in the mounting pads or feet of the machine. Nuts are then used to snug the machine onto the threaded end of the pin or stud, when it is pulled tight against the liner in the slot. Of course, this takes pre-planning in the design of the shop but, once done, offers tremendous convenience compared to jack-hammers and re-cementing.

Figure 3.9 shows T-slots in the working surface of a modern milling machine.

3.3.7 Shaped Rails-and-Ways

There are a number of completely rigid interlocking joints that are used to hold a moving table or other heavy part of a machine tool, such as a milling machine, to the machine bed or frame. These joints allow the table or other part to move

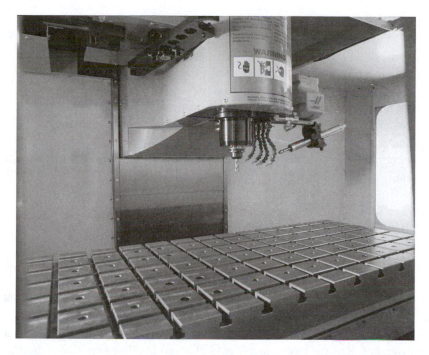

FIGURE 3.9 T-shaped slots (i.e., T-slots) are used on the working tables of machine tools such as milling machines to provide rigid attachments for the quick, easy placement of hold-down tooling employing inverted T-pins. (Photograph courtesy of Haas Automation, Inc., Oxnard, CA; used with permission.)

relative to the fixed bed or frame. Each joint allows motion only in one linear direction, but two joints can be arranged (one over, and at a right angle to, the other) to allow motion in two directions (e.g., x- and y-translation). The joint's design prevents movement in the orthogonal directions for one joint or direction for two joints at right angles to each other. Such joints are known as *rails-and-ways* or simply *ways* or *machine ways*. They typically serve as guides for motion and joints for preventing unwanted separation.

All designs consist of a protrusion on either the moving or the fixed part, typically known as a "rail," which fits with reasonable precision into a similarly-profiled long groove, typically known as a "way."[6] The shapes are chosen such that the protrusion is trapped in the recess. Hence, assembly of the rail into the way must be accomplished with a slide motion along their length. Shapes used for the rail include an inverted *T,* an inverted trapezoid, and a sideways *I*. All are used exclusively in metal.

Figure 7.24 shows an example of shaped rigid ways to hold a traversing unit to the bed of a modern machine tool.

---

[6] Needed movement is made easy and precise using any of several approaches including roller bearings, air-bearings, magnetic-levitation, and others.

3.3.8 Integrally-Threaded Parts

Occasionally, it makes sense to assemble two parts by having them thread together, in which case each needs to be integrally threaded. *Integrally-threaded parts* consist of one each with external and internal threads cast, molded, forged, or machined on. Assembly requires repeated multiple 360-degree twists. Once assembled, integral threads are able to resist shear in all directions and tension that tries to cause part separation. Examples are found in glass (e.g., food jars), ceramics (e.g., socket receptacles for some light-bulbs), and metal (e.g., light-bulb stems and pipes).

3.3.9 Tabbed- and Bayonet-Fittings

These completely rigid interlocking joints require assembly using a two-step, sequenced push-and-twist or, for some, a tilt-and-twist. One or more protruding tabs on one part (usually the one being assembled, known as the "mating part") must be aligned with appropriately matching recess(es) or notch(es) on the surface of the other (usually, fixed) base part, at which point the two are pushed together until the tab(s) reach(es) below an overhanging flange into a recess. At this point, the tabbed part is rotated (with a twist motion) to move the tabs out of alignment with the recess(es) or notch(es). An assembled *tabbed-fitting* is able to resist shear in two orthogonal directions and tension that tries to separate the parts. In a special example, known as a *bayonet-fitting,* rotation is continued until it is prevented at some point by a mechanical stop within the recessed groove. Sometimes reverse rotation is prevented by some type of spring-actuated stop.

A bayonet-fitting can also consist of a short slot cut through the wall of and parallel to the axis of a cylindrical tubular body or part feature and a short-length slot cut in the circumferential direction. A mating part with a projecting tab, pin, or button can be inserted into the axial slot and pushed until the projection seats at the bottom end of that slot. By then rotating the mating part so that the projecting tab, pin, or button runs along the short circumferential slot until it comes to a stop at the end of that slot, the mating part is securely locked to the base part. Such fittings offer secure attachment against accidental disassembly because for the mating part to separate from the base part it would have to undergo a reverse twist, followed by a short pull in the proper sequence and with the proper forces. Having that occur accidentally is unlikely. This added measure of security is the driving motivation behind using combined assembly motions (see Chapter 4, Section 4.5).

Tabbed-fittings are used in glass and ceramics, including well-known examples in fired-clay crock lids-to-bodies. In applications such as a ceramic crock or china sugar bowl, once the tab is aligned over the mating notch, the lid is normally tilted rather than pushed and is then given a twist. When they are used in glass or ceramics, all internal and external radii must be made as generous as possible to prevent concentration of stress. Bayonet-fittings are used in metal, with the best-known example being for the attachment of lenses to single-lens reflex cameras.

3.3.10 Key-and-Slot Joints

Key-and-slot joints consist of one or more protruding feature(s) known as *key(s),* which insert into matching slot(s) or recess(es) in the mating part of the assembly. The most common application is in pre-cast cement or concrete units, in fired-clay units, or in engineered ceramic tiles. By keying with one another, relative in-plane motion is prevented. Assembly requires only a simple out-of-plane push, so accidental or intentional disassembly is possible with an opposite out-of-plane force.

Figure 3.10 schematically illustrates integrally-threaded joints, tabbed fittings, and key-and-recess joints.

Figures 3.11 and 3.12 show examples of integrally-threaded joints in glass and key-and-recess joints in pre-cast cement paving blocks.

Nothing beyond the mating parts of the joint to be created is needed to create a key-and-recess joint. The parts comprising the completely rigid interlocking joints tend to be robust and the joints tend to be strong and reliable.

3.3.10.1 *Naturally-Occurring Rigid Interlocks*

Before leaving this sub-class, it is important to mention that most naturally-occurring interlocks, whether assembled by Nature (e.g., wedged rocks), by animals (e.g., beavers and various birds), or humans, operate by remaining rigid. There are three ways that natural objects can interlock by remaining rigid. First, the shapes of the objects brought together can be such that they nest with one another to cause locking, at least in some directions. The best examples, by far, are the walls of stones found throughout New England in the United States, in particular, but also elsewhere. In New England alone there are more than 240,000 miles of them! Second, natural objects can entangle, such as the sticks in a beaver's dam. Even though they remain rigid, by entangling simply from the complexity of their shapes, interlaced sticks can create a formidable structure. Third, rough surfaces can act, analogous to knurling, to grip and cause joining. This too is found in stone walls.

3.4 INTEGRAL RIGID LOCKING FEATURES

The class of *integral rigid locking features* differs from the preceding class of completely rigid interlocking joint elements (see Section 3.3) in that the parts involved allow attachment and locking to another part or parts, but that(those)

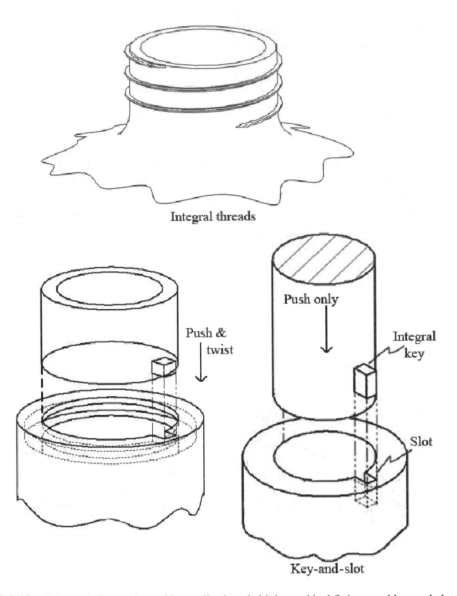

Integral threads

Push &
twist

Push only

Integral
key

Slot

Key-and-slot

FIGURE 3.10 Schematic illustrations of integrally-threaded joints, tabbed fittings, and key-and-slot joints.

other part(s) may or may not contain integral rigid geometric features for attachment and locking. In the class of completely rigid interlocking joint elements, both (or all) parts to comprise the particular joint(s) have integral rigid geometric features. Moreover, for completely rigid interlocking joint elements, both (or all) joints elements are actually parts with a primary function other than simply allowing joining. Some, though not all, parts classified as integral rigid locking features have the primary function of allowing joining and may or

FIGURE 3.11 A photograph showing integrally-threaded joints in glass; here to allow a threaded cap to be attached to a bottle or jar. (Photograph courtesy of Schott North America, Inc., Schott Technologies, Elmsford, NY; used with permission.)

FIGURE 3.12 Key-and-recess joints are used in pre-cast cement paving blocks (or pavers) to permit interlocking. Here, a more elaborate, decorative lizard design (reminiscent of designs by graphic artist M.C. Eschler, 1898–1972) is used to create interlocking among pavers in the entry walk at a restaurant in Kuau Cove, Paia, Hawaii. (Photograph courtesy of the Portland Cement Association, Skokie, IL; used with permission.)

may not, beyond that, have another specific function in the assembly. Another way of highlighting the difference is that with completely rigid interlocking joint elements, the key point is that they are always complete parts, in and of themselves, that can be rigidly interlocked. With integral rigid interlocking features, the emphasis is on the rigid features that allow interlocking between or among parts, not on the parts themselves. The difference is subtle and is mostly one of degree more than substance perhaps, but in many, if not all, cases the subtlety is worthwhile.

As shown in Figure 3.2, there are four major sub-classes within the class of integral rigid locking features: (1) splines and integral keys; (2) knurled surfaces; (3) wedges and Morse tapers; and (4) hinges, hasps, hooks, latches, and turn-buckles, kind of a catchall sub-class. Each sub-class is described below.

3.4.1 Splines and Integral Keys

Splines and *integral keys* are related in that they both have either recessed or protruding rigid features that interlock with some mating part in order to transfer loads in torsion within an assembly in which there is some rotation. The recesses or protrusions ensure that the mating parts of the assembly are locked together to prevent any rotation between those parts. Splines are cylindrical shafts with a radial array of side-by-side recesses or grooves cut into the surface, parallel or at a slight angle to the shaft's axis, and usually only at one end of the shaft. Splines differ from keyways by consisting of multiple recesses or grooves, rather than a single recess or groove. The profile of the grooves in a spline can be round-bottomed *U*'s, somewhat concave-sided *V*'s with a radius at the bottom, trapezoidal, or square, just as they can be for a keyway. The part that mates with and locks with a spline (or a keyway) has similarly profiled raised ridges or teeth (or ridge or tooth). An *integral key* is used with a keyway and has the same profile as the keyway. Splines, having multiple rigid locks, have more bearing surfaces to carry the torsion load in bearing and, so, can transmit much higher torsion loads than integral keys and keyways.

Both splines and integral keys have to be assembled to or with their mating component in the intended assembly using a simple slide motion.

Splines are used predominantly in metal, but they used to be and still can be used in wood. Integral keys are also used in metal but were and can be used in wood.

3.4.2 Knurled Surfaces

A *knurled surface* consists of a pattern of criss-crossed shallow grooves cut perpendicular or at a slight angle off perpendicular to one another. When these grooves are V-shaped (as they usually are), they produce somewhat pointed raised tetragonal square or diamond-shaped prisms. The result is a surface capable of developing excellent traction when it is squeezed (under pressure) to slightly impress into a somewhat softer mating part. As such, the knurled surface is not a part in and of itself but is a rigid feature on a rigid part. The motion needed to accomplish joining between the knurled surface (or part) and the mating part is a simple push.

Knurling[7] is used on metal parts, which can have flat or curved and even round surfaces.

3.4.3 Wedges and Morse Tapers

Wedges are tapered rectilinear prisms or prismatic shapes used to force one part tight against another. To operate, the wedge is pressed or driven into the space or gap between the two parts to be fixed tight with each other. The geometry of the wedge results in a slight incremental movement perpendicular to the direction of insertion, out of the plane of the wedge's inclined or flat surface, as opposed to its end surfaces. Greater insertion produces incrementally greater squeezing or wedging force. The motion required to install the wedge is a simple push.

Wedges can be made of metal, wood, or even the more rigid polymers, and any of these can be used with any combination of metals, wood, ceramics (including cement or concrete), or somewhat rigid polymers, depending on structural demands and environmental factors.

Morse tapers are a special form of wedging in which oppositely-inclined wedges, operating in sets of two, bear against and slide across one another's inclined surfaces at a common interface. When used in two sets of two, they create a gripping action on a normally flat part sandwiched between them. The gripping action (or squeezing force) increases as the sandwiched part is pulled as if to remove it from between the inner wedges. Figure 3.13 schematically illustrates the operation of Morse tapers in a tensile testing machine, for example. Morse tapers are also used as chucks in some machine tools for holding parts. To increase the traction (or friction) against the sandwiched part, the surface of the inner pair of wedges used in Morse tapers is usually knurled (as shown in Figure 3.13).

Morse tapers are made of metal and can be used against softer metals, wood, and polymers.

3.4.4. Hinges, Hasps, Hooks, Latches, and Turn-Buckles

This seeming catchall sub-class within rigid mechanical attachments has more in common within it than might appear, and there is a distinct characteristic that

---

[7] "Knurling" is the process used to create a knurled surface on a part.

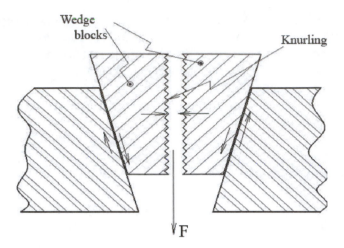

FIGURE 3.13 A schematic illustration of Morse tapers showing their operation in a chuck or grip mechanism for holding test coupons in a tensile testing machine, for example.

differentiates the entire sub-class from other sub-classes within the class of integral rigid interlocking features.

What *hinges, hasps, hooks, latches,* and *turn-buckles* share in common are: (1) they are all used to join and lock parts to which they are attached, usually by mechanical fasteners; (2) they all involve, within themselves, multiple components; and (3) they all operate by remaining rigid (or, at least, having their components remain rigid) throughout the engagement, locking, and service stages of assembly. The distinct characteristic that differentiates the entire sub-class from other sub-classes within integral rigid interlocking features and from other classes of rigid interlocks is that they all operate as supplemental parts with the specific purpose of allowing and enabling joining and locking. In this respect, they could be considered mechanical fasteners, not being, themselves, integral to parts comprising an assembly. In fact, to operate, some of these devices actually require a fastener, beyond those usually needed to attach the device's components to the parts to be interlocked.

The occurrence of such sub-classes that seem to be neither "fish" nor "fowl" is not unusual in classification schemes and taxonomies. Just think of the platypus!

So, let's look at these one by one.

3.4.4.1 *Hinges*

Hinges consist of two plates or leaves[8] joined by a pin about which both plates are free to turn. The plates are usually provided with holes to allow them to be

---

[8] There are one-piece hinges made from flexible polymers, known as "elastic hinges." In these, two heavier, plate-like portions are integrally attached to each other by a thinner, flexible web. The thin flexible web acts as the elastic hinge point.

attached to the structural elements with which they operate using fasteners, typically self-tapping or machine screws, but possibly nuts and bolts or rivets. (While nails are sometimes used to attach hinge plates to wood, it is an improper use of nails, which cannot resist tension well and might pull out.) The hinge permits one member to swing through an angle of up to 180 degrees with respect to the main structure.

There are, in fact, a number of forms and types of hinges including "double-acting hinges," "flush-mount hinges," "torsion-spring hinges," and "piano-type hinges." *Double-acting hinges* consist of three plates, the center one of which is joined at its edges to each of the other two by pins, about which they are free to turn. Each end plate is attached to a structural member. This type of hinge allows near–360-degree rotation of one structural member relative to the other. *Flush-mount hinges* are fit into recesses cut into the parts being hinged, so that the hinge plates do not interfere with allowing those parts to close completely at the hinge's joint. *Torsion-spring hinges* use a spring at the pin joining the two plates to force the hinge to return to a closed position automatically. Well-known examples are the spring-type hinges used on kitchen cabinets. *Piano-type hinges* are typically very long hinges with many, many interlocking loop features. They are designed to operate with long, often heavy, parts, such as the lid of a piano.

Hinges can be made of metal, polymers, or even leather or fabric and are used with metal, wood, and polymers.

3.4.4.2 Hasps

Hasps are devices arranged to accommodate a locking pin or padlock and are typically used on doors or covers. A hasp consists of an eye, which is usually attached to the fixed part of the assembly, and a hinge, one plate of which attaches to the door or cover to be opened and closed, while the other plate swings freely so as to be able to engage with the eye. Once the swinging portion of the hasp fully engages the eye, a locking pin or padlock can be placed through the eye over the swinging plate, thereby preventing the swinging part of the assembly from opening. Sometimes the eye (also known as a "staple") is designed to swivel, to hold the swinging part to the fixed part of the assembly without needing a pin or padlock to keep the assembly closed. A pin or padlock can still be added for additional security against opening. The assembly motion is always a push, although it occurs as a rotation about the fixed hinge's pin.

Hasps are almost always made of metal but can be used with wood or metal.

3.4.4.3 Hooks

Hooks typically operate with eyes and can take many different forms. Some serve as locking devices on doors, covers, etc., and some as anchorages for the attachment of wires, cords, ropes, or cables. Hooks are almost always made of metal but can be used with metal or wood or polymers. The motion to engage a hook with an eye is a simple push, although it is almost always performed around (and, therefore, restrained by) fasteners.[9]

3.4.4.4 Latches

Latches are commonly used for securing doors, covers, or panels in the closed position. They may operate from one side only or from both sides of the door, etc. They always involve one moving and one fixed component, which operate with each other, once engaged, by remaining rigid.

Latches are usually made of metal, but they used to (and still could) be made of wood. They could also be made of a rigid polymer. The motion to cause the engagement and locking of latches often involves more than one of the simple motions, in some specific sequence. By combining simple motions this way, the likelihood of the latch accidentally opening is greatly diminished (see Chapter 4, Section 4.5).

What hinges, hasps, hooks, and latches share in common is that they all require two (or sometimes more) components that operate by remaining rigid throughout engagement, locking, and service *and* they must all be attached to the parts they are to interlock using fasteners.

3.4.4.5 Turn-Buckles

Turn-buckles are interesting in that they operate like a threaded fastener, but they are actually composed of three rigid parts that must operate together for the entire device to operate as a rigid interlocking feature of an assembly. Turn-buckles consist of two rigid rod-like end members with opposite-pitch external threads, that is, one right-handed and one left-handed, and a central connecting piece between these, each end of which contains opposite-pitch internal threads. The outboard ends of the two outer pieces are typically fitted with clevises, loops, or hooks to enable attachment to parts of a structure that are to be tied together by the turn-buckle. By turning the central piece in one direction, the end members pull together to tighten the tie. By turning the central piece in the opposite direction, the end members move farther apart to loosen the tie. Hence, by rigidly attaching each of the end members to two structural parts, the turn-buckle enables those two parts to be tightly held in position or to be supported against outside loads. An example of where turn-buckles are used is to keep supporting cables from becoming slack as the bundled-and-twisted or woven wires making up the cable rearrange or elastically relax. Turn-buckles

---

[9] Actually, most hooks are threaded and, so, can be attached directly to the fixed part with which they are to operate. Such one-piece hinges in polymers are known as "elastic hinges."

are often used on the supporting guidewires for high radio or television transmission towers.

In an alternative design, the internally-threaded end members, each with opposite-pitch threads, can be designed as handles that can be turned on a single rod externally-threaded at each end with opposite-pitch threads. The operation is precisely the same as that design described above.

Turn-buckles operate by repeated twist motions and function much like a threaded fastener in that respect.

Figures 3.14, 3.15, and 3.16 schematically illustrate hinges, hasps and latches, and hooks and turn-buckles, respectively. Figures 3.17 through 3.19 show some examples.

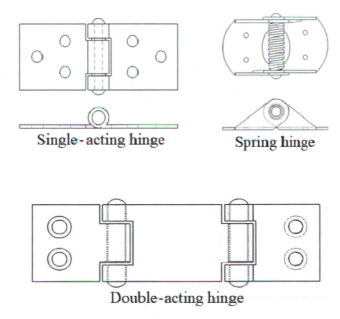

FIGURE 3.14   Schematic illustrations of some major types of hinges.

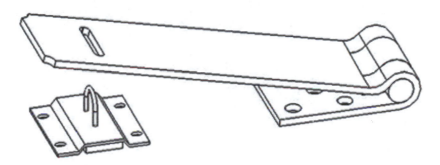

FIGURE 3.15   A schematic illustration of a typical hasp or latch.

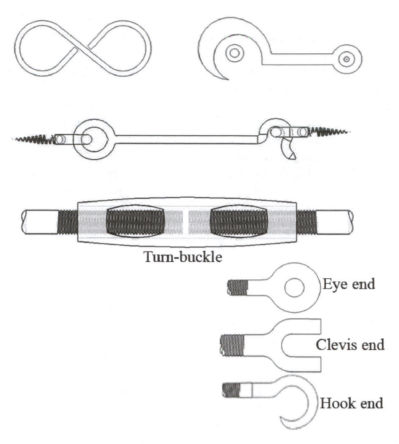

Turn-buckle

Eye end

Clevis end

Hook end

FIGURE 3.16   Schematic illustrations of some hooks and a turn-buckle, including various end attachments for the latter.

FIGURE 3.17   The use of hinges to attach doors in wood cabinets; a conventional two-plate type (left) and a spring type (right). (Photographs courtesy of Thomasville Furniture Industries, Thomasville, NC, and KraftMaid Cabinetry, Middlefield, OH, respectively; used with permission.)

FIGURE 3.18 Photographs showing the use of a latch in wood cabinets or furniture. Note the latch (left) and catch (right) mating parts of the latch, with the catch being designed to spring open elastically. (Photographs courtesy of Thomasville Furniture Industries, Thomasville, NC; used with permission.)

FIGURE 3.19 Photographs showing the use of turn-buckles. In the left-hand photograph, they are used to support the wood roof trusses in Our Lady of the Scared Heart Catholic Church in Boothbay Harbor, ME. (Photograph by the author, Robert W. Messler, Jr.; used with permission.) In the right-hand photograph, while not visible, they are used to keep guide wires properly tensioned on this transmission tower. (Photograph obtained from an unknown public website on radio and TV transmission towers.)

3.5 INTEGRAL RIGID ATTACHMENT FEATURES

Integral rigid attachment features differ slightly, but importantly, from integral rigid locking features in that they enable attachment of the part containing the integral feature but do not, solely from that feature, cause locking to another, mating part. There are four sub-classes of integral rigid attachment features:

- Flanges and shoulders
- Ears and tabs
- Bosses and lands
- Posts

Collars, which can perform the function of flanges or, particularly, shoulders, could be included within this sub-class when, as separate parts initially, they are made integral by some permanent or semi-permanent joining method, but are, instead, included under the sub-class of *rigid collars and connections* (see Section 3.7). Rigid locators or locating features used with integral elastic snap-fit attachments (see Chapter 4) are treated as a separate sub-class in Section 3.8.

3.5.1 Flanges and Shoulders

Flanges and *shoulders* are virtually identical integral geometric features of a part that allow that part to be joined to and locked with another part using either bolts and nuts or clamping rings. Normally found on cylindrical solid or hollow shafts or thick-walled pipes, both flanges and shoulders create a circumferential protruding ridge of diameter larger than the shaft or pipe through which bolts (for example) can be installed or over which a spring-type clamping ring can be applied. *Flanges* are always located at the end of a shaft or pipe, while *shoulders* are located elsewhere along the length of a shaft or pipe. Flanges tend to be used to join one shaft or pipe to another end-to-end but can also be used to attach an end cap (e.g., head) or other part. Shoulders tend to be used to allow the attachment of some other part to a shaft or pipe or as a structural feature for supporting the shaft or pipe.

Flanges and shoulders carry loads in torsion if the shaft rotates, or in shear, perhaps with bending, if not. Flanges, being located at the end of shafts or pipes may carry tensile loads, with bending occurring at their root radii.

Flanges and shoulders are most commonly used with metals but can also be found used with glass or more rigid polymers. They are occasionally used in ceramics. Shoulders, for example, are found on pre-cast cement or concrete pipes.

Figure 3.20 schematically illustrates integral flanges and shoulders, along with a press-fit collar, while Figure 3.21 shows an example of the use of flanges on heavy-walled pipes in a petrochemical refinery.

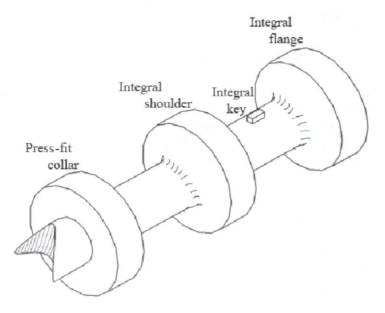

FIGURE 3.20   A schematic illustration showing integral flanges and shoulders, as well as press-fit collars.

FIGURE 3.21   The use of integral flanges for connecting end caps or heads to heavy-walled pipes, here by bolting, in a refinery. (Photograph courtesy of Marathon Ashland Petroleum, LLC, Findlay, OH; used with permission.)

3.5.2 Ears and Tabs

Ears and *tabs* are virtually identical. As the term "ears" suggests, these features protrude from the edge(s) or side(s) of parts. They can be used to attach or secure the part to a fixed base. Ears can be tied down using bolts threaded into internally-threaded parts or embedded inserts or using up-pointing threaded studs on or embedded in the base part. They can also be tied down using rivets or even over-hanging rigid clamps or integral tracks. Tabs tend to be used to interlock parts within a plane, much like key-and-recess or keyed joints described in Section 3.3.

Ears and tabs carry loads in shear, with or without bending, depending on the situation and their specific geometry. They can be found on metal, glass, ceramic, wood, and rigid polymer parts. For the more inherently-brittle materials, the radii at the point of feature attachment to the body of the part should be as generous as possible.

3.5.3 Bosses and Lands

Bosses and *lands* are rigid features that are raised above the plane of the part of which they are a feature. They are normally used to help locate, and possibly lock, another part in place on a planar surface. They do this by providing a rigid feature over which a part with a suitably shaped and sized recess can fit. The motion needed to cause assembly is a simple push. Bosses tend to be round, while lands can be of any planar shape.

Bosses and lands are used in metal, wood, ceramics (including cement and concrete), glass, and rigid polymers. They carry loads in shear or, possibly, when not round, in torsion.

3.5.4 Posts

Posts are rigid features having a relatively high height-to-width or height-to-diameter or aspect ratio. They are typically used to locate and fix the position of another part containing a hole to accept the post. In most applications, posts extend through the part being mounted.

Posts are used in metal, wood, ceramics, glass, and rigid polymers. The motion used to place a part over a post is a simple push. Sometimes the top of the post is plastically deformed at its end to lock the part placed over it in place against separation out-of-plane (see Chapter 5, Section 5.6).

Figure 3.22 schematically illustrates ears and tabs, bosses and lands, and posts. Figures 3.23 and 3.24 show examples of the use of ears and of posts, respectively.

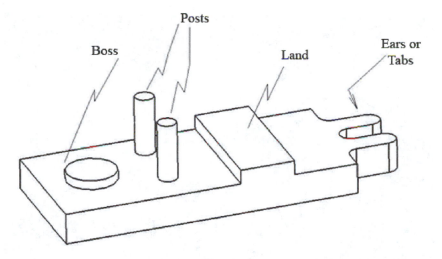

FIGURE 3.22 A schematic illustration showing ears or tabs, bosses and lands, and posts.

FIGURE 3.23 The use of ears on a cast iron machine base for mounting the machine to the shop floor. Also visible on top of the cast unit are the precision machined rigid, shaped rails and ways. (Photograph courtesy of Haas Automation, Inc., Oxnard, CA; used with permission.)

3.6 EMBEDDED RIGID FASTENERS

For some applications it is advantageous, and for some materials and loading situations it is necessary, to embed mechanical fasteners into or permanently attach them onto a part, thereby making them integral with that part. Doing so creates the sub-class *embedded rigid fasteners*. The three major types within this sub-class are:

- Interference-fit pins and keys
- Welded-on, cemented-in, and potted-in studs
- Cast-in and molded-in inserts

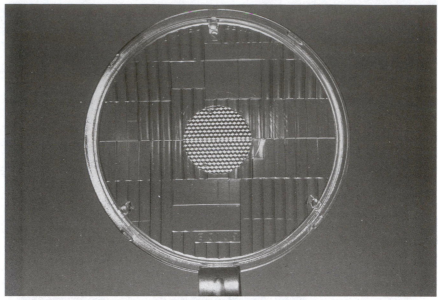

FIGURE 3.24 Molded-in, integral posts are used in glass and in plastic parts to allow mechanical attachment. Here, molded-in posts on glass headlight cover lenses (shown by arrow) are used to aid in beam alignment. (Photograph courtesy of Corning, Inc., Corning, NY; used with permission.)

3.6.1 Interference-Fit Pins and Keys

Interference-fit pins and *interference-fit keys* are embedded into a rigid part to create a rigid feature to be used for further attachment and/or locking of another part. By making the hole for a pin or a slot for a key slightly smaller than the cross-section of the pin or the key, the pin or key has to be forced-fit, thereby creating interference. This interference results in very tight gripping due to the frictional forces developed by the elastic component of the force needed to deform the hole or slot to accept the pin or the key. In addition, the associated compressive residual stress created in the pin or key tends to enhance its resistance to applied tension from shear in tension, bending, or torsion, especially in under-cycling or fluctuating fatigue loading.

Interference-fits can be produced by pressing in *press-fits* or by thermal expansion and contraction in *shrink-fits*. In the former, the pin or key is simply forced into the hole or slot that is too small to accept it without being deformed. In fact, in most cases, pins and keys must be fabricated in materials that will ensure deformation occurs in the base part, around the hole, as opposed to in the pin or key. Occasionally, a softer pin or key is forced into a harder base part to cause deformation to occur in the pin or the key. Examples are soft-metal, for example, Pb, anchor-bolts and threaded sleeves used in cement and concrete. In the case of shrink-fits, the pin or key is made or kept cool relative to the base part containing the hole or slot to receive the pin or key. This can be done by quenching the pin or key into cryogenic liquid nitrogen and/or by torch-heating the base part in the vicinity of the hole or slot. The pin or key is then quickly inserted with relatively little force being needed. Once the pin or key warms back up to room temperature and/or the base part cools back down to room temperature, the pin or key is locked into the hole or slot by the differential shrinkage.[10]

The motion to insert interference-fit fasteners into holes or slots is always a simple push but may involve a slide if the pin or key is guided into place by contact with the base part throughout the assembly process.

3.6.2 Welded-on, Cemented-in, and Potted-in Studs

Either threaded or unthreaded studs[11] can be made integral to a base part by any of a variety of processes. Metal, thermoplastic polymer, and glass studs can be welded onto metals, thermoplastic polymers, or glass, respectively. Metal studs can be either cast into lower-melting metals or cold-cast into ceramics (including cement and concrete) either during part fabrication or later using a compatible ceramic adhesive or mortar. Alternatively, metal studs can be potted into

---

[10] A particularly interesting use of thermal shrink-fitting was when the large (e.g., 16-inch) guns on World War II (WW II) U.S. Navy battleships were designed to allow their steel barrels to be relined with a softer steel sleeve containing the rifling grooves that are used to cause projectiles to spin to impart aerodynamic stability. The idea was to allow the liners to be removed once they became worn and be replaced with new liners. Unfortunately, no liners were ever removed because, more than 50 years after WW II, no one at the U.S. Army arsenal that built the guns at Watervliet, NY, knew how the liners were installed, no less how to get them out!

[11] A "stud," in the context meant here, is a threaded or unthreaded rod or an actual threaded bolt or unthreaded rivet or specially made part used to allow attachment by fastening once the stud is attached to the surface of or embedded into a base part.

metals or ceramics, including cement and concrete, using a polymeric adhesive known as a "potting compound."

Once installed, using a simple push motion, subsequent attachment of a mating part requires a push or a tilt to place the mating part over the stud(s). Threaded studs tend to employ nuts to lock on an attached part. Unthreaded studs tend to have their free end upset plastically to produce a locking "head" or are grooved to accept a retaining clip or are drilled to accept a Cotter pin. In thermoplastics and glasses, this process is known as "thermal staking" (see Chapter 5, Section 5.6). It is not at all uncommon to embed actual headed bolts or rivets (as opposed to using simple threaded or unthreaded rods) head down in the base part. The head then acts to further help lock the "stud" into place.

3.6.3 Cast-in and Molded-in Inserts

Cast-in inserts and *molded-in inserts* are used to create an integral feature in a base part for fastening another part to the base part. Inserts can be internally-threaded receivers or sleeves of any of a variety of types. These are used for anchoring a part using a threaded fastener like a bolt or screw. Alternatively, inserts can be unthreaded solid or split sleeves into which an unthreaded pin can be forced, or a harder, stronger threaded fastener can be used to create mating threads upon installation.

Inserts tend to be made from metal but could be made from wood or polymers. Metal inserts tend to be used with brittle ceramics, including cement and concrete, or soft, easily-deformed polymers, or wood. Wood inserts are occasionally used with cement and concrete. Polymer inserts can be used with metals or ceramics, including cement and concrete. In fact, there are plastic inserts used in gypsum drywall or in cement blocks, for example, to accept screws, which form mating threads during their installation.

Figure 3.25 schematically illustrates interference-fit pins and keys, welded-on, cemented-in, and potted-in studs, and cast-in and molded-in inserts. Figure 3.26 shows an example of welded-on studs.

3.7 RIGID COUPLINGS AND CONNECTORS

There are some parts that have the function of holding other parts of an assembly together, of linking parts of one assembly to another, and/or of supporting and transmitting loads and/or motion in the process. This is often their sole function, and, yet, they cannot be considered to be mechanical fasteners. They also are, for the most part, rigid. That is, they maintain the shape and dimensions under all types of loading for which they are designed.

For lack of any better designation, these types of parts are grouped in this treatment under the sub-class of rigid couplings and connectors. Included under this sub-class are:

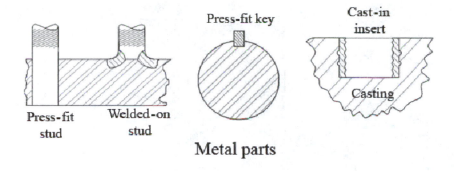

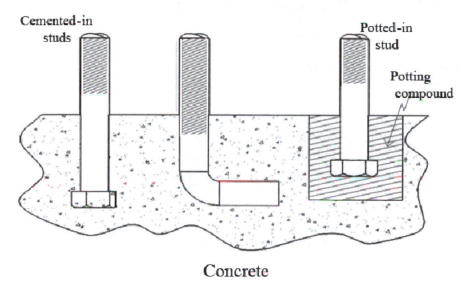

FIGURE 3.25 Schematic illustrations of interference-fit pins and keys, welded-on, cemented-in and potted-in studs, and cast-in and molded-in inserts.

- Collars and sleeves
- Couplings
- Clutches

The last category of parts is rather unique, as will be seen, in that they enable a connection to be made between parts or assemblies intermittently, as desired—that is, on or off.

3.7.1 Collars and Sleeves

Collars are rigid parts used instead of shoulders or flanges to provide a surface for holding other parts on a shaft. Collars differ from shoulders or flanges in that they must be added to rather than being an integral part of the shaft,

FIGURE 3.26 Threaded and unthreaded studs are used to allow rigid mechanical attachment. Here, unthreaded studs that are being percussion resistance welded to steel plates through access holes (left) can be seen down in the recess in a steel cover plate (right). (Photograph courtesy of Nelson Stud Welding, Elyria, OH; used with permission.)

although the degree to which a collar can be made permanent (and, thereby, integral) varies with the means by which it is attached. At one extreme, collars can be made permanent and fully integral by welding them to the shaft. Only slightly less permanent, in reality, are collars that are press-fit or thermal shrink-fitted to a shaft, in which they then act as if they were integral. Least permanent is when collars are attached to a shaft using one or more set screws, with multiple set screws or tapped holes in the shaft to receive the set screw(s) being used for higher loads. An advantage of screw-attached collars is that they are removable far more easily than shrink-fit collars, which are also removable. In all cases, screw-attached collars have lower load-transmitting capability than interference-fit collars, which, in turn, have lower load-transmitting capability than welded-on collars. If done properly (e.g., with sufficiently large welds), welded-on collars can approach the load-transmitting capability of integral shoulders or flanges.

Collars can also be held in place on a shaft by a suitable integral or added key and machined keyway. Such collars have load-transmitting capability close to welded-on collars.

The principal loads being resisted by collars are in the axial direction, although they can also resist torsion loads. Removable collars are especially useful where a fixed collar or shoulder would cause difficulty in assembling or disassembling the shaft.

Sleeves are rigid parts used on a shaft as spacers or for coupling between shafts when only light loads are to be transmitted. Sleeves can be held fast to a shaft using integral keys and keyways or by the wedging action of tapered end members drawn together by bolts and nuts located parallel to the shaft axis. These are known as "compression sleeves."

3.7.2 Couplings

Couplings are used for connecting lengths of shafts or pipes together. There are rigid types (i.e., "screw couplings") that use integral internal threads to tie to externally-threaded shafts or pipes together. There are also rigid types (i.e., "split-sleeve compression couplings") that are split axially into half-cylinders that are then held together by bolts and nuts, and some (i.e., "flanged couplings") that are bolted together at integral flanges at their ends. There is also a type (i.e., a "tapered coupling") used for transmitting somewhat higher torsion loads that consists of a double-taper and held together with bolts and nuts.

There are also so-called *flexible couplings* that are actually composed entirely of rigid components. All are designed to join shafts that are not perfectly aligned, so the shafts are subjected to reversal of stresses during each revolution of the shaft. Various types include (1) multi-jaw couplings, (2) the Torflex™ coupling, (3) the Neidhart™ spring coupling, (4) flexible disk couplings, and (5) various universal joints. The misalignments that can be accommodated may be angular, offset, or combined angular and offset.

Besides these flexible couplings composed of rigid components, there are couplings that are truly flexible. These too are intended to handle misalignments between shafts. There are thin-metal types, braided or woven-metal wire types, rubber and wire-reinforced rubber types, and leather types.

Figure 3.27 schematically illustrates some different types of rigid couplings and a couple of flexible couplings composed of rigid components, including a universal joint. Figure 3.28 shows a rigid coupling used in service.

3.7.3 Clutches

A *clutch* may be considered a coupling that can be separated during operation to start and stop the transmission of power. There are two general types, namely "positive clutches" and "friction clutches." *Positive clutches* are clearly rigid interlocking devices in that the driving and driven members engage each other by means of rigid jaws or teeth, as shown schematically in Figure 3.29. *Friction clutches* are ones in which the driving and driven members are in frictional contact only. The amount of friction is controlled by the amount of pressure of one face of the driving or driven member against the other.

Specific types of positive clutches, listed without description here, are:

- Band clutches
- Sleeve clutch
- Cutoff coupling
- Jawed or toothed clutches (see Figure 3.29)

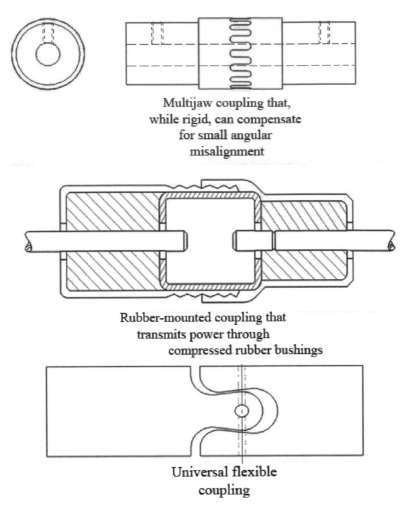

Multijaw coupling that,
while rigid, can compensate
for small angular
misalignment

Rubber-mounted coupling that
transmits power through
compressed rubber bushings

Universal flexible
coupling

FIGURE 3.27 Schematic illustrations of various rigid couplings and a typical flexible coupling comprised of rigid components, known as a universal joint.

- Square-jawed and spiral-jawed contact couplings
- Square-jawed and spiral-jawed sleeve clutches

Specific types of friction clutches, also listed without description here, are:

- Band clutches
- Disk clutches
- Conical clutches

There are also other clutches that use other means for transmitting power, including:

FIGURE 3.28 Typical applications of rigid couplings, here a positive square-toothed type clutch coupling is shown in a machine tool (left) and a coupling with integral flanges that are bolted is shown between a driving and driven unit in a refinery (right). (Photographs courtesy of Haas Automation, Inc., Oxnard, CA, and Marathon-Ashland Petroleum LLC, Findlay, OH, respectively; used with permission.)

- Electromagnetic clutches
- Fluid drive clutches or couplings
- Magnetic-particle "fluid" clutches

Details of the design and operation and application of various types of clutches can be found in design handbooks for engineers, such as that by Avallone and Baumeister (1996) or Shigley et al (2004).

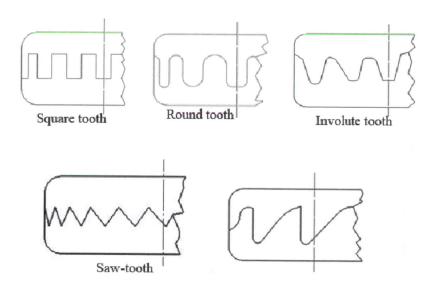

FIGURE 3.29 Schematic illustrations of positive clutch jaws and teeth used on face clutches.

3.8 RIGID LOCATING FEATURES (OR LOCATORS) FOR ELASTIC SNAP-FIT ASSEMBLY

In all joining, the constraint[12] of parts in their appropriate positions, orientations, and alignments is the goal and the most important measure of success. By being properly constrained, the parts of an assembly or structure can operate to resist, sustain, support, and/or transmit loads as required for proper functioning of the assembly and for reliable and safe structural integrity. Constraint is also obviously the most fundamental of the key requirements of integral mechanical attachments. In rigid integral mechanical attachments or rigid interlocks, that constraint comes from the rigid nature of the parts and/or the actual attachment and locking features involved in the joint. As will be seen in Chapter 4 on elastic integral mechanical attachments, typified by so-called "snap-fits," joining requires that a flexible locking feature deflect elastically during part-to-part assembly to allow engagement of the mating part and a base part, followed by the recovery of the locking feature toward its original, non-deflected position. Once this occurs, the physical interference needed to lock the parts together has been achieved.

Bonenberger (2000) defines *constraint features* in elastic snap-fit assembly as "the locking and locating features that actually hold the parts [of an assembly] together." He further properly divides constraint features in snap-fit assembly into two major groups: *locators* and *locks*. As the names imply, "locators" both ensure and facilitate that two parts of the intended assembly, in general, and the integral feature(s) being used to actually accomplish mechanical attachment, specifically, are properly fixed in terms of relative position, relative orientation, and relative alignment. "Locks" physically restrain one part (usually referred to as the "mating part" in the parlance of snap-fit assembly) to another (usually called a "base part"). In snap-fits, as will be seen in Chapter 4, Section 4.1, this is always accomplished using a combination of a "catch" feature on one part and a "latch" feature on the other part.[13]

A significant difference between locators and locks is that locators always operate by remaining rigid (or, at least, relatively rigid compared to locks), while locks operate by being flexible, at least during the engagement stage of assembly. It is because locators or locating features in elastic snap-fit assembly operate by remaining rigid that they are included in this chapter.

Before looking at the various types of rigid locators used in elastic snap-fit assembly, some basics need to be stated. Rigid locating features can be used individually to fix the position and orientation of parts and the alignment of flexible locking features in integrally-snap-fit parts of an assembly *or* they can

---

[12] "Constraint" in the context of joining means the complete prevention or the thoughtful control of relative motion between parts.

[13] A familiar example of a catch-latch pair that is a good analog of those used in snap-fit assembly is found between a door and its jamb.

be used in pairs known as a "locator pair." When used in pairs, rigid locators tend to remove more potential degrees of motion *and* do so more effectively. Also, as pairs, rigid locators become more sensitive to variations in dimensions of and within parts (i.e., to tolerances) and they tend to limit the choices among assembly motions (i.e., push, slide, tip, and twist). Most significantly, to aid the classification of this sub-class of rigid integral mechanical attachments, locators can be logically divided into two groups: (1) those that operate by protruding from a part and (2) those that operate by being recessed into a part. Bonenberger (2000) calls these "protrusion-like" and "void-like" locators.[14]

The major types of rigid locators used in elastic snap-fit assembly, grouped by protruding and recessed types are:

Protruding types:

- Lugs and tracks
- Tabs
- Wedges and cones
- Pins
- Lands[*]
- Catches

Recessed types:

- Holes
- Slots
- Cutouts
- Edges[*]

The operation of the above-listed locators is intuitively obvious from their geometry, so the various types are simply shown schematically in Figure 3.30 without further description. Again, more details can be found in Bonenberger (2000) as well as in an earlier treatment by Genc et al (1998).

3.9   JOINT AND ATTACHMENT FEATURE PERMANENCY

There are those who believe there are times when it is advantageous, if not also necessary, for an assembly or, more commonly, a structure, to be permanent. If that is what is desired, at least critical joints, if not all joints, need to be permanent. When permanence is a requirement in a joint, welding is the joining

---

[14] Bonenberger (2000) also includes a group he calls "surface-like," but these basically employ protruding or recessed features to operate, so do not appear to be fundamentally different, although the designation may offer some additional convenience.

[*] Bonenberger (2000) considers these as "surface-type" locators.

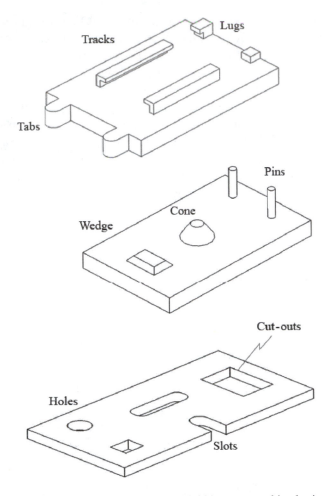

FIGURE 3.30 A schematic illustration showing various rigid locators used in elastic snap-fit assembly.

process of preference (Messler, 2004), although there are clear examples of very old structures held together with mortar, an adhesive.

There are really two questions that need to be asked about the permanency of joints, however. The first is Do we ever *really* want anything to last forever? The designers of the reactor facility at Chernoble, Ukraine, thought so until it became desirable—but impossible—to dismantle it! The second question, irrespective of the answer to the first, is If a joint is to be permanent, why would one ever use a mechanical approach to produce it? The single greatest advantage, and unique attribute, of any mechanical joining process is that it allows intentional disassembly.

So, putting the question of the desirability of permanency of joints aside, there remains the question of the desirability for permanency of integral attachment features used to accomplish what may well be intended to be a permanent joint. The answer to this question depends on, and should actually drive, the selection among the six sub-classes of rigid interlocks (see Figure 3.2).

All of the actual attachment features found in *completely rigid joints, integral rigid locking features,* and *embedded rigid fasteners* are essentially and ought to be permanent. For all of the specific methods within these three sub-classes, it may be desirable to disassemble the joint, but the rigid features themselves should virtually always be preserved. In fact, some of the methods under these three sub-classes were developed and are employed specifically because they allow easy intentional disassembly. Examples are: T-slots (to allow machines to be easily relocated); rigid rails and ways (to allow major components of machines to be easily exchanged); and integral threads and bayonet-fittings (to allow parts to be easily removed). Other examples, for completeness, are: splines; integral keys/keyways; hinges, hasps, hooks, and turn-buckles; and wedges and Morse tapers. A particularly attractive attribute of all types of *embedded rigid fasteners* is that the use of fasteners allows easy intentional disassembly, but by being embedded, the joint can easily be recreated.

Within *integral rigid attachment features,* only certain collars are intended to be removable (i.e., removable collars). Within *rigid couplings or connectors,* removable collars are intended to be removed, all rigid and flexible couplings (of both sub-types) are intended to be removable, as are all sleeves. Clutches are unique in that they are explicitly designed to allow the easy decoupling of the driving and driven members at will, and quickly.

In short, permanency is relative, in the first place, and an annoyance when that permanency of a rigid feature hinders the intentional disassembly of the overall device. The example cited earlier was the desirability of removable collars, versus permanent, integral shoulders or flanges, to allow removal of a shaft.

Figure 3.3 gives the taxonomy of rigid integral mechanical attachments or rigid interlocks in terms of the relative permanence of each sub-class, as well as by methods of rigid feature creation.

3.10 SUMMARY

Rigid integral mechanical attachments or *rigid interlocks* are those types that remain rigid throughout the engagement, locking, and retention in service stages of assembly. By remaining rigid, they have to be designed such that the geometric profile of one part (called the "mating part" when it is the one being added) will cause interference with the profile of another part (called the "base part," as it is normally held fixed while the mating part is added), with externally-applied and internally-generated loads carried by bearing of one against the other. Loads in the parts can be tension, compression, or shear from tension, compression, bending, or torsion.

Rigid interlocks can be classified based on the way the overall joint or actual attachment features creating the joint are intended to operate or, alternatively, by the method by which the actual attachment features are created. Within the former scheme of classification, there are: (1) completely rigid joints; (2) integral locking features; (3) embedded rigid fasteners; (4) integral rigid attachment features; and (5) rigid couplings and connectors. There is also a special subclass known as "rigid locators" that are used in integral elastic snap-fit assembly together with the elastic locking feature(s). Within the latter classification scheme, there are rigid interlocking features that are designed (and fabricated) in the joint elements or parts, permanently added to the joint elements, temporarily added to the joint elements, or detachable or easily removable features of joint elements.

Rigid interlocks can be found that are suitable for use in metals, ceramics (including cement and concrete), glass, rigid polymers, and even composites of all types, provided they are rigid.

Applications of rigid interlocks abound and have been used for centuries, if not millennia. Nothing suggests that will change, unless these fascinating and logistically simple methods for mechanically joining materials and parts proliferate beyond what has ever been seen before. Chapter 13 considers this distinct possibility.

Table 3.1 lists the types of rigid integral mechanical attachments based on their intended function, by major categories, while Table 3.2 lists processes and methods for creating rigid integral mechanical attachments or interlocks.

TABLE 3.1 List of Rigid Interlocks Based on Function (and Materials in Which They Are Found)

Complete Rigid Joints

- Tongues-and-grooves (wood; metal; cement/concrete; ceramics; glass)
  - V-groove (cement/concrete; ceramics; metal seal)
  - trapezoidal-groove (cement/concrete; ceramics)
  - Z-groove or stepped-groove (cement/concrete; ceramics; wood)
  - scored joint (cement/concrete; ceramics; metal)
- Dovetail-and-groove (wood; metal)
- Rabbett (or dado) joints (wood)
- Mortise-and-tenon joints: through- and blind-(wood)
- T-slots and Ts (or tabs) (metal; glass)
- Inverted trapezoid groove (metal)
- I-section groove (metal)
- Bayonet fittings (metal; glass; ceramics)
- Keyed (key-and-slot) joint (cement/concrete; ceramics)
- Integral threads (glass; fired ceramics)

(Continued)

TABLE 3.1 (*Continued*)

Rigid Couplings and Connections

- Collars, integral (also integral rigid attachment features) (metal)
  - welded
  - interference fit (press-fit or shrink-fit)
  - pinned (straight or tapered)
  - set screw
- Collars, removable (also integral rigid attachment features) (metal; wood)
  - set screw (metal)
  - multiple set screws (metal)
  - pegged (wood)
- Coupling, rigid (metal)
  - tapered
  - flanged
  - threaded (and, perhaps, tapered)
- Coupling, flexible (metal; rubber)
- Sleeves (as spacers) (metal; wood; ceramics)
  - compression fit (metal; wood)
  - wedged (wood; metal)
  - keyed (metal; wood)
  - split (and bolted) (metal)
- Clutches (metal)
  - sleeve
  - cutoff
  - jawed or toothed
  - friction
  - conical
  - electromagnetic
  - magnetic fluid

Integral Locking Features

- Splines (metal; wood)
  - square (wood; metal)
  - involute (metal; wood)
  - involute serrations (metal)
  - face-types (metal; wood)
- Integral key (metal) … related to key-and-slot joint in glass
- Knurled surfaces (metal) … on wedges, pins, studs, and shafts
- Hasps (metal)
- Hinges (metal; leather; polymer)
- Turn-buckles (metal)
- Wedges (wood; metal)
- Morse tapers (metal)

(*Continued*)

TABLE 3.1 (*Continued*)

Integral Rigid Attachment Features

- Shoulders (shrink-fit) (metal)
- Ears and tabs (metal; glass; polymers)
- Bosses (metal; ceramics; polymers)
- Lands (metal; ceramics; glass; polymers)
- Posts (metal; polymers; glass; ceramics)
- Collars, integral (also rigid couplings and connections) (metal)
  - welded
  - interference-fit (press-fit or shrink-fit)
  - pinned (straight or tapered)
  - set screw
- Collars, removable (also rigid couplings and connections) (metal; wood)
  - set screw (metal)
  - multiple set screws (metal)
  - pegged (wood)
- Rigid locator (or locating) features … used with snap-fits (polymers; metal)

Embedded Rigid Fasteners

- Press-fit (Interference-fit) pins (metal; wood)
- Press-fit keys (metal; wood)
- Welded threaded studs (metal)
- Welded unthreaded studs (metal)
- Cemented-in (or potted-in) studs (threaded) or posts (unthreaded) (cement/concrete; ceramics)
- Molded-in studs (threaded) (polymers)
- Molded-in inserts (threaded) (polymers)
- Cast-in inserts (threaded) (metals; cement/concrete; ceramics)

TABLE 3.2   List of Rigid Interlocks Based on Methods of Feature Creation (and Materials in Which They Are Found)

Naturally-occurring Features

- Interlocking or nesting shapes (stone)
- Entangling shapes (sticks)
- Gripping surfaces (stone)

Designed-in Features

- Tongues-and-grooves (wood; metal; cement/concrete; ceramics; glass)
  - V-groove (cement/concrete; ceramics; metal seal)
  - trapezoidal-groove (cement/concrete; ceramics)
  - Z-groove or Stepped-groove (cement/concrete; ceramics; wood)
  - scored joint (cement/concrete; ceramics; metal)

(*Continued*)

TABLE 3.2 (*Continued*)

Designed-in Features—Cont'd

- Dovetail-and-groove (wood; metal)
- Rabbett (or dado) joints (wood)
- Mortise-and-tenon joints. Through- and Blind-(wood)
- T-slots and Ts (or tabs) (metal; glass)
- Inverted trapezoid groove (metal)
- I-section groove (metal)
- Bayonet fittings (metal; glass; ceramics)
- Keyed (key-and-slot) joint (cement/concrete; ceramics)
- Integral threads (glass; fired ceramics)
- Rigid locator features (used with elastic snap-fit attachments) (polymers)

Permanently Added Features

- Shoulders (shrink-fit) (metal) ... could be temporary
- Ears and tabs (metal; glass; polymers)
- Bosses (metal; ceramics; polymers)
- Lands (metal; ceramics; glass; polymers)
- Posts (metal; polymers; glass; ceramics)
- Collars, integral (also rigid couplings and connections) (metal)
  - welded
  - press-fit
  - shrink-fit ... can be temporary
- Sleeves (as spacers) (metal; wood; ceramics)
  - compression fit (metal; wood)
- Press-fit (Interference-fit) pins (metal; wood)
- Press-fit keys (metal; wood)
- Welded threaded studs (metal)
- Welded unthreaded studs (metal)
- Cemented-in (or potted-in) studs (threaded) or posts (unthreaded) (cement/concrete; ceramics)
- Molded-in studs (threaded) (polymers)
- Molded-in inserts (threaded) (polymers)
- Cast-in inserts (threaded) (metals; cement/concrete; ceramics)

Temporarily Added Features

- Shoulders (shrink-fit) (metal) ... could be permanent
- Collars, integral (metal)
  - shrink-fit (can be permanent)
  - pinned (straight or tapered)
  - set screw

(Continued)

TABLE 3.2 (*Continued*)

Temporarily Added Features—Cont'd

- Collars, removable (metal; wood)
  - set screw (metal)
  - multiple set screws (metal)
  - pegged (wood)
- Coupling, rigid (metal)
  - tapered
  - flanged
  - threaded (and, perhaps, tapered)
- Coupling, flexible (metal; rubber)
- Sleeves (as spacers) (metal; wood; ceramics)
  - compression fit (can be permanent) (metal; wood)
  - wedged (wood; metal)
  - keyed (metal; wood)
  - split (and bolted) (metal)
- Hasps (metal) (can be easily removed)
- Hinges (metal; leather, polymer) (can be easily removed)
- Turn-buckles (metal) (can be easily removed)
- Wedges (wood; metal) (can be easily removed)

Detachable, Removable Features

- Clutches (metal)
  - sleeve
  - cutoff
  - jawed or toothed
  - friction
  - conical
  - electromagnetic
  - magnetic fluid
- Splines (metal; wood)
  - square (wood; metal)
  - involute (metal; wood)
  - involute serrations (metal)
  - face-types (metal; wood)
- Integral key (metal) … related to key-and-slot joint in glass
- Knurled surfaces (metal) … on wedges, pins, studs, and shafts
- Hasps (metal) (can be difficult to remove, but are temporary)
- Hinges (metal; leather; polymer) (can be difficult to remove, but are temporary)
- Turn-buckles (metal)
- Wedges (wood; metal)
- Morse tapers (metal)

(*Continued*)

TABLE 3.2   *(Continued)*

Rigid Locator Features (used with Elastic Snap-fit Attachment) (polymers)

- Lugs and tracks
- Tabs
- Wedges and cones
- Pins
- Catches
- Lands
- Holes and slots
- Cut-outs

REFERENCES

Avallone, E.A., and Baumeister, T., III., *Mark's Standard Handbook for Mechanical Engineers,* 10th ed., McGraw-Hill, New York, NY, 1996, 8-34-8-37 and 8-37-8-39 ("Dry and Viscous Couplings" and "Clutches").

Bonenberger, P.R., *The First Snap-fit Handbook: Creating Attachments for Plastic Parts,* Hanser Gardner Publications, Inc., Cincinnati, OH, 2000.

Genc, S., Messler, R.W., Jr., and Gabriele, G.A., "A systematic approach to integral snap-fit assembly," *Research in Engineering Design,* 10, 1998, 84–93.

Messler, R.W., Jr., *Joining Materials and Structures: From Pragmatic Process to Enabling Technology,* Elsevier/Butterworth-Heinemann, Burlington, MA, 2004.

Shigley, J.E., Mischke, C.R., and Brown, T.H., *Standard Handbook of Machine Design,* McGraw-Hill, New York, NY, 2004.

4 ELASTIC INTEGRAL MECHANICAL ATTACHMENTS OR INTERLOCKS

4.1 HOW ELASTIC INTERLOCKS WORK

An increasingly popular method of mechanical joining is "snap-fit fastening." While there are, indeed, fasteners that operate with what has come to be known as a "snap-fit" (see Section 4.2), most of the time when people speak of snap-fit fastening, the term is actually a misnomer because fasteners (i.e., supplemental devices specifically designed for joining parts together, without any other function) are not usually involved. Rather, elastic integral mechanical attachment features or elastic interlocks are.

As described in Chapter 2, Section 2.3, *elastic integral mechanical attachments* or *elastic interlocks* function by having a geometric feature on one part in a mating pair be designed so that it can and does elastically deflect when it comes into contact with a relatively more rigid geometric feature on the mating part. Once deflection of the elastic feature occurs and the insertion of the mating part, which is normally the one being moved, into the base part, which is normally the one that remains fixed or stationary, reaches a certain point, the designs of the deflecting and rigid features are such that some portion of the elastic feature clears some portion of the rigid feature and is able to recover at least partially, if not completely. When this recovery takes place, there is usually a distinctive "snap" that can be both heard and felt. This interesting and unique characteristic of these interlocks enables built-in quality assurance that successful part-to-part engagement *and* locking has occurred. In the recovered position, the detailed geometry of the engaging features act to lock the two parts

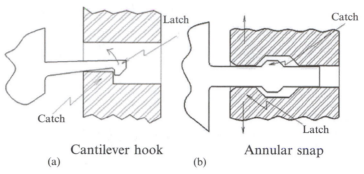

Cantilever hook	Annular snap
(a)	(b)

FIGURE 4.1 A schematic illustration of the role of a "catch-latch" pair of details for the operation of elastic snap-fit locking features, the hook as the latch and a ledge as the catch in a cantilever hook snap-fit feature pair (a) and the rigid ring as the catch and the flexible annulus as the latch in an annular snap feature pair (b).

together. Thus, the two features operate as a "catch" and a "latch,"[1] with one deflecting and the other remaining at least relatively rigid. In "catch-latch" pairs involved in *all* elastic interlocks, there is usually no particular difference as to which part does what. The concept of elastic snap-fits always involving a catch-latch pair was described and classified by Genc et al (1998a).

Figure 4.1 schematically illustrates how typical elastic integral snap-fit features operate as "catch-latch" pairs.

As will be seen in Section 4.2 on the sub-classification of elastic interlocks, there are actually many types. All rely on the elastic deflection of at least one part or part feature in a mating pair, but some primarily function to accomplish joining and secondarily, if at all, as parts with another primary function (carrying or transferring loads or forces, allowing motion, causing actuation, etc.). As such, the latter operate more like mechanical fasteners than like integral mechanical attachments. Pure fasteners that themselves operate by relying on elastic deflection and recovery, including the associated audible and tactile "snap," will be included in the sub-classification scheme but will not be described or discussed further. Suffice to say that there are such fasteners that can accurately be referred to as "snap-fit fasteners" (Messler, 1993). Most are clevis-like rivets or pins that are inserted into pre-prepared holes by squeezing some elastically deformable feature until the fastener is fully in place, and then releasing the squeezing force to allow the fastener to elastically recover, at least partially, and lock the assembly.

---

[1] The terms "catch" and "latch" are best exemplified by the features that allow a door to click into place in a jamb. A feature that can somehow deflect is caused to deflect by a relatively rigid feature on the mating part. For a door, the protruding feature is called the "latch" and the recessed feature is called the "catch." In general, either the catch or the latch can be elastic or rigid, so long as one behaves elastically and the other rigidly. Also, the catch or latch can, in the example of a door, be located on the door, as long as the mating feature is located on the jamb. For most doors, the latch deflects (as it is backed by a spring), and the catch (usually a cutout in a metal plate) is rigid.

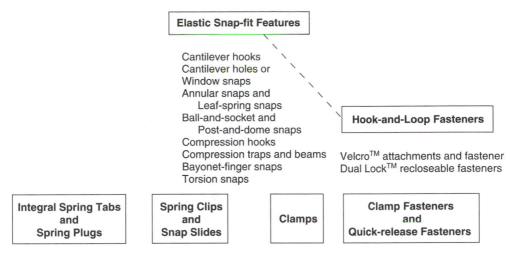

FIGURE 4.2 Taxonomy for elastic integral mechanical attachment methods.

4.2 SUB-CLASSIFICATION OF ELASTIC INTERLOCKS

Using the elastic deflection and at least partial recovery of parts or integral features of parts to accomplish assembly and locking is an old, diverse, and widespread method of mechanical joining, in general, and of integral mechanical attachment, in particular. Including approaches that actually employ true, fairly conventional fasteners,[2] Figure 4.2 gives a proposed taxonomy that attempts to reflect technical, if not familial, relationships. In fact, other than the close similarity between typical elastic snap-fit features and hook-and-loop fasteners, there is very little familial relationship. The sole similarity is the dependence of design features which operate to allow and enable attachment and locking via elastic deflection and at least some recovery; possibly with some remaining, residual stress.

The largest sub-class, by far, is also the newest application of elastic interlocks, namely so-called "plastic snap-fit features" or, simply, "plastic snap-fits." Contrary to the use of the term "plastic" in reference to integral mechanical attachments elsewhere in this book (e.g., Chapter 5), the term "plastic" here refers to the fact that these elastic integral features are used extensively (though no longer exclusively!) in parts made from plastics. Examples of the use of plastic

---

[2] True, conventional fasteners that operate using elastic deflection and at least partial recovery include: certain clevis-like (bifurcated) or two-piece tubular rivets, various clevis and spring pins, many eyelet/grommet pairs, virtually all retaining rings and clips, and some washers (e.g., spring-washers, Belleville washers, and curved and conical washers). Examples exist made of metal and made of more rigid polymers (Messler, 1993, 2004; Parmley, 1989).

snap-fits abound from parts of automobiles and aircraft to parts of computers and computer peripherals (keyboard, mouse, speakers, printers, scanners, etc.) to parts of children's toys and adults' appliances, furniture, and lawn and garden equipment. The variety of types will be described in Section 4.3.

So important and successful have elastic integral snap-fit features become in the assembly of parts fabricated from plastics, particularly parts injection molded from thermoplastics, that extension of the approach to other materials, most notably metals, but also to various reinforced composites, was inevitable (Goldsworthy and Johnson, 1994; Messler and Genc, 1998). Performance enhancement has occurred with plastic snap-fits by employing combined assembly motions for more secure part-to-part locking (Section 4.5) and through the use of certain designed-in enhancements (Section 4.6).

Closely related to "plastic snap-fits" are so-called *"hook-and-loop"* attachments.[3] While mostly found made from thermoplastic polymers, the possibility of analogs in higher-performance thermosetting polymers, polymer-matrix composites, and even metals has not gone unnoticed, unexplored, or unexploited (Messler and Genc, 1998). The actual features responsible for allowing joining are really just miniature versions of many of the types of features found in larger-scale plastic snap-fits. Hook-and-loop attachments are described in Section 4.7.

The remaining sub-classes of elastic integral mechanical attachments or elastic interlocks span a wide spectrum of sizes, geometric forms, materials of construction, applications, and degree of integration within a part or parts of the actual feature(s) responsible for allowing interlocking, as opposed to operating more as fasteners. In fact, the various sub-classes are sometimes difficult to relate to one another, but all share the common trait that they accomplish joining through the elastic deflection and at least partial recovery of one part or feature of a part with another. The various sub-classes, from most to least integral to a part, are:

- Integral spring tabs
- Spring plugs
- Snap slides and snap clips
- Clamp fasteners
- Clamps
- Quick-release fasteners

There are almost certainly other devices (e.g., toggles and certain types of latches) that could be included under the classification of elastic interlocks, but what is covered in this chapter thoroughly covers the operating concepts, if not the full diversity of specific embodiments.

---

[3] Hook-and-loop attachments are sometimes called "hook-and-loop fasteners" because their sole function, as parts, is to allow joining. Nothing else!

4.3 ELASTIC INTEGRAL SNAP-FITS USED IN ASSEMBLY OF PLASTIC PARTS

With the appearance of products of all types made from plastics after World War II, new methods for joining plastic parts became necessary. As is usually the case, initial approaches simply extended what was being used with metal,[4] the material of choice prior to World War II, namely, mechanical fastening. Self-tapping screws (later modified for plastics from the types used with sheet-metal), machine screws and bolts with internally-threaded parts or with nuts, and, occasionally, rivets (later modified specifically for use with plastics, including some types of "pop rivets"). The problem with all of these mechanical fastening methods is that they create severe point or concentrated loading and highly localized stresses. Such concentration of stress easily leads to stress relaxation in polymers, even at relatively low temperatures, because of their inherent viscoelastic[5] strain behavior. As a result of stress relaxation (or "cold flow"), tightened screws or bolts and nuts become loose as the underlying material strains and relaxes with time. As a result of this same viscoelastic strain behavior, fasteners that operate by carrying shear by bearing against the side of the fastener hole, rivets, screws, and bolts, as well as keys, pins, and other fasteners, elongate the hole and lose their effectiveness of joining. Because of the softness of most polymers compared to most metals, the threads produced by self-tapping screws tear out in time or under heavy loading. The only answer to these problems has been to customize fasteners for use with plastics. Head and nut bearing faces are made larger to spread loading, load-spreading washers are used, threaded inserts are used with screws (to preclude tear-out), and fastener holes are lined with sleeves to reduce the bearing stress somewhat.

As a further and generally more appropriate response to these problems, the use of load-spreading adhesive bonding emerged and proliferated. To a lesser extent, thermal bonding (actually a welding process for thermoplastics) emerged. But, joints produced by both adhesive bonding and thermal bonding proved difficult to inspect for defects and even more difficult, if at all possible, to repair.

Finally, it occurred to designers that by designing into parts geometric features that explicitly enabled and allowed interlocking between mating parts, several problems could be overcome at once. First, the need for supplemental

---

[4] A similar thing happened with the appearance of reinforced plastics and other composites, when rivets were used to assemble parts made from composite materials in the aerospace industry, based more on experiential knowledge and facility than on purely technical rationale.

[5] "Viscoelastic strain behavior" means that, after responding elastically, and perhaps also plastically, to a load immediately upon its application, a material continues to strain with time, albeit at a decreasing rate. Likewise, upon unloading, the material immediately recovers the elastic component of strain and then continues to recover some additional strain with time.

fasteners was virtually eliminated, precluding problems both of logistics in production manufacturing and with the often labor-intensive preparation of suitable fastener holes. Second, by taking advantage of the inherent property of polymers to deflect easily under the application of a force and to recover (at least partially), catch-latch mechanisms for attachment could be created. Integral snap-fits were born! Of course, what appeared as an unexpected by-product—and bonus—was an embedded system for quality assurance (i.e., listening for and/or sensing the force response associated with the snap's recovery). Beyond these advantages are: partial or complete elimination of fixtures and/or tools for assembling parts, since snap-fit joints tend to self-align, self-key, and, thus, self-fixture; elimination of fastener fallout, with the attendant dangers posed to young children who often put such small objects in their mouths and choke; and security against unwanted disassembly without specific knowledge of where integral attachment features are located within the interior of a closed assembly and, perhaps, the requirement for specialized tools.

There are almost as many different designs of snap-fits as there are designers. That's both the good news and the bad news. It's good news because it suggests there should be a design suitable for the assembly of almost any type of parts for any type of situation, including: prevention of unwanted movement in selected directions, ease of insertion of one part into another during assembly, strength of retention against unintentional disengagement, and so on. It's bad news because without some standardization of designs, new designers, unfamiliar with snap-fits from previous experience, may be too intimidated to try them in a design. The really good news is that there have been a relatively small and manageable variety of snap-fit feature types for which there is now fairly significant information on application, at least as guides.[6] The following major types (as sub-classes) of snap-fits will be described[7]:

- Cantilever hooks
- Cantilevered holes or window snaps
- Annular and leaf-spring snaps
- Ball-and-socket or post-and-dome snaps
- Compression hooks and L-shaped and U-shaped hooks
- Compression traps and beams
- Bayonet-and-finger snaps
- Torsion snaps

---

[6] In fact, in a paper by Genc et al (1997), an attempt was made to enumerate the number of possible designs based on some key factors relating to the fundamental geometry of the parts being assembled.

[7] The particular contribution of descriptions from Dr. Dean Q. Lewis, former graduate student leader of the Integral Fastening Program at Rensselaer Polytechnic Institute, Troy, NY, after 1999, is gratefully acknowledged (Lewis, 2005).

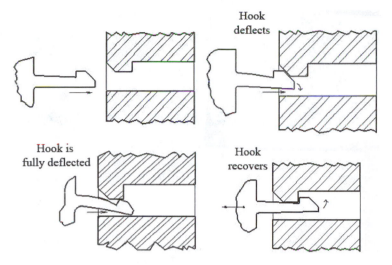

FIGURE 4.3 A schematic illustrating the key steps in the process of assembling cantilever hook snap-fits.

4.3.1 Cantilever Hooks

The most familiar of snap-fit features is surely the basic *cantilever hook*. The latch in this snap-fit locking pair[8] is composed of an elastic beam with a trapezoid-shaped head located at its free end referred to as the "hook." The body of the cantilever hook, that is, the beam portion, is integral to the part at its root. The hook-shaped latch of the cantilever hook snap-fit is paired with a catch that can be a raised ledge protruding from the wall of a mating part, a hole in the wall of a mating part, or even the edge of the mating part itself. The elastic beam portion of the latch feature may be tapered so that the cross-section at the base is larger than just beneath the head or hook, further facilitating beam deflection. The cross-section itself may be rectangular, trapezoidal, or curvilinear (e.g., a circular arc).

Cantilever hook features should be designed (as with all snap-fit pairs) so that the distance between the latch and catch will result in the latch being fully engaged with and locked to the catch when the parts are fully mated. As the active features on a mating part and a base part are brought into closer proximity so that contact first occurs between them, the catch feature will contact the hook along its angled face and, thereby, cause the beam to deflect back as the insertion motion continues. Once the head has passed the recessed edge of the catch, the elastic stress in the deflected cantilevered beam will cause it to "snap" into place.

Figure 4.3 shows the key steps in the process of assembling cantilever hook snap-fits.

---

[8] Recall from Chapter 3, Section 3.8, Bonenberger (2000) recognized that snap-fits fall into two broad categories: locating features (or locators) and locking features. The former are almost always rigid integral features and serve to help guide and fix the position of a mating part on a base part. The latter are always elastic integral features and serve to actually hold parts together once they are fully and properly engaged one with the other.

FIGURE 4.4 The use of a cantilever hook snap-fit for securing the cover of the battery compartment for a TV hand-held remote control unit by sliding the cover until the hook (left) engages the unit's housing (middle), is deflected, and then recovers and locks the cover in place (right). (Photographs taken by Sam Chiappone for the author, Robert W. Messler, Jr.; used with permission.)

When the snap-fit attachment is designed to be permanent and to be disengaged only with a special tool, the mating edges of the cantilever's head and the catch should be nearly perpendicular to the elastic beam. In this case, if access is allowed during design by placement of a molded-in hole in a plastic part, the snap can be disengaged by physically causing the elastic beam to deflect back so that the latch and the catch disengage. If a sloped edge (without a re-entrant angle) is used for either the cantilever hook's head or the mating catch's edge, a sufficient force in the separating direction will cause the two to disengage.

The insertion and retention forces for the cantilever hook snap are dependent upon the physical dimensions of the elastic beam (i.e., length, width and depth, cross-sectional shape, and taper) as well as on the angles and height of the head or hook. The material properties of the feature (e.g., flexural modulus, yield strength, fracture strength) are, obviously, also important, but are fixed by the material used to make the parts.

Figure 4.4 shows the use of a cantilever hook snap-fit on the sliding cover for the battery compartment of a hand-held TV remote control unit.

Cantilever hooks offer easy push assembly, reasonably secure locking for preventing motion in controlled directions by proper feature placement and orientation, and flexibility to resist shock loads (e.g., from dropping).

4.3.2 Cantilevered Holes or Window Snaps

A similar feature to the cantilever hook type of snap-fit is the *cantilevered hole feature,* also known as a *window snap feature*. The major difference from the cantilever hook is that the elastic beam member of the locking pair contains a hole or window instead of a protruding head or hook. Most designs will have a rectangular cross-section for the elastic beam portion, but a curvilinear cross-section could also be employed. The mating catch for this type of latch must be a feature that will cause the elastic beam to deflect back during assembly, and then once the hole clears and aligns with a similar-shaped protrusion on the more rigid catch, the beam will recover and the parts will be locked or latched. Both the assembly and the disassembly processes for the cantilevered hole-type snap are similar to the cantilever hook-type snap. The insertion and retention forces for this type of snap are, as they are for the cantilever hook, dependent upon the physical dimensions of the beam (e.g., length, width and depth, moment of inertia), as well as by the angles and height of the catch.

The cantilevered-hole snap-fit also operates with a simple push and is capable of resisting unwanted movements by proper placement (i.e., location and orientation) of the latch-catch pairs. Otherwise, the choice between this and the cantilever-hook snap-fit is largely based on convenience in the molding of parts and, possibly, clearance issues in the assembly (see Section 4.6).

Figure 4.5 schematically compares the cantilever hook and cantilevered hole snap-fits as mating pairs.

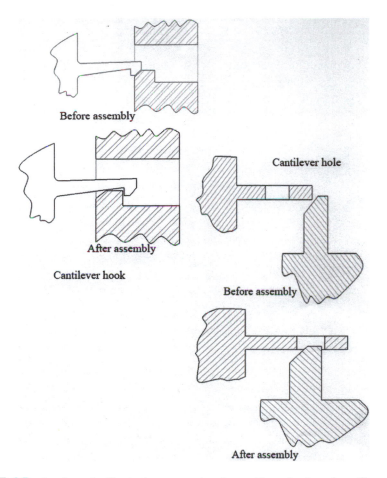

FIGURE 4.5 A schematic illustration comparing the cantilever hook and cantilevered hole snap-fits, shown before and after assembly in each case.

4.3.3 Annular and Leaf-Spring Snaps

A series of cantilever-hook features, each with a curvilinear cross-section, placed in a circular array can act as the latch features for engaging a circular hole-shaped catch. When the cantilever-hook features are connected to form a continuous ring, the resulting feature is called an *annular snap*. Because this type of feature can no longer act as a true cantilevered beam, a smaller amount of deflection can occur. If the inner part in the mating pair for an annular snap has protrusions (usually in the form of a raised ring), then it tends to act more rigidly, so elastic deflection must take place in the outer annular body by a hoop stress. If, on the other hand, the protrusions are freer to deflect elastically, by being on a continuous or split hollow ring, for example, the outer catches can be more rigid. In either case, locking arises from the development of hoop stresses.

Once engaged, an annular snap can provide a strong, secure hold because of its ability to tolerate externally applied or internally generated loads in any direction orthogonal to the insertion direction. A simple push assembly motion is all that is needed to engage an annular snap. The annular snap feature will also limit motion in the retention direction. Rather uniquely among snap-fits, the annular snap-fit allows rotation about the centerline of the annular feature.

In order to disengage this type of feature pair, enough force must be applied in the disassembly direction to cause enough deflection for the inner feature to pass back out of the outer feature. The insertion and retention forces are dependent upon the angles of the mating surfaces on the latch and catch features, as well as on the inner and outer dimensions of the features (i.e., the wall thickness of each).

Figure 4.6 schematically illustrates a solid-core annular snap-fit feature pair in which the core behaves rigidly and the annulus behaves elastically. The reverse tends to be true for hollow-core annular snaps.

Sometimes it is necessary that an integral mechanical attachment has a profile that does not protrude much from the surface of mating parts. For such situations, a *leaf-spring snap* can be employed along an edge of a part by creating

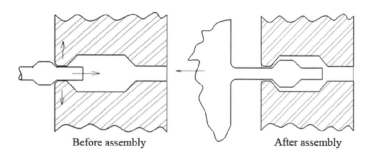

Before assembly   After assembly

FIGURE 4.6 A schematic illustration of a solid-core annular snap-fit feature pair in which the solid core remains rigid while the annulus deflects elastically.

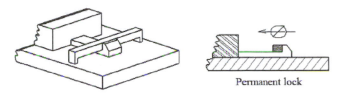

Permanent lock

FIGURE 4.7 A schematic illustration of a typical leaf-spring snap-fit. Here the right angle at the hook catch to beam junction makes the lock permanent.

a thin slot parallel to the free edge. If the thickness of the created band of material between the slot and the edge of the part is thin enough, it will be able to flex and act as a leaf spring. By placing a protruding head at either the midpoint of the leaf-spring feature or on the opposite mating part's surface, and a matching hole for engagement on the other part in the mating pair, an integral attachment can be created. As the two surfaces mate, the leaf-spring latch should deflect inward toward the slot as the protruding head pushes the surface back. The protruding head should then snap into the hole to cause engagement and locking.

Only a significant force in the disassembly direction or physically moving the leaf spring with a tool can result in disengagement, so locking security is quite good. The physical dimensions of the leaf spring that affect insertion and retention forces are its length, width, and depth, along with the dimensions and angles of the protruding head. A right or re-entrant angle on the hook tends to make the attachment permanent, while an angled face on the hook allows disassembly.

Figure 4.7 schematically illustrates a typical leaf-spring snap-fit.

4.3.4 Ball-and-Socket or Post-and-Dome Snaps

An attachment feature similar to the annular snap is the *ball-and-socket* or *post-and-dome snap*. This type of locking pair also has an inner and an outer feature, but, in this case, the inner feature is usually a solid spherical ball mounted on the end of an upstanding beam or post. This ball mates into a spherical socket that has an inner diameter that is approximately the same as the outer diameter of the ball. The opening to the socket through which the ball end of the post must pass is slightly smaller than the diameter of the ball. This demands that the ball elastically deflect the socket open with a hoop stress to allow full engagement. Once the ball is fully in the socket, the opening recovers elastically and locking of the feature pair has been completed. The inner feature can be considered to be the rigid latch, as neither the ball nor the post is intended to deflect or otherwise deform during assembly. The socket can be considered the elastic catch because it needs to expand to allow the ball to pass through its opening during engagement, and then recover to cause locking.

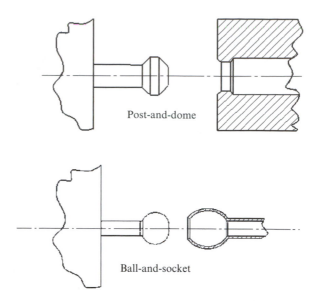

Post-and-dome

Ball-and-socket

FIGURE 4.8 Schematic illustrations of a couple of common post-and-dome snaps, the lower one of which is a ball-and-socket snap.

Once mated, this type of feature pair will restrict displacement motion but allow rotational motion around the centerline of the post and some limited rotation about the two axes orthogonal to this axis. Disengagement for this type of feature pair can only occur when enough force is applied in the disassembly direction to cause the socket to expand enough to release the ball end of the post. The diameter of the ball and of the socket, as well as the wall thickness of the socket, will determine the insertion and retention forces. Retention force can be enhanced by designing the socket to have re-entrant angles and no sloping surface to aid extraction of the ball.

Figure 4.8 schematically illustrates a couple of common post-and-dome snaps, one of which is commonly, and understandably, referred to as a "ball-and-socket" snap. Hip and shoulder joints in humans, and many other animals, employ a ball-and-socket configuration.

4.3.5 Compression Hook and L-shaped and U-shaped Hooks

A traditional cantilever hook feature will experience tensile forces while under retention. By modifying the design of the cantilever hook slightly, it can be made to experience compressive forces and provide a strong snap in a small feature. The compression hook snap engages in a similar way to the cantilever hook in that the elastic beam portion deflects away as the angled face of the head passes over and passed the catch. But, since the beam is under compression to keep the parts mated, either it will need to be physically moved away from the catch or enough force will need to be applied to buckle the beam in order to cause disengagement.

This type of snap is commonly used for small electrical connections, including telephone jacks. The cross-section of the beam portion of the latch feature may be tapered. The cross-sectional shape, as well as the physical dimensions of the beam and the engaging head, are factors that determine the insertion and retention forces.

Figure 4.9a schematically illustrates a compression hook.

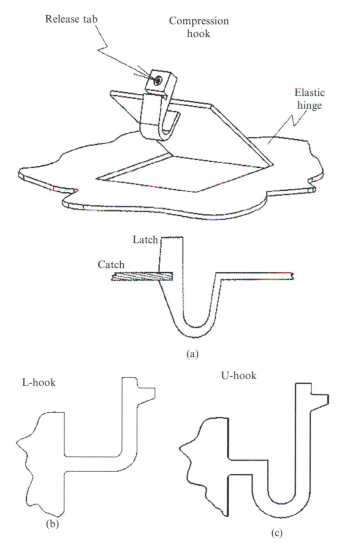

FIGURE 4.9 Schematic illustrations of the operation of a compression hook (a), and an L-shaped hook (b) and a U-shaped hook (c).

FIGURE 4.10 The use of a U-hook snap-fit for securing the cover of the battery compartment of a DVD hand-held remote control unit by engaging rigid hinges at the left end of the cover (in the photo) and tilting it downward until the U-hook (right) engages the edge of the opening in the unit's housing (middle) and locks into place after the deflected hook recovers elastically (right). (Photographs taken by Sam Chiappone for the author, Robert W. Messler, Jr.; used with permission.)

Besides changes to the physical dimensions of the snap feature in a compression hook, their base conditions also play a part in their performance. The *L-shaped* and *U-shaped hooks* are variations of the compression hook type of integral snap-fit.

The *L-shaped hook* (shown in Figure 4.9b) extends at a right angle out of the wall of a part to which it is integral and has a 90-degree bend in the beam portion to create an assembly direction that is parallel to the wall's surface.[9] In this way, the mating part will slide across the base part during assembly, which can sometimes facilitate the assembly process by allowing features to be mated to pre-align (Section 4.6). The *U-shaped hook* (shown in Figure 4.9c) also extends at a right angle out of the surface of a part but then has a 180-degree bend in its beam portion. Referring to Figure 4.9b and c, it can be seen that the U-shaped hook still allows the mating parts to be assembled in a direction parallel to the wall to which the U-hook is integral, but the extra bends give this snap feature much more elastic compliance or flexibility.

The different end conditions associated with L-shaped and U-shaped hooks also have an effect on the insertion and retention forces as they provide more flexibility in the snap. The insertion force tends to be "softened" more than lessened, and the retention force tends to be lowered somewhat; however, unintentional disengagement from impact, for example, tends to be reduced because the bent beams offer more compliance, absorbing some of the shock impulse in the flexible beam. Both types also help greatly with tolerance issues that are often associated with the manufacture of the molded plastic parts in which snap-fits are so commonly used (see Section 4.6).

Figure 4.10 shows a U-hook used to lock a hinged cover to the battery compartment of a DVD remote control unit to the unit's housing.

---

[9] In most instances, assembly motions are perpendicular to the placement of the base of an integral snap-fit feature.

4.3.6 Compressive Traps and Beams

The concept of a plug (such as a split clevis) that is inserted into a hole and then expands by recovering elastically on the opposite side so that it cannot be removed is the idea behind a *compressive beam snap*. Many variations of this snap exist, but the general form consists of a rigid post with flexible members protruding from it to act as the latch features in a locking pair. When the post is inserted through a hole (which serves as the catch feature in the locking pair), these flexible members are forced to elastically deflect back toward the post, allowing the post plug to pass through the hole. Once fully engaged, the deflected members recover to their original, pre-deflected positions. Together, they then prevent the post plug from passing back through or being withdrawn from the hole.

To disengage this type of snap without permanently damaging either it or the mating part's catch, some type of tool is necessary to compress the relaxed latch members close enough to the post to allow the plug portion to be withdrawn from the hole. Since there are many different ways that the latches in such snaps can be implemented, the insertion and retention forces are dependent on the design details, dimensions of the latch members, and the orientations of the latch members. Compressive trap-type snap-fits all share the sometimes advantageous characteristic that they tend to resist disengagement the more they are forced to disengage improperly.

Figure 4.11 schematically illustrates a couple of designs for a compressive trap and one design for what is known as a *compressive beam snap-fit*, which operates virtually the same way.

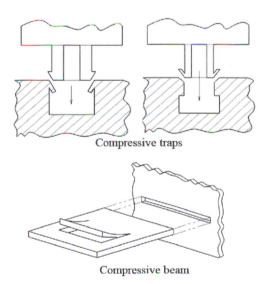

Compressive traps

Compressive beam

FIGURE 4.11 Schematic illustrations of two different compressive traps (top) and a compressive beam (bottom) snap-fit locking pair.

4.3.7 Bayonet-and-Finger Snaps

Occasionally, features from different types of snap-fits can be combined to create a different type of snap with different beneficial attributes. The *bayonet-and-finger snap* is an example of this, being a cross between a cantilever hook and a compressive trap integral attachment feature.

In a bayonet-and-finger snap, two "fingers" or trap features project out of one of the parts to be mated, and a cantilever hook or "bayonet" feature projects from the other part so that it must pass between the two fingers (Figure 4.12). The fingers must be designed so that the space between their tips is approximately the width of the beam portion of the bayonet below the head. As the two features of a locking pair are brought together, the fingers deflect back to allow the head of the bayonet to pass between them, after which they elastically recover and act as supports to keep the features and parts attached.

The finger that engages the head of the bayonet can move slightly with the head, keeping the engagement intact, and would need to either buckle or be physically moved by a tool in order to release the bayonet. The finger behind the head of the bayonet applies pressure on the bayonet to help prevent it from cantilevering away from the engagement finger.

A high (10–20 times) ratio of retention force to insertion force can be obtained with this type of locking pair. In order to disengage this type of snap, a tool capable of separating the fingers enough to allow the bayonet's head to be released would be needed. Additional strength can be obtained by designing the bayonet to have a head that protrudes in both directions, rather than in only one direction, so that both fingers will be engaged simultaneously. The physical dimensions of the bayonet (e.g., length, width, and depth of the beam portion, as well as dimensions and angles of the head) and the physical dimensions of the fingers (e.g., length, width, and depth) and the angle at which they protrude, all influence the insertion and retention forces for this snap.

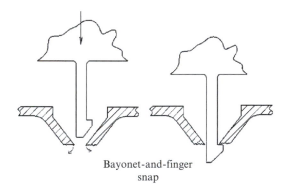

Bayonet-and-finger
snap

FIGURE 4.12   A schematic illustration of a bayonet-and-finger snap, shown before and after assembly.

Torsional snap locks

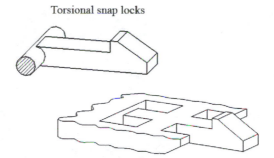

FIGURE 4.13   Schematic illustrations of two different designs for torsional snaps.

The snaps discussed above are the most commonly used examples, but this list is not exhaustive. Many other types of integral attachment snap-fit features have been designed and applied with success. Specific needs can be addressed with unique designs that may find only limited use. All, however, are based on elastic beams.

One more example is noteworthy, that is, torsional snaps.

4.3.8   Torsional Snaps

If it is desired that the typical pushing and pulling forces and motions that parts undergo not cause the snap-fit to disengage, a *torsional snap* can be used. Here, the latch feature in the mating part pair is engaged and disengaged with a twisting motion of its elastic beam, instead of a cantilever motion. The catch feature can be designed to impose a twist to the latch during assembly, but a tool would be needed to cause the torsional motion on the latch to cause it to disengage. A familiar example of torsional snaps is found in some plastic medicine bottles. A couple of designs for torsional snaps are shown schematically in Figure 4.13.

Figures 4.14 through 4.19 show applications of various elastic integral snap-fit features in plastic parts and assemblies.

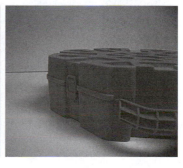

FIGURE 4.14   The use of a cantilevered hole type snap-fit to serve as the easy to operate, but still effective latch for the lid of a child's tool kit. The toolkit is shown having the cover-securing latch snapped together. (Photograph courtesy of KNex Industries, Inc., Hatfield, PA; used with permission.)

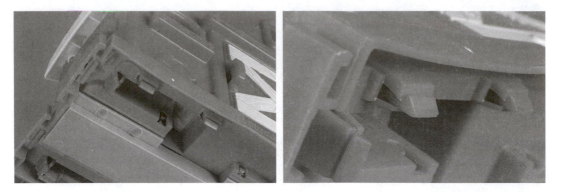

FIGURE 4.15 Multiple bayonet-and-finger-snaps (overview at left and close-up at right) are used in the assembly of the barn to the base of the Little People Farm™ shown in Figure 2.9. (Photograph courtesy of Fisher-Price; used with permission.)

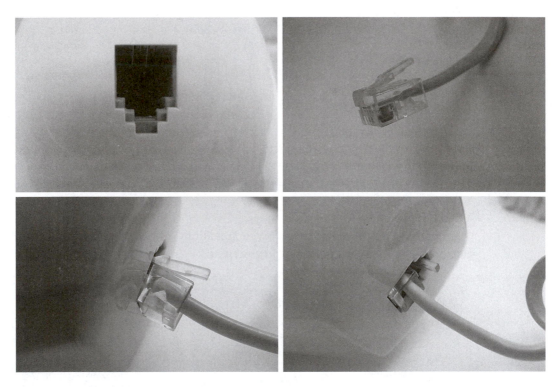

FIGURE 4.16 Photographs showing the use of a compression trap snap-fit at the end of the coiled wire that connects to a telephone receiver and wall jack. The receptacle in the end of the telephone (upper left) is shaped to accept the compression trap connector fitting on the end of the coiled wire (upper right) in only one orientation. As the compression trap connector enters the receptacle (lower left), the spring on the snap-fit is deflected to allow engagement. Once the connector is fully inserted, the deflected spring recovers elastically and locks the connector into the receptacle (lower right). (Photographs taken by Sam Chiappone for the author, Robert W. Messler, Jr.; used with permission.)

FIGURE 4.17 Photograph showing the use of an annular snap-fit in a child's toy. The cut-away annulus (left) is elastically deflected by the relatively more rigid post to allow engagement and locking (right). (Photograph courtesy of Fisher-Price, East Aurora, NY; used with permission.)

FIGURE 4.18 Another type of annular snap uses molded-in serrations to create a pine tree–like shape that aids in interlocking in an annulus simply from compression forces from elastic spring-back and from friction. (Photograph courtesy of Fisher-Price, East Aurora, NY; used with permission.)

4.4 DESIGN ANALYSIS FOR SNAP-FITS

By now it should have become apparent that integral snap-fits are a very simple, economical, and rapid way of joining two different components to produce an assembly. The common characteristics of all integral elastic snap-fit locking features are as follows:

- All types have in common a protruding portion on one component that engages a depression or undercut in the mating component, one acting as a latch and the other as a catch.

FIGURE 4.19 Ball-and-socket snap-fit joints allow children to easily assemble and disassemble parts of toys intended for this educational and entertainment purpose (top). Details of one such joint can be seen in the lower left and lower right images. (Photographs courtesy of Fisher-Price, East Aurora, NY; used with permission.)

- The catch-latch pair of integral locking features operates during assembly by either one deflecting elastically and the other remaining relatively rigid.
- After the joining operation, the integral locking joint features should return to a stress-free state.
- The resulting joint may or may not be separable, depending on the detailed shape of the undercut and the protruding feature.
- The force required to cause assembly (i.e., the insertion force) is typically low and is always 2–20 times lower than the retention force holding the assembly together.
- The force required to separate the components varies greatly depending on the design of the mating features—for a particular material of construction.
- Since they operate elastically, design analysis is based on elastic beam theory (Pilkey, 2002).

In the previous section (4.3), it was apparent that there are a wide range of design possibilities for snap-fit joints. Calculation principles have been derived (analytically, numerically, or empirically from experimental data, depending on the feature complexity) for many types including: cantilever hooks (Luscher, 1996); post-and-dome (Nichols and Luscher, 2000); compression hook (Lewis et al, 1997); bayonet-and-finger (Lewis et al, 1997); and annular snaps (Bayer Polymers Division, 1998).

Because polymers[10] normally exhibit high levels of flexibility, they are particularly well suited materials for elastic snap-fit assembly. Not surprisingly, a number of major polymer manufacturers have prepared excellent design manuals or guides for plastic snap-fit joints (Allied Signal Plastics; BASF; Bayer Polymers Division; DSM Engineering Polymers; Dupont Polymers; GE Plastics; Hoechst Celanese/Ticona). These have tended to focus on three major categories of snap-fits, and calculation principles have been rather well developed for these three, including:

1. *Cantilever snap joints,* for which the load is mainly flexural
2. *Torsion snap joints,* in which shear stresses carry the load
3. *Annular snap joints,* which are rotationally symmetrical and involve multiaxial stresses

Table 4.1, taken from the fine design guide by Bayer's Polymer Division (formerly Miles-Mobay) (Pittsburgh, PA), gives equations for calculating permissible deflection[11] y and deflection force P for cantilever snap features with various beam cross-sections. Figure 4.20 gives the geometric factors K (upper portion of the figure) and Z (lower portion of the figure) required for ring segments in Table 4.1.

The most comprehensive discussion of design using integral elastic snap-fit features appears in the seven-part series by Messler et al, (1997–1998).

4.5 COMBINING ASSEMBLY MOTIONS FOR SNAP-FIT ASSEMBLY SECURITY

Anyone who has been around long enough knows that if something can go wrong, it will go wrong.[12] To those involved with joining, this appears as if something can be disassembled intentionally, it can disassemble accidentally. We have all experienced this, but let me give one personal example specific to elastic integral snap-fits.

---

[10] While commonly known as "plastics" in the vernacular, materials scientists and engineers tend to refer to these materials as polymers.

[11] "Permissible deflection" derives from the permissible strain for the polymer used in the snap feature, which relates to a yield point for the material, and on the shape of the beam.

[12] This is popularly known as "Murphy's Law" and has a corollary that states, "And, when something does go wrong, it will do so at the worst possible time." While not a law of physics, it surely seems to be a perverse fact of life.

TABLE 4.1 Equations for dimensioning cantilevers (From Snap-fit Joints for Plastics—A Design Guide, 1996, Bayer Polymer Division, Bayer Corporation, Pittsburgh, PA [formerly Miles-Mobay Corporation]; used with permission.)

Shape of cross section / Type of design	A — Rectangle	B — Trapezoid	C — Ring segment	D — Irregular cross section
(Permissible) deflection — 1 Cross section constant over the length	$y = 0.67 \cdot \dfrac{\epsilon \cdot l^2}{h}$	$y = \dfrac{a+b_{(1)}}{2a+b} \cdot \dfrac{\epsilon \cdot l^2}{h}$	$y = K_{(2)} \dfrac{\epsilon \cdot l^2}{r_2}$	$y = \dfrac{1}{3} \cdot \dfrac{\epsilon \cdot l^2}{c_{(3)}}$
(Permissible) deflection — 2 All dimensions in direction y, e.g., h or Δr, decrease to one-half	$y = 1.09 \cdot \dfrac{\epsilon \cdot l^2}{h}$	$y = 1.64 \dfrac{a+b_{(1)}}{2a+b} \cdot \dfrac{\epsilon \cdot l^2}{h}$	$y = 1.64 \cdot K_{(2)} \dfrac{\epsilon \cdot l^2}{r_2}$	$y = 0.55 \cdot \dfrac{\epsilon \cdot l^2}{c_{(3)}}$
(Permissible) deflection — 3 All dimensions in direction z, e.g., b and a, decrease to one-quarter	$y = 0.86 \cdot \dfrac{\epsilon \cdot l^2}{h}$	$y = 1.28 \dfrac{a+b_{(1)}}{2a+b} \cdot \dfrac{\epsilon \cdot l^2}{h}$	$y = 1.28 \cdot K_{(2)} \dfrac{\epsilon \cdot l^2}{r_2}$	$y = 0.43 \cdot \dfrac{\epsilon \cdot l^2}{c_{(3)}}$
Deflection force — 1, 2, 3	$P = \overbrace{\dfrac{bh^2}{6}}^{Z} \cdot \dfrac{E_s \epsilon}{l}$	$P = \overbrace{\dfrac{h^2}{12} \cdot \dfrac{a^2+4ab_{(1)}+b^2}{2a+b}}^{Z} \cdot \dfrac{E_s \epsilon}{l}$	$P = Z_{(4)} \cdot \dfrac{E_s \epsilon}{l}$	$P = Z_{(4)} \cdot \dfrac{E_s \epsilon}{l}$

Subscript numbers in parenthesis designate the note to refer to.

Symbols

y = (permissible) deflection (= undercut)

ϵ = (permissible) strain in the outer fiber at the root; in formulae: ϵ as absolute value = percentage/100 (see Table 2)

l = length of arm

h = thickness at root

b = width at root

c = distance between outer fiber and neutral fiber (center of gravity)

Z = section modulus $Z = \dfrac{I}{C}$, where I = axial moment of inertia

E_s = secant modulus (see Figure 16)

P = (permissible) deflection force

K = geometric factor (see Figure 10)

Notes

(1) These formulae apply when the tensile stress is in the small surface area b. If it occurs in the larger surface area a, however, a and b must be interchanged.

(2) If the tensile stress occurs in the convex surface, use K_2 in Figure 10; if it occurs in the concave surface, use K_1 accordingly.

(3) c is the distance between the outer fiber and the center of gravity (neutral axis) in the surface subject to tensile stress.

(4) The section modulus should be determined for the surface subject to tensile stress. Section moduli for cross-section shape type C are given in Figure 11. Section moduli for other basic geometrical shapes are to be found in mechanical engineering manuals.

Permissible stresses are usually more affected by temperatures than the associated strains. One preferably determines the strain associated with the permissible stress at room temperature. As a first approximation, the computation may be based on this value regardless of the temperature. Although the equations in Table 1 may appear unfamiliar, they are simple manipulations of the conventional engineering equations to put the analysis in terms of permissible strain levels.

Geometric factors K and Z for ring segment (Shape C in Table 1)

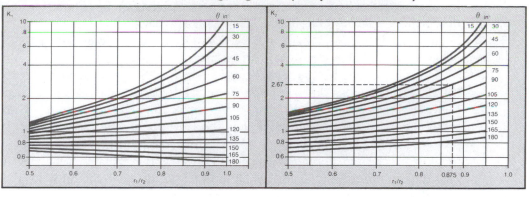

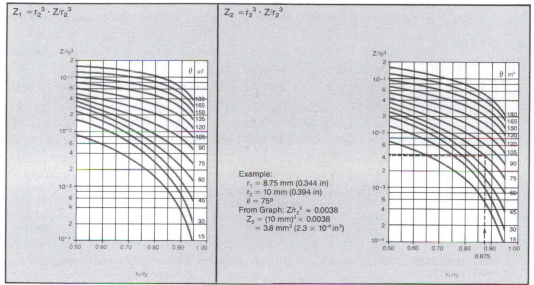

Example:
r_1 = 8.75 mm (0.344 in)
r_2 = 10 mm (0.394 in)
θ = 75°
From Graph: $Z/r_2^3 \approx 0.0038$
Z_2 = (10 mm)$^3 \times 0.0038$
= 3.8 mm^3 (2.3 × 10^{-4} in^3)

FIGURE 4.20 Plots for determining geometric factors K (top) and Z (bottom) required the design of cantilever hooks consisting of ring segments. (From *Snap-fit Joints for Plastics—A Design Guide*, 1998, Bayer Polymer Division, Bayer Corporation, Pittsburgh, PA [formerly Miles-Mobay Corporation]; used with permission.)

Years ago, I had a wristwatch that had a stainless steel linked wristband that latched with a thin sheet-metal hinged piece with two small tabs or prongs at its free end. These were angled inward, toward the wearer's wrist, and were designed to contact a round pin on the catch portion of the band. They did so by deflecting elastically a small amount until they cleared the pin, at which point there was a distinct and characteristic snap that ensured that the clasp was fully engaged and locked in place.

One day during a lecture on joining, ironically, I clapped my hands together fairly vigorously to make a point about impact loading on joints, and when I did, the clasp of my watchband sprang open and the watch slid up my hand, nearly flying off! What had happened, which I immediately turned from an accident of a fortuitous and timely lesson, was that the impact from clapping my hands together generated an equal and opposite reaction force (ala Newton!) that, being opposite the direction of the simple push motion (albeit around a hinge) used to lock the clasp, unlocked it. I went on to point out that such behavior in snap-fit assembly was an ever-present danger.

Sometime later, I bought a new, highly-esteemed Swiss-made watch. It too had a wristband that latched with an integral elastic snap-fit on a clasp. However, besides being made from stainless steel and 18k yellow gold, it had a small clip-like cover that hinged around the same pin used as the catch for the pronged, latching clasp. The clip had to be rotated out of the way of the clasp to allow the clasp to close with a snap and then had to be rotated back over the closed clasp to secure it against being opened accidentally should it snag on something or be jarred by an overly vigorous owner.

Two—actually three—lessons were learned by me from this fine Swiss watch: first, elastic integral snap-fit assembled joints are prone to accidental disengagement by a force of sufficient magnitude acting in a direction opposite the simple assembly direction. Second, elastic integral snap-fit assembled joints can be made secure against accidental disengagement by designing in a security feature.[13] There are, in turn, two fundamental approaches for designing-in security. The first, to be described here, is to combine more than one simple assembly motion. The second, to be described in Section 4.6, is to add certain additional features that enhance the performance of the snap-fit in one or more of several ways, including enhanced security against accidental disengagement.

The third lesson was that "You get what you pay for" in watchbands, not just watches.

If one thinks of a child-proof plastic container for medicines or toxic substances, most have caps or lids that require a specific combination of relatively simple actions and/or motions to be performed in a very specific sequence or simultaneously and with the right forces. As a familiar example, many of the caps on over-the-counter pain remedies (such as aspirin) require that they be pushed down (against a resisting spring-action) and turned once they come up against a fixed stop. This is all accomplished by proper design placement and sizing of various protruding and recessed features in the cap and or bottle's mouth. Another example appears in the covers to battery compartments on hand-held telephones, calculators, cameras, and toys. To remove the cover requires a slight downward push and a simultaneous slide in the proper direction. What both of these combinations do is drastically reduce the probability that such coordinated (i.e., simultaneous or sequential) actions could occur accidentally, say, by dropping the assembly.

So, by combining two or more simple assembly motions into a precise sequence or to be performed simultaneously, accidental disengagement is quickly reduced to a near-zero probability.

---

[13] Women and men know the principle of security locks from their common use on the clasps of fine jewelry.

4.6 SNAP-FIT FEATURE ENHANCEMENTS

It has been said that "Good designs work and great designs give pleasure." The difference is often more attention to detail than anything else. The designer thinks about not only what the design must do, and makes sure it will do it, but also about how to produce the actual device or structure to ensure it will be correct and how to ensure that the user uses it to its fullest potential. In the design of plastic parts that will be assembled using integral snap-fit features, these worthy goals are often ensured through the use of *enhancements*. While they may not be essential to the design of a snap-fit assembly, enhancements complete the snap-fit system by adding robustness and user-friendliness to the part and to the assembly portion of the manufacturing process and to product use. According to Bonenberger (2000), *enhancements* in snap-fit design are the third component required to form a mechanical attachment between parts beyond the proper selection and arrangement of locators and locks, which have already been discussed elsewhere.

Enhancements can and probably ought to be used at several stages in the evolution and production of a snap-fit assembly, including: (1) for supporting the actual assembly of mating and base parts; (2) for achieving optimum, beyond proper, snap-fit performance; (3) for supporting the physical activation and use of snap-fits; and (4) for supporting the manufacturing of detail parts of a snap-fit assembly. Each of these, and more, is covered thoroughly by Bonenberger (2000) in his fine book *The First Snap-fit Handbook: Creating Attachments for Plastic Parts,* as well as other references on plastic part design (Tres, 2000). What will be given here is a short overview.

4.6.1 Enhancements to Assembly

It is possible, and worthwhile, to employ certain features and attributes[14] that specifically support the assembly of a product using snap-fits. Doing so improves both the consistency and the efficiency of the assembly process. Possible enhancements for assembly are of three types: (1) physical features that provide guidance for locating the mating part to the base part to reduce the actual process of assembly to only the final simple motion needed to achieve engagement and locking; (2) the attribute of clearance that assures that once a mating part is brought to a base part nothing will interfere with its engagement and locking; and (3) the attribute of feedback to verify to the assembly operator or device (e.g., robot) that assembly and locking have occurred properly. Guidance, clearance, and feedback are three enhancements to assembly worth considering a little more.

Guidance is achieved using separate and distinct physical guides or using the added detail of a pilot or pilots on snap-fit feature locating and/or locking

---

[14] "Attributes" are to designs and processes what desirable properties are to materials.

features. *Guides* are added physical details or features that help the assembly process by simplifying the coarse or gross movements needed to bring mating parts to base parts to even begin their assembly. Most often, guides stabilize the mating part to the base part. Examples of common guides are integral pins or posts that extend from the mating and base parts to help align the parts for proper nesting and the locking features for proper attachment, or, alternatively, guiding extensions on the locking features themselves. Bevels, tapers, reduced sections at free ends are all examples of guiding extensions. *Pilots* are another form of details added to locking features that help them align with one another upon initial contact. Pilots are usually reduced sections on posts or pins but can take other forms.

Clearance needs to be dealt with by the designer at the point that locating and locking features are being placed on parts, and even earlier as the parts are being designed to perform their needed function(s). Walls, stiffeners, support posts, and all other features required for structural integrity and for assembly must be considered when two or more parts are to be joined. Details of one part cannot, obviously, physically interfere with details of another part if they are to mate properly. Likewise, actual locator and lock features must be selected (by type), located, and sized to preclude interference during assembly.

Feedback to verify proper assembly uses a combination of visual, audible, and tactile signals, including the distinctive "snap" that gives snap-fits their name. Beyond actually adding visual guides or cues, and audible and tactile "snaps," the designer can and should be sensitive to the "feel" during assembly. Feel, whether to be sensed by a human operator or an automated device (through sensors), is somewhat more esoteric. It involves controlling the force-deflection signature for the snap-fits. A familiar example of feel in assembly is how one senses whether a nut is being placed onto a bolt correctly, or not. If the nut is slightly off-angle, the external and internal threads engage incorrectly in what is known as "cross-threading." This can be felt by the lack of smoothness and eventual sticking of the nut and bolt.

4.6.2   Enhancements for Performance

The performance of snap-fit features in integral attachment to have them not only work properly but also work optimally can be enhanced through the use of physical guards and/or retainers and/or through part and/or feature compliance. *Guards* are physical details designed and fabricated into mating and base parts that protect sensitive locking features from damage. *Retainers* are physical details that provide lock and locking strength and lock performance. *Compliance* is a design attribute that allows the integral attachment features to accommodate dimensional variations in parts so that parts will still assemble without binding or with too much looseness or sloppiness. The designer can develop such compliance through some combination of local yielding in details, elasticity in details, or, if absolutely necessary, by using isolating materials such as soft (e.g., felt or rubber) fillers, O-rings, cushions, pads, and so on.

Another valuable enhancement for the performance of snap-fit assemblies is the use of *back-up locks*. These extra devices provide a locking alternative in the event the intended lock features cannot provide adequate locking, or fail in service. Most often, back-up locks involve provisions for threaded fasteners (to be installed later, if required), use of special push-in fasteners, or metal spring clips. Many designers frown on the use of such back-up locks because they seem to suggest lack of confidence in the snap-fits. However, for critical assemblies, better to be safe than sorry!

Figure 4.21 schematically illustrates some of the enhancements discussed earlier.

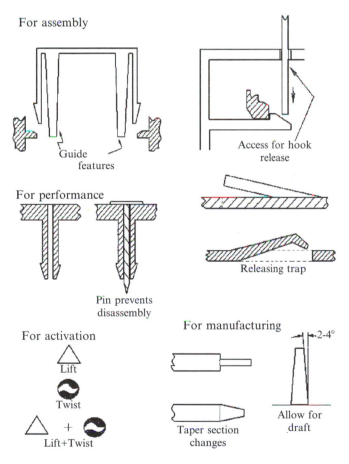

FIGURE 4.21 Schematic illustrations of various enhancements used in snap-fit assembly, for example, for guiding assembly or facilitating disassembly, for enhancing performance, for informing the user how to activate the snap, or for facilitating manufacturing by molding of plastic.

4.6.3 Enhancements for Activating and Using Snap-Fits

One of the advantages of integral snap-fits in the assembly of parts made from plastics is the clean look they provide from outside the assembly. Freedom from obvious fasteners can be purely aesthetic but can also be necessary for proper functionality. Two examples are clearance between nesting assemblies and aerodynamic smoothness. However, when integral snap-fits are used, one of the problems encountered with them is that they cannot be seen—and thus located—from outside the assembly. If they cannot be seen, it is very difficult to disassemble the assembly, which might be the intent, that is, to make the assembly tamper proof! Another problem is, even if they can be located, they cannot be disengaged. This is where some combination of mechanical and informational features can help support attachment usage and, particularly, disassembly for some essential purpose (e.g., repair or ultimate disposal).

On many snap-fit–assembled devices, molded-in, embossed, or adhesive-labeled written or symbolic information is provided to identify the presence and type of motion needed to operate a hidden snap-fit feature. Figure 4.22 shows some major examples.

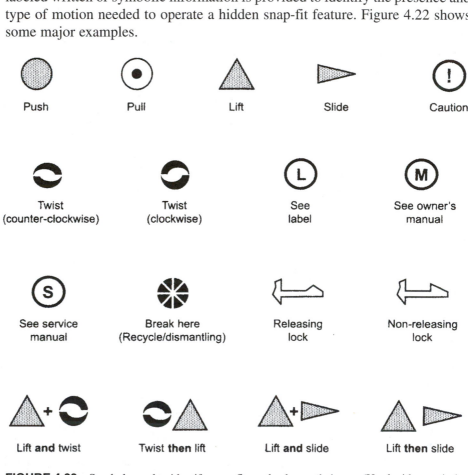

FIGURE 4.22 Symbols used to identify snap-fits and enhance their use. (Used with permission of the publisher from P.R. Bonenberger's fine detailed reference *The First Snap-fit Handbook: Creating Attachments for Plastic Parts*, Hanser Gardner Publications, Inc., Cincinnati, OH, 2000.)

Mechanical assists can also be used to make easier the release of locked features. Typical *assists* include: finger tabs; lock release tabs; recesses for finger pulls; tool access holes; and manufacturer-provided lock-release push-pins.

4.6.4 Enhancements for Snap-Fit Manufacturing

There are a number of ways in which plastic parts that are to be assembled using integral snap-fits can be designed to facilitate the manufacture of what usually turn out to be more geometrically complicated detail parts. The specialized nature of the manufacturing processes used with plastics/polymers, most notably, injection molding, warrants that the interested reader refer to any of several excellent references on plastic parts design for manufacture (Tres, 2000; GE Plastics).

4.7 HOOK-AND-LOOP ATTACHMENTS

Anyone who has walked through a field in the autumn has undoubtedly encountered "thistle burrs." These are the dried seed-bearing pods of the thistle plant, which ensures the perpetuation of the species by having the pods attach themselves to the clothes of an unaware human or the fur of a passing animal. The burrs are small (typically the size of a gumdrop or jellybean), seed-filled hollow spheres bristling with fine spikes with small hooks at their ends. When these hooks encounter any material that consists of woven fibers or of curly hair or fur, they snag that fiber or fur, latching on to travel with the person or animal until they are brushed off, to deposit there seeds and re-generate a new plant.

From Nature's example of the thistle burr were born synthetic *hook-and-loop attachments*. The best known examples are Velcro™ made by Dupont (Wilmington, DE) and Dual Lock™ Recloseable Fasteners made by 3M (Minneapolis/St.Paul, MN). Velcro™ is available in a wide variety of forms (adhesive-backed tapes, double-sided hooks and loops), shapes (sheets, strips, circular or square pads, etc.), colors, and coarseness/fineness (which corresponds to strength of attachment). All employ a two-dimensional (2-D) array of hooked polymeric spikes, analogous to the thistle burrs' spikes. When pushed to intermesh with another piece of Velcro™ or a material (such as a woven fabric), these hooks interlock with other hooks or with naturally-occurring loops in the fabric. This interlocking between hook and loop features allows attachment of one part with bonded-on Velcro™ to another or to a suitable fabric. Velcro™ fasteners are well known for their use in wind-resistant closures on ski jackets and on some athletic shoes in lieu of laces.

Dual Lock™ Recloseable Fasteners by 3M differ somewhat in design, but not in the fundamental concept of interlocking between appropriately shaped elastically-deformable features. Designs exist that consist essentially of a close-packed, orderly 2-D array of posts with free ends that have been thermally

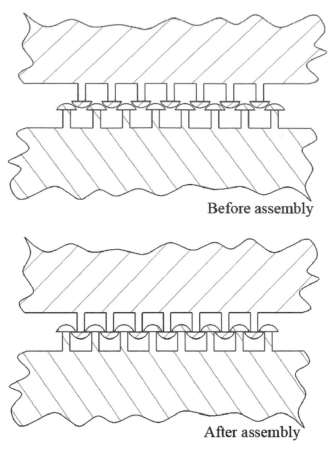

Before assembly

After assembly

FIGURE 4.23 A schematic illustration of mushroom-shaped features such as those used in 3M's Dual Lock™ Recloseable Fasteners; before and after assembly.

(or otherwise) plastically shaped to create a bulbous feature. When pressed face-to-face, two such arrays interlock when the small-scale ball-ended posts interlock.

Figure 4.23 schematically illustrates a typical design for 3M's Dual Lock™ fasteners to show the concept by which they interlock.

Depending on the stiffness of the plastic from which the hooks-and-loops or miniature interlocking ball-ended posts are made, together with their heights, height-to-diameter ratio (as a determining factor in resistance to deflection), and density in the 2-D array, insertion and retention forces can vary considerable, but to surprisingly high levels. These types of interlocks work particularly well to resist shear in the plane of the array and pretty well for pure tension. They work less well to resist separation by a peeling motion.

A description of hooks-and-loops as one type of integral micro-mechanical interlock was presented by Messler and Genc (1998).

Hook-and-loop attachments are widely used for attaching interior fabrics (e.g., head-liners) in automobiles, as well as for attaching some interior and

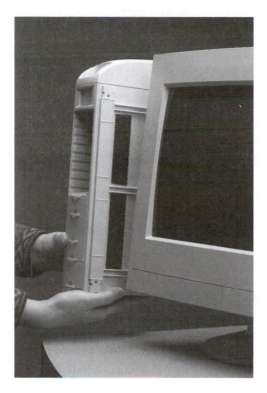

FIGURE 4.24 A photograph showing the use of hook-and-loop attachment, here to help hold the side-unit up against the monitor of a computer system. (Photograph courtesy of 3M, St. Paul, MN; used with permission.)

even exterior trim to the automobile. They are also increasingly used in closures for garments and in architecture.

Figure 4.24 shows an example of the use of hook-and-loop attachment.

4.8 OTHER ELASTIC ATTACHMENT METHODS

If ever a section should be titled "Miscellaneous," this is it! Section 4.3 describes what are surely the best known, most obvious, and most rapidly growing class of elastic integral mechanical attachments, that is, integral (elastic) snap-fits. Section 4.7 describes a close, sometimes smaller-scale cousin, that is, hook-and-loop attachments. What's left to describe is a somewhat rag-tag group of seemingly unrelated methods of accomplishing attachment using the elastic behavior of features. While all rely on elastic behavior to operate, sometimes the devices are elastic features that are integral to one or both parts to be joined, but with joining there is no telltale snap. Sometimes the devices operate like a plug with inherent springiness to create a joint, but there is no telltale snap. Many times devices operate with the aid of an actual spring or by themselves operating as a spring, but there is no telltale snap. And, finally, there are some devices that are actually fasteners, pure and simple. However, while operating to cause joining by being actual separate parts, whose sole purpose is to accomplish joining, they operate in a way that permits quick disassembly, and there is no telltale snap.

So, what will be described from here on are the following specific and, sometimes, general elastic methods:

- Integral spring tabs
- Spring-plugs
- Snap slides and spring clips
- Clamp fasteners
- Clamps
- Quick-release fasteners.

4.8.1 Integral Spring Tabs

Chapter 5, Section 5.8, describes formed tabs as integral features punched and formed into sheet-gauge metal parts. Most of the time, these features are created after the sheet-metal parts are assembled, and they are always created by plastic deformation. For this reason, they are classified as, and are described under, plastic integral interlocks. However, there are some that, once formed plastically, can be used to join or re-join a mating part to a base part containing the integral formed tabs strictly because of the elastic behavior of those tabs, that is, the formed tabs act as spring locks. For completeness, as much or more than for precision of classification, these *integral spring tabs* are presented here.

A variety of types of integral spring tabs are shown schematically in Figure 4.25 and will be presented again in Chapter 7, Section 7.2, on integral attachment of sheet-metal parts.

With suitably designed and fabricated integral spring tabs, a mating part can be slid under the integral tabs in a base part, provided those tabs can be elastically deflected open enough to let the mating part slide into place. Once the mating part is in place, the sprung tabs are allowed to relax back to their pre-deflected position and hold the mating part to the base part with the resulting spring clamping force and the associated frictional force that arises therefrom.

4.8.2 Spring Plugs

There is a small sub-class of elastic attachments known as a "spring plug". One particular example of a *spring plug* is a pressure connection used almost exclusively to hold an electrical wire in place in a terminal connector. As shown

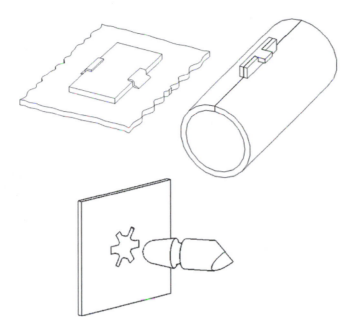

FIGURE 4.25 A schematic illustration of some formed tabs that operate as integral spring tabs.

schematically in Figure 4.26, the design consists of a funnel-like opening in a barrel-section made of very thin-gauge roll-formed sheet or from a thin-walled tube. When a wire, for example, is pushed into the funnel-like opening and through the reduced neck of the funnel, the wire forces the thin collar of metal to expand outward elastically under a hoop stress. Normal elastic spring-back (recovery) of the elastically-expanded neck grips the wire and holds it by the friction that develops.

Another type of spring is shown in Figure 4.25, as a spring-tab attachment. This design consists of a hole punched into a thin sheet of material (usually metal, but possibly plastics) whose interior is lined with teeth. When a round-ended rod is pushed into the hole so that it forces the teeth to push back elastically (away from the entering rod), in trying to spring back, the teeth grip the

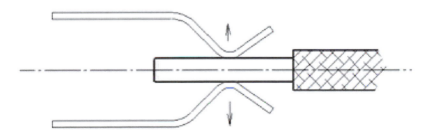

FIGURE 4.26 Schematic illustration of a typical spring plug.

rod. The grip can be dramatically increased if the rod contains a recessed groove back from its tip, and the teeth are caused to relax back into this groove once the rod is pushed into the toothed hole far enough. The resulting joint is semi-rigid. The device clearly depends on the springiness of the teeth.

4.8.3 Snap Slides and Spring Clips

Snap slides, such as shown schematically in Figure 4.27, consist of a clevis clip with a V-shaped opening and a recess or undercut located back from the ends of the clevis clips prongs. When such a clip is subjected to a sliding motion to mate with a round pin containing a recessed groove, or, alternatively, a shouldered stud, the two pieces can be used to join parts together with the resulting locking force. Motion is prevented parallel to the pin's axis and parallel to the length of the clevis clip (i.e., the clip's prongs). While the clevis clip and pin are

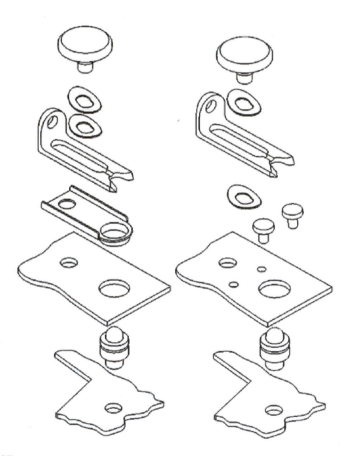

FIGURE 4.27 Schematic illustrations of two approaches for using snap slides for assembly.

often separate pieces used entirely to allow joining of other parts, they can be integral to the parts being joined. In either case, they operate by elastic deflection and recover to create interlocking.

Spring clips, such as those shown schematically in Figure 4.28, are obviously leaf-spring–like parts for serving as electrical terminals, in one case (upper portion of the figure), and for pushing into holes in sheet-metal parts to hold another part or serve as an attachment point for some other part, in the other case (lower portion of the figure). In any case, joining results from the elastic deflection and recovery between two portions of a part (for the former case) or between two parts (for the latter case).

4.8.4 Clamp Fasteners

A wide variety of *clamp fasteners* are available that lock in place with a simple quarter turn of a rotating member. Most of these work against a spring, so that some axial pressure is required while turning the rotating member. Usually, heads are provided to aid turning using a screwdriver slot, a wire ring, or knurled-, hexagonal-, or wing-shaped projections. In most cases, the clamp fastener makes use of a cam action to lock the rotating member in place under spring pressure. The spring can be of either the helical-wire type or of the sheet-metal type.

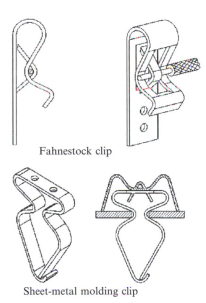

Fahnestock clip

Sheet-metal molding clip

FIGURE 4.28 Schematic illustrations of some spring clips; the Fahnestock clip for securing stripped electrical wires (top) and a sheet-metal clip for securing molding (bottom).

4.8.5 Clamps

Clamps are specially-designed devices to attach one part to another using an elastic squeezing force of some origin. Many require the use of fasteners, such as nuts and bolts or screws to pull split clamp-body elements together. Without going into much detail, because of the sheer variety of clamps, and for the fact that details abound in design handbooks for mechanical engineers (Avallone and Baumeister, 1996; Juvinall and Marshek, 1991; Shigley et al, 2004), there are the following specific types: (1) hose clamps (for joining hoses to nibbles or other fittings); (2) conduit clamps (for securing sheet-metal or plastic conduit or tubing together); and (3) wire and cable clamps (for splicing or otherwise connecting wires or cables to one another or to another part). Figure 4.29 shows some examples of clamps for hoses, conduits, and wires or cables.

4.8.6 Quick-Release Fasteners

There is a special group of fasteners, virtually always involving some type of rotating member that resembles (if it is not actually) a screw, that is intended to allow the quick engagement or intentional disengagement of parts of which

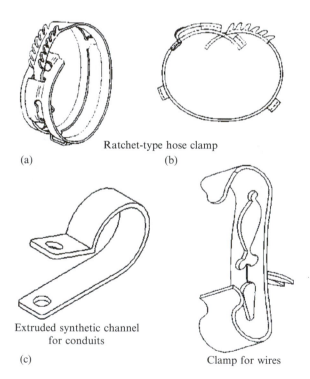

Ratchet-type hose clamp

(a)   (b)

Extruded synthetic channel
for conduits

(c)

Clamp for wires

FIGURE 4.29 Schematic illustrations of examples of clamps for hoses (a), conduits (b), and wires or cables (c).

they are a part or with which they are used to hold those parts together. The key characteristic that differentiates *quick-release fasteners* from other fasteners is that they fully engage or disengage with a partial twist rather than requiring several turns. Because the highly specialized nature of these devices, and because they are really fasteners, even though they all operate using the principle of elastic interlocking, details will be left to the interested reader to find in standard handbooks on machine design (Avallone and Baumeister, 1996; Juvinall and Marshek, 1991; Shigley et al, 2004).

Figure 4.30 schematically illustrates some typical quick-release fasteners that operate by the elastic deflection and recovery of integral design features.

4.9 SUMMARY

Integral mechanical attachments can operate by deflecting elastically during assembly, with some degree of recovery to lock the assembled parts together. Such attachments are known as *elastic integral mechanical attachments* or *elas-*

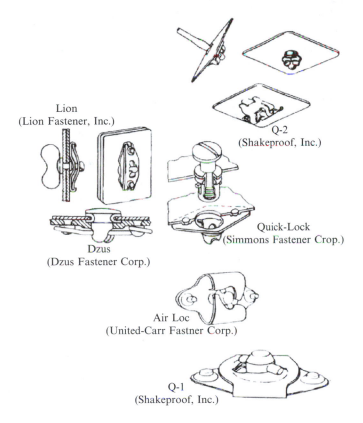

FIGURE 4.30 Schematic illustration of some typical quick-release fasteners, all of which rely on elastic deflection of a design features to operate.

tic interlocks. Because they are inherently highly elastic materials, polymers or plastics are ideally suited to assembly using elastic interlocks; however, metals are another possibility. Elastic integral mechanical attachments used to assemble parts made from plastics are commonly known as *snap-fit features* or simply *snap-fits*. The name derives from the characteristic telltale audible and tactile "snap" that occurs when a deflected feature in a pair of locking features recovers to complete the locking action.

Snap-fits come in a tremendous variety of types, each developed to satisfy a certain design need. The most common types are: (1) cantilever hooks; (2) cantilevered holes or window snaps; (3) annular and leaf-spring snaps; (4) ball-and-socket or post-and-dome snaps; (5) compression hooks and L- and U-shaped snaps; (6) compression traps and beams; (7) bayonet-and-finger snaps; and (8) torsion snaps.

The analysis of snap-fit features for a design has been reasonably well developed for the major types, treating them as elastic beams and using elastic beam theory to calculate permissible deflection and expected/required deflection force.

The security of snap-fits can be increased by employing a combination of simple assembly motions, in sequence or simultaneously, and performance as well as manufacture can be enhanced with what are known as "enhancements."

Hook-and-loop attachments are smaller-scale cousins of snap-fits, with their origin undoubtedly being the burrs found in nature.

Other methods for accomplishing joining using elastic attachments are: (1) integral spring tabs; (2) spring-plugs; (3) snap slides and spring clips; (4) clamp fasteners; (5) clamps; and (6) quick-release fasteners. While they differ more than they appear alike, all rely on the elastic deflection and at least partial recovery of some part in their design to enable and cause attachment and locking.

As a group, elastic interlocks are a valuable approach for accomplishing mechanical joining and assembly. Table 4.2 lists the major processes and methods for achieving elastic integral mechanical attachment, along with the materials in which these processes or methods tend to be found.

TABLE 4.2 List of Major Methods and Processes for Achieving Elastic Integral Mechanical Attachment (and Materials Where Used)

Integral Snap-Fit Attachments

Cantilever hooks (polymers; metal; wood)

Cantilevered holes/window snaps (polymers; metal)

Annular snaps (polymers)

Leaf-spring snaps (polymers; metal)

Ball-and-socket (polymers; metal; bone)

TABLE 4.2   (*Continued*)

	Hook-and-Loop Attachments
Post-and-dome (polymers; metal)	
Compression hooks (polymers)	
– straight	Velcro™ attachments/fasteners (polymers)
– L-shaped	Dual Lock™ recloseable fasteners (polymers)
– U-shaped	
Compression traps and beams (polymers)	
Bayonet-and-finger snaps (polymers; metal)	
Torsion snaps (polymers)	

Other Elastic Attachment Methods

Integral spring tabs (metal)

Spring plugs (metal)

Snap slides (metal)

Spring clips (metal)

Clamps (metal)

  – hose clamps

  – conduit clamps

  – wire and cable clamps

Quick-release fasteners

Plastic Integral Mechanical Attachments

Co-formed Features and Interlocks	Conformed (or Conforming) Features and Interlocks
Stakes/Staking	Crimps/crimping
Metal clinches/clinching	Hems/hemming
– Tog-L-Locs™	Formed tabs/sheet-metal forming
– Lance-N-Locs™	Indentation-type joints
Formed tabs/Sheet-metal forming	Beaded-assembly joints
	Folded joints
	Roll-and-press joints

Other Methods

Metal stitches/metal stitching

REFERENCES

Allied Signal Plastics, *Modulus Snap-fit Design Manual,* Allied Signal Plastics, Morristwon, NJ, 1997.

Avallone, E.A., and Baumeister, T., III., *Mark's Standard Handbook for Mechanical Engineers,* 10th ed., McGraw-Hill, New York, NY, 1996, 8-34-8-37 and 8-37-8-39 ("Dry and Viscous Couplings" and "Clutches").

BASF, *Snap-fit Design Manual,* BASF.com.

Bayer Polymers Division, *Snap-fit Joints for Plastics—A Design Guide,* Bayer Corporation, Pittsburgh, PA (formerly Miles-Mobay Corporation), 1998.

Bayer Material Science, *Thermoplastic Part and Mold Design Guide,* bayerplastics.com.

Bonenberger, P.R., *The First Snap-fit Handbook: Creating Attachments for Plastic Parts,* Hanser Gardner Publications, Inc., Cincinnati, OH, 2000.

DSM Engineering Polymers, *Design Guide,* dsm.com.

Dupont Polymers, *Dupont Engineering Polymers-Product Information Guide,* Dupont Polymers Department, Wilmington, DE.

GE Plastics, *GE Engineering Thermoplastics Injection Molding Processing Guide,* The General Electric Company, Pittsfield, MA. Also, geplastics.com

Genc, S., Messler, R.W., Jr., and Gabriele, G.A., "Enumerating possible design options for integral attachment using a hierarchical classification scheme," *Journal of Mechanical Design,* 119(2), 1997, 178–184.

Genc, S., Messler, R.W., Jr., and Gabriele, G.A., "Selection issues for injection molded integral snap-fit locking features," *Journal of Injection Molding Technology,* 1(4), 1998a, 217–223.

Genc, S., Messler, R.W., Jr., and Gabriele, G.A., "A systematic approach to integral snap-fit assembly," *Research in Engineering Design,* 10, 1998, 84–93.

Goldsworthy, W.B., and Johnson, D.W., *Pultruded Joint System and Tower Structure Made Thereof,* U.S. Patent 5,285,613, 1994.

Hoechst Technical Polymers, *Designing With Plastics-The Fundamentals,* Design Manual TDM-1, Ticona LLC, Summit, NJ (now a division of Celanese AG, formerly Hoechst Celanese Corporation), 1996.

3M (Dual Lock™ Recloseable Fasteners).

Lewis, D.Q., "A Design Analysis Method for Improved Drop-test Survival of Products with [Snap-fit] Plastic Housings," Ph.D. thesis, Rensselaer Polytechnic Institute, Troy, NY, December 2005.

Lewis, D.Q., Knapp, K.N., and Gabriele, G.A., "An investigation of the compression hook integral attachment feature for injection molded parts," *Journal of Injection Molding Technology,* 1(4), 1997, 224–234.

Lewis, D.Q., Wang, L., and Gabriele, G.A., "A finite element investigation of the bayonet-and-finger integral attachment feature for injection molded parts," *Journal of Injection Molding Technology,* 1(4), 1997, 235–341.

Luscher, A.F., "An investigation into the performance of cantilever hook type integral attachment features," *Proceedings of the 1996 ASME Design Engineering Technical Conferences and Computers in Engineering Conference (DETC-96/DAC-1127),* August 18–22, 1996, Irvine, CA, 1–9.

Messler, R.W., Jr., *Joining Materials and Structures: From Pragmatic Process to Enabling Technology,* Elsevier/Butterworth-Heinemann, Burlington, MA, 2004.

Messler, R.W., Jr., and Genc, S., "Integral micro-mechanical interlock (IMMI) joints for composite structures," *Journal of Thermoplastic Composite Materials,* 11(5), 1998, 200–215.

Messler, R.W., Jr., Genc, S., and Gabriele, G.A., "Integral attachment using snap-fit features: A key to assembly automation," *Journal of Assembly Automation,* Part 1: "Introduction to integral attachment using snap-fit features" 17(2), 1997, 140–152/Part 2: "Bringing order to integral attachment-level design" 17(2), 1997, 153–162/Part 3: "An attachment-level design methodology" 17(3), 1997, 239–248/Part 4: "Selection of locking features" 17(4), 315–328/ Part 5: "Constraining parts in integral attachments using locating features" 18(1), 1998, 58–74/Part 6: "Evaluating alternatives for design optimization" 18(2), 1998, 151–165/Part 7: "Testing a conceptual design methodology with a case study" 18(3), 1998, 223–236.

Nichols, D.A., and Luscher, A.F., "Numerical modeling of a post-and-dome snap-fit feature," *Research in Engineering Design,* 12, 2000, 103–111.

Pilkey, W.D., *Analysis and Design of Elastic Beams: Computational Methods,* John Wiley & Sons, Inc., New York, NY, 2002.

Shigley, J.E., Mischke, C.R., and Brown, T.H., *Standard Handbook of Machine Design,* McGraw-Hill, New York, NY, 2004.

Tres, P.A., *Designing Plastic Parts for Assembly,* Hanser Gardner Publications, Inc., Cincinnati, OH, 2000.

5 PLASTIC (FORMED-IN) INTEGRAL MECHANICAL ATTACHMENTS OR INTERLOCKS

5.1 HOW PLASTIC (FORMED-IN) INTERLOCKS WORK

There are applications for which mechanical joining is needed and preferred over processes that accomplish joining by creating atomic-level bonding (e.g., adhesive bonding or welding). Furthermore, for some of these applications, it is preferable to use integral features of the parts themselves for accomplishing the needed attachment and interlocking, but, for one reason or another, it is not feasible or practical to create those features in the detail parts during their fabrication prior to their assembly. For these applications, forming in (i.e., creating) the locking features once the parts of the intended assembly are in their proper relative positions, orientations, and alignments is preferable. For such applications, plastic formed-in integral mechanical attachments or interlocks are the answer.

Plastic (formed-in) integral mechanical attachments or *plastic interlocks* possess two differentiating characteristics compared to rigid and elastic integral mechanical attachments or interlocks. First, while their use for joining is planned by the designer, no particular geometric features are generally designed into the detail parts and, absolutely, no features capable of accomplishing actual locking are designed in or created prior to full part assembly. At most, in some forms of plastic interlocking, certain geometric features are designed into the detail parts that will allow a plastic interlock to be created by an additional and quite specific fabrication or processing operation. The second characteristic that distinguishes plastic interlocks from rigid or elastic interlocks is that the mating part being brought into appropriate intimate contact with the base part must be subjected to a specific *processing* operation that causes the mating and base part or parts to interlock. For both rigid and elastic types of integral attachment features, locking occurs as soon as the parts of the intended assembly are fully engaged due to their pre-existing geometry, not to any formed-in change.

In all cases, what causes actual interlocking of one part to another is the forcing of material from one part into natural or created recesses in the other part. Figure 5.1 schematically illustrates how a softer material can be formed into pre-existing recesses in a part made from a harder material or, alternatively, how material, if suitably ductile, in both parts can be moved during the same operation to create interlocking protruding and recessed features. In fact, softer mate-

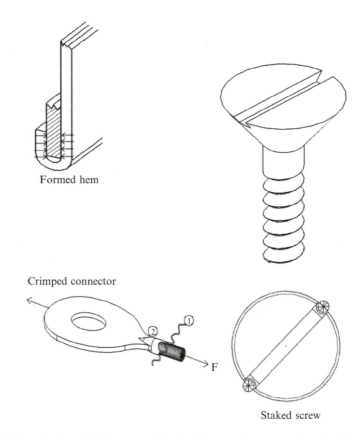

Formed hem

Crimped connector

Staked screw

FIGURE 5.1 A schematic illustration showing how a softer material can be formed into pre-existing recesses (as in crimping of a soft-metal terminal connector around electrical wires), or over pre-existing protrusions, in a harder material, how mating protrusions and recesses can be formed simultaneously in identical materials (as in the staking of a screw in a part housing), and how a portion of a softer-material part can be formed over and around a portion of a harder-material part (as in hemmed joints).

rial can also be forced to form around a pre-existing protrusion in a harder material, in which case a recess is being formed in the softer material. In all cases, physical interference arises between the newly-formed protrusion(s) and recess(es). While it is not impossible that cold welding or micro-welding[1] occurs between metals forced into intimate contact, with the formation of actual metallic bonds, such atomic-level bonding is not necessary or common.

Once interlocking features are created, several forces act together, in some combination, to keep parts for which those features are integrally joined together. The first, most obvious force is the bearing force that arises when the protruding feature pushes against the walls of the recess, as shown in Figure 5.1. This force, of course, is a reaction force that only arises when one part tries to move relative to the other. The protruding and recessed features being referred to are often macroscopic but can be, and usually are also, microscopic. The microscopic interlocks occur as the protruding and recessed asperities that always exist on the surface of real materials interfere with one another under shear forces. This microscopic interference is the origin of friction and is the second force that operates to hold parts or portions of parts forced together plastically to remain together. This "friction" is, in turn, both the result of physical interference between asperities under a shear force and the result of micro-welding. Micro-welding results in actual atomic-level bond formation and is able to resist tensile forces that attempt to pull one part away from another. Not surprisingly, the resisting force is generally of quite a small magnitude. The origin of friction on the microscopic scale is the always present peaks and valleys of asperities.

What is not usually involved as an active force or mechanism in plastic interlocks is actual inter-diffusion that occurs between the materials comprising the parts being joined mechanically. While actual atom movement and exchange by diffusion in the solid state surely occurs between two clean metals that have any solid-state solubility in one another, the amount of diffusion is normally very small at room temperature because the diffusion rate is so slow. Significant inter-diffusion between metals does not become significant or occur in reasonable time-frames, unless and until the temperature exceeds about 50–60% of the absolute melting point of the metals involved.[2]

The third force that contributes to holding together two parts joined by the formation of a plastic interlock is the elastic recovery force. This is the elastic component of the force that caused plastic deformation, commonly seen as "spring-back" in forming. The elastic recovery force may apply a squeezing

---

[1] In "cold welding" or "micro-welding," atomically clean metal can be caused to form a weld through the creation of actual metallic bonding, by forcing metal comprising one part into intimate contact with the metal comprising the other part. Once in intimate contact, atoms of metals form bonds.

[2] "Absolute melting point" is the melting point of a pure metal, or the temperature at which an alloy begins to melt (i.e., its solidus temperature) on an absolute temperature scale, either degrees Kelvin (analogous to degrees Celsius) or degrees Rankine (analogous to degrees Fahrenheit).

force to the surface of any and all protruding features if newly-formed features are oriented properly.[3] This squeezing force increases the friction force.[4]

5.2 SUB-CLASSIFICATION OF PLASTIC (FORMED-IN) INTERLOCKS

Plastic integral mechanical attachments or plastic interlocks can logically be divided into two sub-classes based on whether the interlocking protruding and recessed features created are formed simultaneously or whether either the pro-truding or the recessed features pre-existed and the mating interlocking features were created over, into, or around these. The former, simultaneously formed interlocks are aptly called "co-formed features." The latter are aptly called "conformed" or "conforming features."

Co-formed plastic interlocks tend to be found in mating parts composed of the same or similarly-behaving materials from the standpoint of ease of plastic defor-mation (i.e., ductility and yield strength). The best known example of co-formed interlocks is "staking" (see Section 5.3). Other, generally more recently developed examples are various proprietary integral attachment techniques generically referred to as "metal clinching" (see Section 5.4), such as BTM Corporation's Tog-L-Loc™ and Lance-N-Loc™. The best known and most widely used examples of conformed or conforming plastic interlocks are *crimping* and *hemming* (see Section 5.6). There are also some techniques that tend to operate more like con-forming than like co-formed interlocks, including thermal staking (see Section 5.7) used in some assembly applications involving glass, possibly with glass-to-metal seals. Thermal staking can also be used for assembling parts made from thermo-plastics or for joining parts made from other materials to thermoplastics.

An old method of using plastic interlocks is that of formed tabs (see Section 5.8). Depending on precisely how these are made, *formed tabs* can be consid-ered either co-formed or conforming types of plastic interlocks. In any case, they were widely used in the manufacturing of what were called "tin-type" toys but were actually often thin sheet-gauge steel toys. The use of formed tabs is still used today in the manufacturing of many products made from sheet metal, particularly sheet-gauge steel.

A list of the major sub-classes of plastic integral mechanical attachments or plastic interlocks thus includes:

- Setting and staking
- Metal stitching and metal clinching
- Indentation-type and beaded-assembly joints

---

[3] It is possible for elastic recovery to cause newly-formed features to separate from one another, too, if the features are not oriented properly. After all, the elastic recovery force is a relaxation force. As examples, a protrusion forced plastically to fit in a recess will expand slightly due to elastic recovery after insertion and help hold the protrusion in the recess.

[4] The force of friction, $F = \eta N$, where η is the coefficient of friction (normally <1.0) and N is the force act-ing normal to the surfaces in contact.

Plastic Integral Mechanical Attachments

Co-formed Features and Interlocks	Conformed (or Conforming) Features and Interlocks

Stakes/staking
Metal clinches/clinching
   - Tog-L-Locs™
   - Lance-N-Locs™
Formed tabs/sheet-metal forming

Crimps/crimping
Hems/hemming
Formed tabs/sheet-metal forming
Indentation-type joints
Beaded-assembly joints
Folded joints
Roll-and-press joints

Other Methods

Metal stitches/metal stitching

FIGURE 5.2 A taxonomy of plastic integral mechanical attachments.

- Crimping and hemming
- Thermal staking
- Formed tabs

Figure 5.2 gives the author's proposed taxonomy for plastic integral mechanical attachments or plastic interlocks.

Sections 5.3 through 5.8 describe the major methods for producing plastic interlocks.

5.3 SETTING AND STAKING

Setting involves the spreading of metal by plastic deformation applied to the driven end of hollow or tubular rivets and eyelets (Messler, 2004; Parmley, 1989). It may be done with a single blow, with repeated blows, or with a sustained squeeze using a tool normally provided with a reduced-section pilot for centering the punch on the rivet or eyelet. The shape of the setting tool, at an undercut end, determines the form of the driven part, being round, flat, or specially shaped. Examples of setting applied to a variety of rivets and eyelets are shown schematically in Figure 5.3.

One-piece solid rivets too are set or "upset" to produce a second head or "foot" on the end that protrudes through to the back-side of a stack-up of parts. And, finally, some two-piece, high-strength/high-shear rivets are set by swaging one piece onto or into the other. The process of locking the two parts together is identical to what is described in Section 5.5.

While rivets and eyelets are fasteners not integral attachments, most are set using plastic deformation following fastener installation, so they bear some resemblance to plastic interlocks.

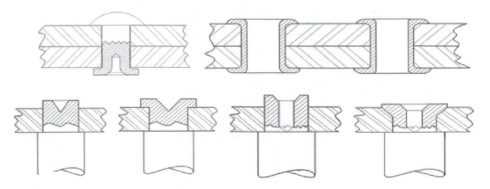

FIGURE 5.3 Schematic illustrations of various forms for setting applied to rivets and eyelets.

Staking is, without question, a method of creating plastic interlocks using formed-in integral features. Staking involves plastically deforming the metal of assembled parts in such a way that they cannot loosen or come apart under operating conditions. This deformation co-forms a protruding locking feature and a recessed locking feature in mating parts, simultaneously. In most cases, these formed-in locking features are small compared to the size of the parts involved.

In its most commonly seen form, staking is applied to an inner cylindrical part at one or more points around the interface between it and a close-fitting outer part. A sharp, hard tool forces metal at points along the interface to simultaneously form a protrusion and a perfectly-mating recess that result in interlocking against movement in one or more directions of translation and/or rotation. Staking is frequently applied to the head of a screw or to the face of a nut at the edge of the screw's threaded body. With flat-headed screws, staking locks the screw in place after tightening. With nuts, staking locks the nut in place after tightening.

Rectilinear parts can also be staked using a similar approach that spreads metal in the protruding end of one part where it extends through a recess in another part.

Figure 5.4 schematically illustrates staking.

Another form of staking is "ring staking." *Ring staking* is done with a circular tool having a projecting ring on its face. This projecting ring is shaped so as to force metal in one of the mating parts of an assembly either inward or outward to cause locking against removal. Such staking is sometimes done to compensate for wear that has caused a part to loosen.

Examples of ring staking are shown schematically in Figure 5.5.

5.4 METAL STITCHING AND METAL CLINCHING

"Metal stitching" is a term used for hundreds of years to refer to ways in which certain difficult-to-weld metals were repaired, if not joined originally. However, in the context of integral mechanical attachment, it is sometimes used in a

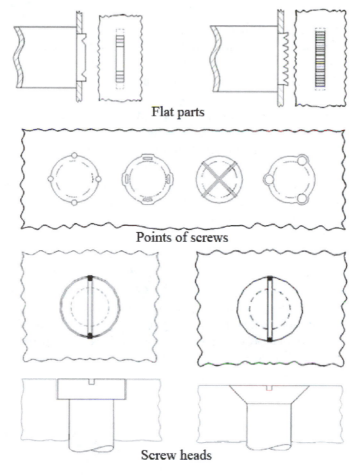

FIGURE 5.4 Schematic illustrations of various forms of staking.

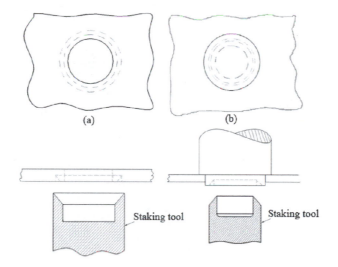

FIGURE 5.5 A schematic illustration of ring staking.

broader sense. In integral mechanical attachment, *metal stitching* refers to any of several predominantly proprietary processes or methods by which two or more layers of sheet-gauge metal are joined by co-forming interlocking features. A more appropriate term is actually "metal clinching." Before describing these more modern plastic integral mechanical attachment metal clinching methods, however, a brief description of traditional "metal stitching" is warranted.

"Metal stitching," also known as "cold metal stitching," is a hundreds-of-years-old method by which inherently brittle, hard-to-weld metals or alloys, most notably various cast irons, had cracks repaired or, in times preceding oxy-fuel and electric arc welding, were assembled. The severe sensitivity of such metals and alloys to thermal shock on heating or on cooling generally precluded fusion welding as a means of repairing cracks that arose either during the casting process itself (due to thermal shock and/or residual stresses) or in service (due to stress, especially from shock loads). In lieu of welding, a number of lines of side-by-side holes were drilled at right angles to and across the crack (see Figure 5.6, Step 3). These holes were subsequently extended to produce slots (see Figure 5.6, Step 4) that would accept pre-formed softer, tougher (usually steel) keys that rigidly interlocked into the slots (see Figure 5.6, Step 5). By spanning the crack, these keys rigidly lock together the portions of the part on either side of the crack and provided restored structural integrity. Sometimes additional holes were drilled along the crack itself and were tapped to receive special threaded studs. These

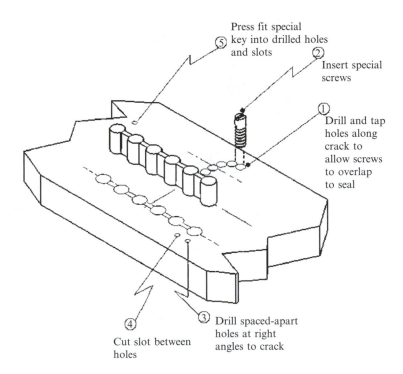

FIGURE 5.6 Schematic illustration of the steps involved in "cold metal stitching" to repair a cracked casting.

studs were then tied together with another part or by wire wrapping. Depending on the application, once fully installed, the bridging keys could be ground flush to match the surface of the casting to become virtually invisible (see Figure 5.6, Steps 1 and 2). Figure 5.7 shows a couple of "cold metal stitching" repairs of cast parts.

Metal clinching refers to any of several approaches by which one piece of sheet-gauge metal is forced into, over, or around another to create discrete points of attachment. Most processes involve a combination of drawing and interlocking. Older methods of metal clinching tended to punch in features that resembled tabs cut along two opposite sides or along three orthogonal sides and integrally "hinged" at the remaining uncut side(s). Newer methods tend to produce buttons that push through multiple layers without cutting the materials involved. The buttons are formed in completely and solely by plastic deformation.

Prime examples of a modern method of metal clinching are the patented Tog-L-Loc™ and Lance-N-Loc™ sheet-metal joining systems developed and marketed by BTM Corporation (Marysville, MI). In both of these methods, which differ only in the shape of the co-formed plastic interlock produced, upper layers of a stack of sheet-gauge metal are literally plastically deformed to produce an indentation that passes all the way through the stack-up to create a locking feature.

FIGURE 5.7 Photographs showing a totally fractured (top left) cast iron press frame repaired using "cold metal stitching" (top right). Another elaborate repair of the cast iron crown of a large machine is also shown (bottom) in which details of the repair technique are even more clear. (Photographs courtesy of Metalock Engineering UK, Ltd., Coventry, England; used with permission.)

A specially-designed tool set is used (see upper portion of Figure 5.8). Each layer of sheet metal is locked with other layers by the physical interference that results from the co-formed protrusions and precisely mating recesses. Resulting joints are said by the manufacturer to be "strong, consistent, fatigue resistant." The joining system is also said to offer a "leak proof alternative to spot-welding, threaded fasteners and rivets." The lower portion of Figure 5.8 schematically shows the Tog-L-Loc™ and Lance-N-Loc™ systems from a number of perspectives. From the schematics in the lower portion of Figure 5.8, it is clear that what prevents the layers of thin sheets from separating out of plane is the use of re-entrant angles in the co-formed protruding and recessed features. Interlocks such as these work better in materials with high elastic recovery forces, such as metals, but can, not surprisingly, also be created in thermoplastic polymers.

Figure 5.9 shows how the Tog-L-Loc™ joining system is used to join galvanized steel "drill-mill" pads to a galvanized steel automobile space-frame for subsequent attachment of composite outer skins using fasteners.

A number of other systems for producing plastic interlocks between sheet-gauge metals via co-forming exist and continue to emerge. A major driver of late

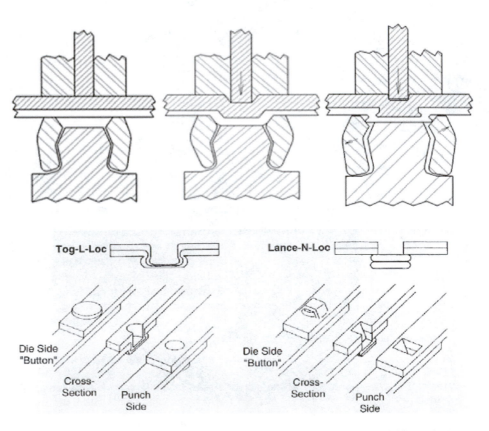

FIGURE 5.8 Schematic illustration of the process by which a Tog-L-Loc™ is made (upper portion) along with details of the Tog-L-Loc™ and Lance-N-Loc™ sheet-metal joining systems (lower portion). (Used with permission of BTM Corporation, Marysville, MI.)

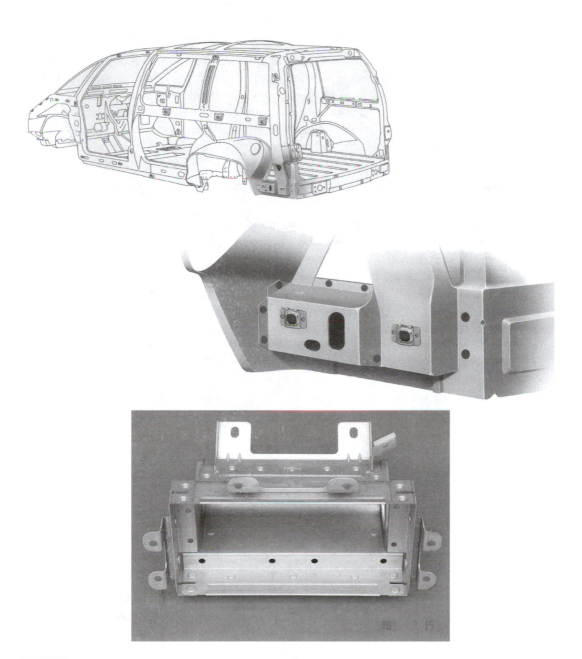

FIGURE 5.9 The general use of Tog-L-Locs™ and Lance-N-Locs™ in automobile assembly (top) and an example of their use in the sheet-metal housing for the air-bag system (bottom). (Photographs courtesy of BTM Corporation, Marysville, MI; used with permission.)

is the desire to produce some aluminum-intensive automobiles in order to reduce structural weight and, thereby, improve fuel efficiency. Considerable process development has occurred at aluminum producers as well as at companies involved in or supporting aluminum fabrication. Aluminum alloys, being highly formable, are ideal for the mechanical clinch forming of buttons that lock sheet-gauge skins, inner-skin doublers or stiffeners, space-frame components, and so on (Green, 1997). One interesting example is Tempress Technologies' (Kent, WA) hyper-pressure pulse bonding process. In it, ultrahigh-pressure jet pulses (actually impulses) from compressed water bond aluminum components by causing plastic flow and mechanical interlocking at the interface between aluminum or other metal components (Kolle', 1994). Pressures as high as 7 GPa can be achieved with such pulse jets.

5.5 INDENTATION-TYPE AND BEADED-ASSEMBLY–TYPE JOINTS

Before moving on to methods for producing plastic interlocks by conforming one part to another, it is necessary to look at two related methods that, as similar as they appear, actually transition from using co-formed to using conformed interlocking features.

In *indentation-type joints,* one sheet-gauge metal part contains a pre-existing recess, normally in the form of a dimple. A second sheet-gauge part placed on top of this dimpled part is then pressed by some means to plastically form its surface down into the dimple(s). Once the second part is forced down into the dimple(s), the two parts are locked together against translation and/or rotation in the plane of their common interface. Alternatively, in a co-formed approach, the two parts can be simultaneously deformed to produce a dimple and mating protrusion.

Figure 5.10 (top right) schematically illustrates an indentation-type conformed plastic interlock between two sheet-metal-gauge tubes.

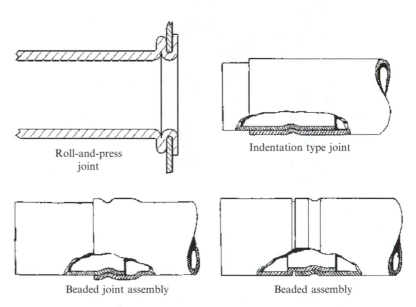

Roll-and-press
joint

Indentation type joint

Beaded joint assembly

Beaded assembly

FIGURE 5.10 Schematic illustrations of indentation-type (top right), two forms of beaded-assembly joints (bottom left and right) and "roll-and-press" joint (top left) in sheet-gauge metal components; here, tubes.

In *beaded-assembly joints,* either one sheet-gauge metal part, such as a tube, can either be forced down into a pre-existing groove or the two stacked sheets (or concentric tubes) can have male and female beads co-formed. In either case, a plastic interlock is created that resists axial motion between the tubes, for example. One common approach to producing beaded joints is to use hydrostatic pressure in what is generically known as "hydro-forming."

Figure 5.10 (bottom left and right) schematically illustrates a beaded-assembly joint in which an outer tube is caused to conform with a pre-grooved or pre-beaded inner tube.

Both indentation-type and beaded-assembly joints are found in metals but could, conceivably, be used in thermoplastic polymers as well. In metals, forming could be done cold or hot, depending on the particular metal. In thermoplastics, forming would always be done hot (i.e., above the glass transition temperature for the particular thermoplastic).

Figure 5.10 also shows what is known as a "roll-and-press joint" at the top left.

5.6   CRIMPING AND HEMMING

Crimping and *hemming* are similar processes in that both join metal or thermoplastic parts by plastically deforming one part over and/or into another part while they are in contact in order to cause interlocking. Both processes rely on one part conforming closely to the other. Parts are held together against unwanted movement by a combination of macroscopic (albeit sometimes small-scale) physical interference and friction from interlocking microscopic asperities being squeezed under a force from the elastic recovery component in the material used in the parts.

In *crimping,* an outer piece is crushed, squeezed, or otherwise plastically deformed around another to prevent subsequent relative movement between the two. To work, the outer part must be made from a material that is easily deformed, or, in some cases, a soft, malleable metal is sandwiched between the crimped and fixed parts to provide better compliance and result in more intimate contact. As an example, twisted or braided fine metal wires are commonly locked into a soft metal terminal connector by crimping the connector body down onto and around the wires for electrical assemblies.

In *hemming,* a linear joint is formed by plastically folding one piece of sheet metal over and around another to create an immovable seam through the resulting conforming and interlocking features produced. A well-known example of hemming is in automobiles with steel body panels. Outer, decorative skins are folded around inner structural skins using hemmed joints. Occasionally, hemmed joints in automobiles are subsequently spot welded to provide additional strength during collisions.

Figure 5.11 schematically illustrates crimping and hemming plastic interlocks.

Figure 5.12 shows a typical example of hemming used to join decorative outer skin panels to structural stiffening inner panels of a modern automobile,

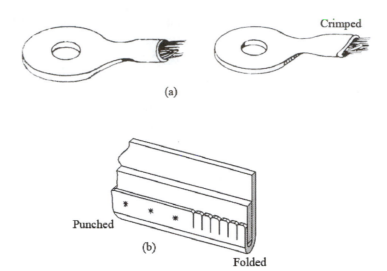

FIGURE 5.11 A schematic illustration showing some examples of crimping (a) and of hemming (b). (Reproduced from *Joining of Materials and Structures,* Robert W. Messler, Jr., 2004, with permission from Butterworth-Heinemann/Elsevier, Burlington, MA, publisher.)

while Figure 5.13 shows a typical hemming press (a) and robot used for accomplishing roller hems (b).

To work properly, care must be taken during design and production process planning to ensure that the outermost piece in either crimping or hemming assembly does not elastically spring back from the part(s) trying to be gripped.

5.7 THERMAL STAKING

The process of *thermal staking* is similar in both objective and principle to the process of setting or plastic upsetting described in Section 5.3. Like setting, which is used to spread metal at the driven end of hollow or tubular rivets or eyelets, and upsetting, which is used to form a second "head" or a "foot" on a solid rivet, thermal staking plastically deforms material at the end of a solid post that extends through, over, or around another part. By plastically deforming the end of such a post extending from a base part, the mating part is locked in position, orientation, and alignment, and unwanted movement is prevented. Unlike setting, however, which can be performed when the material is cold or hot, thermal staking is always performed when the material is hot and pliable or plastic. Furthermore, it is used extensively on materials that only plastically deform while hot. The two examples are glasses and thermoplastic polymers.

Another unique characteristic of thermal staking is that it is used to create a plastic interlock on the free end of an integral feature of the base part, not on the driven end of a fastener.

FIGURE 5.12 A photograph showing sheet-metal hemming in a modern automobile's assembly; here a front door. (Photograph courtesy of General Motors Corporation, Detroit, MI; used with permission.)

Normally, integral posts[5] in glass or thermoplastic polymer base parts are thermally staked. Occasionally, thermal staking is used to do more than just form a locking "head" or "foot" at the free end of a post. Sometimes, particularly with glass base parts, thermal staking forms an integral feature (like a post) on a base part completely around or over a mating part, often fabricated from a different material (i.e., not a glass). The best known example is the creation of metal-to-glass seals such as those described in Chapter 11, Section 11.3.

Figures 5.14 and 5.15 show some examples of thermal staking schematically and in practice, respectively.

---

[5] "Integral posts," in the context of this treatment on integral mechanical attachment, are features that extend out of the plane of a part and typically have a length- or height-to-width or -diameter ratio that is larger than 1:1. As seen in Chapter 3, Section 3.8, posts are rigid locating features.

(a)

(b)

FIGURE 5.13 Photographs of a typical hemming press (a) and a robot used for accomplishing roller hemming (b). (Photographs courtesy of General Motors Corporation, Detroit, MI; used with permission.)

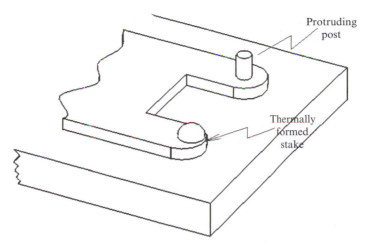

FIGURE 5.14 A schematic illustration of thermal staking in thermoplastic or, alternatively, in glass.

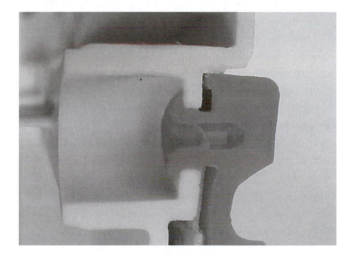

FIGURE 5.15 A photograph showing thermal staking of a thermoplastic part in a child's toy. Such staking absolutely prevents disassembly without destroying the part(s). (Photograph courtesy of Fisher-Price, East Aurora, NY; used with permission.)

5.8 FORMED TABS

Perhaps no method of accomplishing joining using plastic integral mechanical attachments or interlocks is better known than *formed tabs*. Widely used in assembling thin-gauge sheet-metal into parts and parts into assemblies, formed tabs have been used since people first created useful objects from thin metal. Ancient people used formed tabs for producing both utilitarian and decorative metal ware from hammered copper, bronze, and iron. With the dawn of the Industrial Revolution in England in the late 18th century, metal-working, aided by steam-powered machines, rapidly expanded, and formed tabs, as well as other plastic interlocking techniques such as bead-assembled (see Section 5.5) and hemmed joints (see Section 5.6), were the most common means by which sheet-gauge material was joined. At various times since then, right up to present times, the use of formed tabs has waxed and waned but never disappeared as an important method of joining. Following World War I, and again following World War II, in the United States, for example, so-called "tin-type" toys became the rage. Actually fabricated from thin-gauge steel, like that used in so-called "tin cans," virtually all assembly was done using formed tabs.

Formed tabs refer to one of two generic features found in thin-gauge sheet-metal. First, they refer to protruding "ears" (see Chapter 3, Section 3.5) located at various points around the perimeter of initially flat stamped parts. Alternatively, they refer to flaps punched out around most of their perimeter at locations in the interior of an initially flat stamped part, but still being attached to the base sheet at one edge to act as an integral hinge. In either case, these ear-like protrusions or "tabs" can then be folded or otherwise formed up and around another part to hold that part to the base part to which the tabs are integral. Once folded or formed so as to lock the mating part to the base part, unwanted movement between the two is prevented by suitable tab location and orientation. Separating forces are resisted by the formed tab having to be bent back from its formed position. This approach may be known from the way in which diploma certificates or family portrait-photographs, for example, are held behind protective glass in a picture frame. Thin soft-metal tabs located around the inside perimeter of the frame on its back side are folded out of plane to allow the diploma or photograph to be inserted, followed by a backing piece of cardboard, and then the tabs are pressed tight against the backing cardboard to hold everything in place.

In other applications, assembly is further enabled by having the tabs pass through a narrow slot during part-to-part engagement, and then the tab, once it fully extends through the slot, is folded or formed plastically to lock the parts together. A familiar example of this approach is the use of tabs and slots in the cardboard boxes that contain dry cereal or crackers. By inserting the tab into the slot, the cover of the box remains closed to keep the contents fresher and secure from "critters."

By judiciously choosing where to locate and how to orient tabs on a base part, and how to form them to trap a mating part, it is possible to obtain a wide variety of constrained and/or allowed directions of motion in translation or rotation.

Figure 5.16 schematically illustrates some formed tabs used in thin-gauge sheet metal, while Figure 5.17 shows an example of the use of formed tabs in office furniture. Figure 1.13 shows formed tabs used to assemble an old "tin-type" toy.

Chapter 7 describes and portrays a great variety of formed tabs used in sheet-metal fabrication and assembly.

Formed tabs can also be created in thermoplastic using heat to form and set the tabs. However, such use is not at all commonplace.

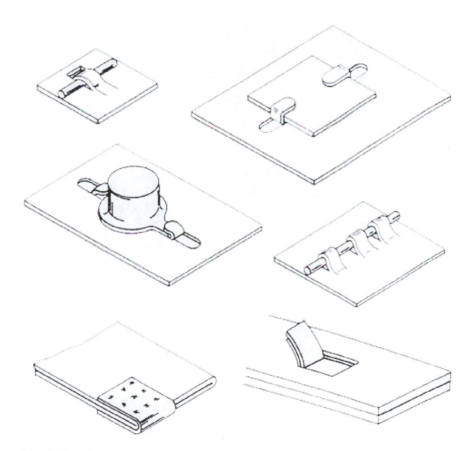

FIGURE 5.16   Schematic illustrations of formed tabs such as are commonly used in assemblies of formed thin-gauge sheet-metal parts.

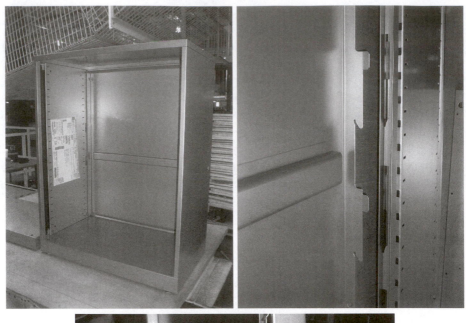

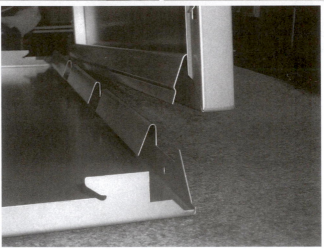

FIGURE 5.17 The use of formed tabs is common in the assembly of formed sheet-metal part such as found in certain office storage cabinets and file cabinets. Here, a metal storage cabinet (top left) has all five punched and formed sheet-metal panels (i.e., top, two sides, back, and bottom) that slide together and interlock using a variety of formed tabs, slots, and so on. Only two screws are used in the final step to complete the assembly. Details are shown in the photographs at the top right and bottom. (Photographs courtesy of Steelcase, Inc., Grand Rapids, MI; used with permission.)

5.9 SUMMARY

Plastic integral mechanical attachments or *plastic interlocks* are created between parts following their assembly. Actual production of interlocking features by plastically deforming a portion of one part over, around, or into another locks those parts together against unwanted motion. The force responsible for locking one part to another is the combination of macroscopic physical interference (leading to load transfer in bearing), with some degree of microscopic interference, in the form of friction, from the interference and interaction of microscopic asperities on the surfaces of the interacting parts. This friction force is enhanced by any elastic recovery component of the stress used to create the plastic interlock(s) that might be acting.

Basically, all plastic interlocks are created either by co-forming interlocking features between parts of an intended assembly or by causing one part to conform to another and create locking. Within these two generic approaches are found several methods of creating plastic interlocks, including: (1) setting (actually used with rivets and eyelet fasteners) and staking; (2) various types of metal stitching, typified by BMT Corporation's patented Tog-L-Loc™ and Lance-N-Loc™ sheet-metal joining methods; (3) indentation-type and beaded-assembly joints; (4) crimping and hemming; (5) thermal staking; and (6) formed tabs.

Not surprisingly, plastic interlocks can only be employed with materials that can be easily plastically deformed and still provide a resisting force once formed. Metal is the most common material in which plastic interlocks are found, but various types can also be created in the more rigid thermoplastic polymers and in glasses of all types.

Table 5.1 lists the major processes and methods for creating plastic integral mechanical attachments or interlocks.

TABLE 5.1 List of Plastic Integral Mechanical Attachments or Plastic Interlocks (and Materials in Which These are Found)

Co-Formed Attachment Features or Interlocks

Staked interlocks (metal/thermoplastic polymers)

Clinched metal (e.g., Tog-L-Loc™ and Lance-N-Loc™) (metal)

Formed tabs (metal)

Conformed (or Conforming) Attachment Features or Interlocks

Crimps (metal/thermoplastic polymers)

Hems (metal; thermoplastic polymers)

Formed tabs (metal/thermoplastic polymers)

Indentation-type joints (metal)

Beaded-assembly joints (metal)

Folded joints (metal/thermoplastic polymers)

Roll-and-press joints (metal)

Other Methods

Metal stitches/staples (cast iron)

REFERENCES

Green, R.C., "Clinching is a cinch," *Forming and Fabrication Magazine,* February 1997.

Kolle', J.J., "Development and applications of hydrostatic pulse generation," *Mechanical Engineering,* 116(5), 1994, 81–85.

Messler, R.W., Jr., *Joining of Materials and Structure: From Pragmatic Process to Enabling Technology,* Elsevier/Butterworth-Heinemann, Burlington, MA, 2004.

Parmley, R.O., *Standard Handbook of Fastening & Joining,* 2nd edition, McGraw-Hill, New York, NY, 1989.

6   INTEGRAL MECHANICAL ATTACHMENT CLASSIFICATION REVISITED

6.1   COMPARISON OF METHODS: RELATIVE ADVANTAGES AND DISADVANTAGES

It is as true in joining, generally, and in integral mechanical attachment, specifically, as it is anywhere else that "nothing is perfect." But, it is also true that "some things are better than others" and "it is important to use the right tool for the right job." With these in mind, it is worthwhile to revisit the classification of integral mechanical attachment in the context of joining, overall, and of the various methods available within integral mechanical attachment, in particular. The place to begin is with relative advantages and disadvantages.

Table 1.2 in Chapter 1 summarizes the general advantages of mechanical joining compared to the two other major processes by which materials and structures can be joined, that is, adhesive bonding and welding. Most important, mechanical joining uses only mechanical forces to hold the parts of an assembly or structure together, relying on the physical interference of one part with another to resist unwanted movement or motion in selected directions of translation and/or rotation. The other two primary joining processes, adhesive bonding and welding, rely on chemical and purely electrostatic (i.e., physical) forces and atomic-level bonding between the materials comprising parts to hold those parts together. Both offer a permanency of joining that may seem attractive, but not without its drawbacks. Within mechanical joining, integral mechanical attachment, as opposed to mechanical fastening, offers some specific advantages that are also summarized in Table 1.2.

TABLE 6.1 Relative Advantages and Disadvantages of Rigid, Elastic, and Plastic Integral Mechanical Attachments

Rigid	Elastic	Plastic
Advantages:		
1. Capable of carrying highest loads/tolerate highest stresses/ best retention		1. Capable of carrying modest loads/tolerate high loads/good retention
2. Can create permanent joints	2. Joints can be semi-permanent	2. Joints tend to be permanent
		3. Simplest feature geometry
4. Simple, low-force assembly	4. Simple, low-force assembly	
	5. Lowest per unit cost for features	
6. Normally easiest to intentionally disassemble	6. Sometimes can be intentionally disassembled easily	
7. Design analysis is straightforward		7. Design analysis is reasonable
8. Best for inherently rigid materials	8. Best for elastic, semi-rigid materials	8. Best for plastic, semi-rigid materials
9. High retention force	9. High retention-to-insertion force ratio	9. High retention force
	10. Tend to tolerate dimensional variations	10. Most tolerant of dimensional variations
Disadvantages:		
	1. Capable of carrying only modest loads/tolerate only low stresses/fair retention	
3. Feature geometry can be complex	3. Feature geometry is most complex	
		4. Most complicated assembly/ requires forming after assembly/assembly forces are the highest
5. Per unit feature can be high		5. Per unit feature cost can be high
	6. Intentional disassembly can be complicated	6. Difficult to impossible to intentionally disassemble
	7. Design analysis is complicated	
10. Intolerant of dimensional variations		
	11. Accidental disengagement is the most likely	

Note: Numbers in table indicate correspondence between/among attachment types.

Conceding that nothing is perfect, Table 1.3 summarizes the general potential disadvantages of mechanical joining compared to adhesive bonding and welding, including the issue of lack of joint permanency. It also summarizes the specific potential shortcomings of integral mechanical attachment within mechanical joining compared to mechanical fastening.

Using the classification of integral mechanical attachments based on the character or operation of the features responsible for allowing and enabling joining presented in Section 2.3 and shown in Figure 2.3, integral mechanical attachments were divided into rigid, elastic, and plastic types. Table 6.1, presented here, summarizes the relative advantages and disadvantages of the rigid, elastic, and plastic types. For simplicity, each advantage, under each type, is numbered. Advantages bearing the same number under the different types mean that these types share this advantage, at least to a significant (albeit, somewhat different) degree. Disadvantages listed under any particular type of integral attachment feature bearing the same number as an advantage for another type of integral attachment feature allow the different types to be compared one to the other.

Without rehashing the entire contents of Table 6.1, suffice to say here, in summary:

- Designed-in rigid interlocks are capable of carrying the highest loads or, more correctly, tolerating the highest stresses, and they offer the best retention against unwanted separation in directions for which they are to resist separation.
- Formed-in plastic interlocks can, depending on the material in which they are produced, tolerate stresses as high as most rigid interlocks. (It must be recognized that materials capable of being plastically deformed to allow the creation of plastic interlocks, tend to be inherently less strong than brittle materials suited solely to rigid interlocks.)
- Joint permanency tends to follow the ability of an interlock to tolerate stress in the separation direction(s). However, when a joint is permanent, it cannot be disassembled for any reason by any means that does not damage the parts comprising the joint.
- The geometric complexity of features tends to be greatest for the elastic interlocks, followed by the rigid interlocks. The geometry of plastic interlock features is almost always quite simple.
- Rigid interlocks, followed by elastic interlocks, require the lowest forces to accomplish assembly of parts, while creation of plastic interlocks, which must be formed-in after parts are assembled, requires the most force to create the joint.
- The cost of creating the actual locking features is highest for rigid and plastic interlocks because whatever the labor intensity needed is, it recurs for each joint or joint element. Which of these two has the higher cost depends on the specific complexity of the lock feature's geometry and size. The labor intensity and skill-level required to produce some plastic interlocks (e.g., hems) can be high. The cost of elastic interlock features tends to be low

because, being established by molding dies, for example, the labor cost is non-recurring and is amortized over the number of parts produced.

- Rigidly interlocked joints are almost always the easiest to intentionally disassemble. Elastically-interlocked, snap-fit joints can be difficult to intentionally disassemble, unless provision to do so is taken into account with the inclusion of certain enhancements. Plastically-interlocked joints are the most difficult to disassemble and can be impossible to disassemble without damaging the parts involved.

- The analysis required during the design stage is most straightforward and is almost always the least complicated, for rigid interlocks, followed closely by plastic interlocks. The analysis required during the design stage to create properly functioning elastic interlocks is definitely the most involved. Prior experience with elastic interlocks is a major advantage for designing new elastic interlocks.

- There is a direct correlation between the inherent properties of the material in which integral mechanical attachments are to be used and the character of the interlock, that is, rigid, elastic, or plastic. (More will be said about this in Section 6.3.)

- Because they are formed-in after parts are assembled, plastic interlocks are the most tolerant of always-present variations in the dimensions of the parts involved. Elastic interlocks can be designed to tolerate some dimensional variations by considering certain enhancements during the design stage. Rigid interlocks can be quite intolerant of dimensional variations in the parts involved in the joint.

- Elastic interlocks are, by far, the most prone to accidental disengagement in planned retention directions, although tolerance can be improved with the addition of certain enhancements.

6.2 CLASSIFICATION OF INTEGRAL MECHANICAL ATTACHMENT METHODS

Section 2.2 discussed and Figure 2.1 presented a taxonomy that included all joining processes. This taxonomy, as is usually the case, shows familial relationships between the various processes and methods. Section 2.3 discussed and Figure 2.3 presented a classification scheme specifically for integral mechanical attachments based on the character and operation of the features that allow and enable joining. It was here that division into rigid, elastic, and plastic types appeared. Section 2.4 discussed and Figure 2.4 presented an alternative classification scheme based on the method by which the features used to allow and enable joining are created in the parts of an intended assembly.

Table 6.2, presented here, presents a classification of all of the major methods of integral mechanical attachment discussed in Chapters 3, 4, and 5 for rigid, elastic, and plastic types, respectively. In fact, at the highest level, the

TABLE 6.2 Classification of Major Methods of Integral Mechanical Attachment

Feature Deformation for Joint Creation		
None (Rigid Interlocks)	**Limited temporary (Elastic Interlocks)**	**Permanent (Plastic Interlocks)**
Designed-in Features:	**Designed-in Features:**	**Designed-in Features:**
Tongues-and-grooves	Cantilever hook snaps	Formed tabs (some)
Dovetail-and-grooves	Cantilevered holes snaps/window snaps	
Rabbets/dados	Annular snaps	
Dovetail-fingers	Leaf-spring snaps	
Mortise-and-tenons	Post-and-dome snaps	
T-slots	Compression hook snaps	
Rails-and-ways	Compressive traps	
Integral threads	Compressive beams	
Tabbed fittings	Bayonet-and-finger snaps	
Bayonet fittings	Torsion snaps	
Keys-and-slots	Velcro™ hooks-and-loops	
Splines	3M Dual Locks™	
Integral keys	Integral spring tabs	
Flanges		
Shoulders		
Ears and tabs		
Bosses and lands		
Posts		
Rigid locators (in snap-fits)		
Added Features:		
Press-fits		
Shrink-fits		
Welded-on studs		
Cemented-in studs		
Potted-in studs		
Cast-in inserts		
Molded-in inserts		
Collars		
Sleeves		
Detachable Features:		
Wedges/Morse tapers	Spring slides	
Hinges	Spring clips	
Hasps	Clamp fasteners	
Hooks	Clamps	
Latches	Quick-release fasteners	
Turn-buckles		
Rigid couplings		
Flexible/linked couplings		
Clutches		

(Continued)

TABLE 6.2 *(Continued)*

Feature Deformation for Joint Creation		
None (Rigid Interlocks)	**Limited temporary (Elastic Interlocks)**	**Permanent (Plastic Interlocks)**
Designed-in Features:	**Designed-in Features:**	**Designed-in Features:**
	Processed-in Features:	
Knurled surfaces	Setting	
		Staking
		Metal clinching
		Indentation-type joints
		Beaded-assembly joints
		Crimping
		Hemming
		Thermal staking
		Formed tabs (some)

table is divided into these three types for which no feature deformation is required to create a joint (for rigid interlocks), limited temporary deformation in the form of deflection is required to create a joint (for elastic interlocks), or permanent deformation is required to create a joint (for plastic interlocks). Below each type, specific methods are grouped under headings as to whether the features that allow and enable joining are designed-in (prior to detail part fabrication), added (following detail part fabrication), or are processed-in (following the actual assembly of pre-fabricated detail parts). Also included is a group for methods that allow the easy detachment of detail parts in a joint. No further discussion is needed.

After looking at Table 6.2, the interested reader can search out details on a specific attachment method by referring to the Index.

6.3 CORRELATIONS BETWEEN JOINT MATERIALS AND ATTACHMENT METHODS

Section 2.8 discussed the rationale for various integral mechanical attachment methods to be identified with specific classes or types of materials. Without repeating Section 2.8 here, suffice to say:

- Inherently brittle materials are most suited to integral mechanical attachment, specifically, and mechanical joining, in general, using rigid interlock methods. Obviously, plastic interlocks cannot be created in inherently brittle materials, and elastic interlocks could be prone to damage during deflection due to the intolerance of brittle materials to localized stress concentration associated with discrete points of attachment or fastening.

- Materials that have a relatively high elastic limit and a fair tolerance of strain to fracture are most suited to integral mechanical attachment using elastic interlocks, in general, and snap-fits, in particular. On the other hand, materials must not be too flexible, or elastic snap-fits, in particular, are prone to accidental disengagement from either elastic relaxation or flexure.
- Only materials that can be plastically deformed can be considered for integral mechanical attachment using plastic interlocks. However, too much plasticity (as in soft metals and in elastomeric polymers) can prevent formed-in locking features from holding under load/stress.

Table 6.3 presents a correlation between the material comprising joint elements and the possible methods by which parts might be joined by integral mechanical attachment. In this table, an "X" indicates that examples are absolutely known to exist or have existed for this combination of material and method. A "?" indicates that no specific example is known to exist or to have ever existed, but the feasibility for success seems likely. A "–" indicates that the material and method are definitely incompatible or highly unlikely to be compatible.

The materials included in Table 6.3 are: ductile metals (being ones that can be plastically deformed by reasonable means); brittle metals (being ones that cannot generally be shaped by plastic deformation at practical, if any, temperatures); engineering or engineered ceramics (being ones that have been either highly refined or synthesized for desired properties); cement (including concrete); thermoplastic (TP) and thermosetting (TS) polymers (which are not too elastomeric); and fiber-reinforced plastics (or polymer-matrix), metal-matrix, ceramic-matrix composites, and wood (a natural composite). Integral mechanical attachment methods are divided by rigid, elastic, and plastic types, and every method discussed in Chapters 3, 4, and 5 is included.

6.4 SUMMARY

Following the detail description and discussion of integral mechanical attachment as a process for joining parts using only the mechanical forces that arise from the physical interference of one part with another and of the specific classes, that is, rigid, elastic, and plastic, it was considered worthwhile to revisit the classification of these important methods for joining once again.

Upon reflection by the author and the reader, relative advantages and disadvantages of rigid, elastic, and plastic integral mechanical attachments were considered and summarized in Table 6.1. The goal of the table is simplicity for making comparisons.

Following detailed discussion of the many specific major methods of integral mechanical attachment presented in Chapters 3, 4, and 5 for rigid, elastic, and plastic types, respectively, all major methods were classified again in a

TABLE 6.3 Correlation Between Joint Materials and Integral Mechanical Attachment Methods

	Metals		Ceramics		Glass	Polymers		FRP	Composites		Wood
	Ductile	Brittle	Engrd.	Cement		TP's	TS's		MMC	CMC	
					Rigid:						
Tongues-and-grooves	–	–	X	X	X	X	X	X	?	X	X
Dovetail-and-groove	–	–	–	–	–	–	–	–	–	–	X
Rabbets/Dados	–	–	–	–	–	–	–	–	–	–	X
Dovetail-finger	–	–	–	–	–	–	–	–	–	–	X
Mortise-and-tenon	–	–	–	–	X	–	–	–	–	–	X
T-slots	X	X	–	X	X	?	?	?	?	–	?
Rails-and-ways	X	X	–	–	–	–	–	–	?	?	?
Integral threads	X	X	X	–	X	X	X	?	?	?	X
Tabbed fittings/bayonets	X	X	–	–	X	X	?	?	?	?	?
Keys-and-slots	X	X	X	–	X	?	?	?	?	?	X
Splines	X	–	–	–	–	–	?	–	?	–	?
Integral keys	X	X	X	–	X	?	?	?	?	?	X
Knurled surfaces	X	?	–	–	–	–	–	–	?	–	–
Wedges/Morse tapers	X	X	–	–	–	–	–	–	?	–	X
Hinges	X	–	–	–	–	X	–	?	–	–	–
Hasps	X	–	–	–	–	?	?	?	–	–	–
Hooks	X	–	–	–	–	X	X	?	?	–	?
Latches	X	–	–	–	–	X	X	?	?	–	?
Turn-buckles	X	–	–	–	–	–	–	–	–	–	–
Flanges and shoulders	X	X	X	X	X	X	X	X	X	X	X
Ears and tabs	X	X	X	X	X	X	X	X	X	X	X
Bosses and lands	X	X	X	X	X	X	X	X	X	X	X
Posts	X	X	X	?	X	X	X	X	?	?	X
Press-fits	X	–	–	–	–	?	?	?	?	–	X
Shrink-fits	X	X	–	–	–	–	?	?	?	–	X
Welded-on studs	X	–	X	–	X	X	–	–	?	–	–
Cemented-in studs	–	–	X	X	–	–	–	–	–	?	–

	C1	C2	C3	C4	C5	C6	C7	C8	C9	C10
Potted-in studs	X	X	X	X	–	–	–	–	?	–
Cast-in inserts	X	X	X	X	–	–	–	–	?	–
Molded-in inserts	X	X	X	–	X	X	X	–	–	–
Collars	X	X	X	–	X	?	?	?	?	X
Sleeves	X	–	–	–	X	?	?	?	?	–
Rigid couplings	X	–	–	–	?	?	?	?	?	–
Flexible couplings	X	–	–	–	?	?	?	?	?	–
Clutches	X	–	–	–	–	–	–	–	?	–
Rigid locators in snap-fit	X	–	–	–	X	X	X	X	?	–
Elastic:										
Cantilever hook snaps	X	–	–	–	X	X	X	X	?	X
Cantilevered hole snaps	X	–	–	–	X	X	X	X	?	X
Annular snaps/Leaf-spring snaps	?	–	–	–	X	X	?	?	?	–
Post-and-dome snaps	?	–	–	–	X	X	?	?	?	–
Compression hooks	?	–	–	–	X	X	?	?	?	–
Compressive traps and beams	?	–	–	–	X	X	?	?	?	–
Bayonet-and-finger snaps	?	–	–	–	X	X	?	?	?	–
Torsion snaps	?	–	–	–	X	X	?	?	?	–
Velcro™ hook-and-loop	?	–	–	–	X	?	?	?	–	–
3M Dual Lock™ recloseables	?	–	–	–	X	?	?	?	?	–
Integral spring tabs	X	–	–	–	X	X	X	?	?	–
Spring slides and clips	X	–	–	–	–	–	X	–	?	–
Clamp fasteners	X	–	–	–	–	–	?	–	?	–
Clamps	X	–	–	–	?	?	?	?	?	–
Quick-release fasteners	X	–	–	–	X	X	X	?	?	–

TABLE 6.3 (*Continued*)

	Metals		Ceramics			Polymers		FRP	Composites		Wood
	Ductile	Brittle	Engrd.	Cement	Glass	TP's	TS's		MMC	CMC	
					Plastic:						
Setting	X	–	–	–	–	X	–	?	?	–	–
Staking	X	–	–	–	–	X	?	?	?	–	–
Metal clinching	X	–	–	–	–	–	–	–	–	–	–
Indentation-/Beaded-assembly	X	–	–	–	–	X	–	?	?	–	–
Crimping	X	–	–	–	–	X	–	?	?	–	–
Hemming	X	–	–	–	–	X	–	?	?	–	–
Thermal staking	–	–	–	–	X	X	–	?	–	–	–
Formed tabs	X	–	–	–	–	?	–	?	?	–	–

slightly different manner in Table 6.2. Specifically, this classification considered whether each major method had its features designed into detail parts prior to their fabrication, added following detail part fabrication prior to or during assembly, or processed in following assembly of the detail parts into their intended arrangement. An additional group was added to include all major methods that readily allow intentional disassembly of the joint or detachment of the lock.

Finally, the relationship between the material used to fabricate detail parts and the method used to join those parts using integral attachment features was reconsidered and summarized in Table 6.3. This table presents an easy reference for the designer to consider viable options for integrally attaching parts of any material.

It is now time to consider how specific types of materials can or should be joined to advantage using integral mechanical attachments. This is done in Chapters 7 through 12.

7 METAL ATTACHMENT SCHEMES AND ATTACHMENTS

7.1 PROPERTIES OF METALS THAT FACILITATE INTEGRAL MECHANICAL ATTACHMENT

Among the 111 elements listed in current versions of the periodic table (see Appendix A), only 17 are absolutely not metals; that is, they are nonmetals.[1] Of the others, 86 are definitely metals and another 8 are considered metalloids, by which is meant that they sometimes exhibit properties like metals and sometimes properties like nonmetals. Of the 86 metals, those above around atomic number 94 (plutonium, Pu) have half-lives that are measured in very small fractions of a second, so they are never encountered on Earth. Admittedly, not all of the other 69 occur in great abundance or are used in making actual physical objects, but they are metals and, as such, share many common characteristics. When all is said and done, there are probably about 35 or so that are used to make things at all, and most things one is likely to encounter are made from one or more of 27 metallic elements.[2] The point is

---

[1] Among the 17 nonmetals in the periodic table, 11 are gases at standard conditions of pressure and temperature (i.e., 1 atm and 25˚C), including 6 inert gases. One nonmetallic element (Br) is a liquid at standard pressure and temperature, and five (C, P, S, Se, and I) are solids. Chemists tend to think of C as a nonmetal, while materials scientists and engineers tend to think of it as a metalloid because, in the form of diamond, it is a semiconductor.

[2] As either relatively pure metals or as alloys (i.e., solutions of metals), the following metallic elements (from lowest to highest atomic number) are used: Be, Mg, Al, Ti, V, Cr, Mn, Fe, Co, Ni, Cu, Zn, Zr, Nb, Mo, Pd, Ag, Sn, Hf, Ta, W, Re, Pt, Au, Pb, Bi, and U.

that metals are abundant[3] and are both technologically and economically important. So, let's consider why, and then we'll know why it is important to be able to join them to produce useful objects, devices, assemblies, and structures.

A *metal* is defined by the ASM International (Materials Park, OH) as "an opaque, lustrous elemental chemical substance that is a good conductor of heat and electricity and, when polished, a good reflector of light." A more sophisticated and basic definition used by material scientists centers around the type of bonding that exists between the atoms comprising the pure elemental species and around the characteristics of the outermost electrons associated with these atoms. The bonding (known as "metallic bonding," which isn't terribly enlightening!) is said to be delocalized (as opposed to what occurs in covalent bonding), in that outermost electrons are shared among many near-neighbor atoms in a solid aggregate in which there is long-range order to the arrangements of the atoms.[4] A solid possessing such long-range order is said to be crystalline.

Besides having physical properties (e.g., melting point, boiling point, density, Poisson's ratio, coefficient of thermal expansion) that arise from details of the type of bonding, metals have mechanical properties (e.g., strength, hardness, stiffness, or modulus of elasticity) and electrical properties (e.g., conductivity and/or resistivity) that, besides arising from the type of bonding, often exhibit a dependency on directions in the crystal. All metals except mercury (Hg) are solid at room temperature, most (except Cs, Ga, K, Sr, Na, In, Li, Sn, Cd, Pb, and Zn) melt above 600°C (1,000°F), and many (e.g., W, Ta, Mo, Hf, Re, Os, and Ir) melt at very high temperatures (>2,000°C or 3,600°F). Almost all (except Li, Na, and K) are more dense than water (at 1 g/cm^3 or 1 kg/m^3) and than most of the solid nonmetallic elements. And, as stated earlier in the ASM definition, all metals conduct both heat and electricity as a result of their outermost electrons being able to move freely from one atom's core[5] to another's. Finally, and characteristically for all metals, electrical conductivity decreases (or, contrarily, resistivity increases) as temperature increases.

---

[3] In fact, the crust (or lithosphere) of the Earth, which varies from 6 to 20 miles (10–32 km) in thickness, consists of mostly nonmetallic minerals containing Si, O, H, N, C, S, Al, Fe; predominantly in the form of granite, and a relatively small fraction of metals. The lower (inner) portion of the crust consists of mostly basalt and igneous rocks. The molten core of the Earth, on the other hand, is believed to be almost entirely metallic, probably mostly Fe and Ni.

[4] "Long-range order" means that it is possible to predict with certainty (ignoring the occurrence of defects in crystals) that an atom will exist at a point in three-dimensional space for integral multiples of what are known as "unit lengths" in each of three directions for any of seven basic crystal systems of unit lengths and included angles (i.e., "crystal systems").

[5] In this context, an atom's "core" consists of its proton- and neutron-containing nucleus and all electron shells except the outermost valence shell.

Most interesting and most relevant for the purposes of joining, metals exhibit some interesting mechanical properties, where mechanical properties are the response of a material to a force or stress and/or a strain. It is mechanical properties that drive the selection of materials for structural applications, including mechanisms, and it is very often the preservation, if not optimization, of mechanical properties that drives the need for and selection of a specific process for joining (Ashby, 2004).

Another way in which metals are reasonably characterized (with some exceptions, such as liquid Hg and a few inherently brittle metals, like Cr) is that they are strong, hard, and ductile or malleable. By "strong" is meant that metals (and other materials said to be strong) are able to resist the application of a force without fracturing and breaking. By "hard" is meant that metals (and other materials said to be hard) are able to resist scratching, indentation, or penetration. By "ductile" is meant that metals (and other materials said to be ductile) are able to be compressed or stretched and have their shape changed without breaking or fracturing. "Malleable" literally means capable of being formed or shaped. It is these properties, usually, but not always, in combinations, that make metals particularly attractive for use in creating all types of physical objects, parts, and, structures. More details on the mechanical properties of metals (and other materials) can be found in any good basic textbook on materials science and engineering (see the bibliography at the end of this chapter).

The reasons for metals being particularly attractive for their inherent strength, hardness, and ductility or malleability are as follows: first, physical objects, parts, and structures must virtually always resist applied forces or loads and often must do so with the smallest size and lowest weight as is possible to achieve. Succeeding at this requires that the material of construction be strong. Second, in assemblies of parts in which one part of the assembly moves relative to another part, neither part should be worn away by the inevitable friction forces at the points or areas of contact. Even in assemblies of parts or in structures composed of multiple, smaller structural elements, it is usually essential that one part not indent or penetrate or even scratch any other part(s) that it touches. Succeeding at this requires that the material of construction be hard. Third, since the objects or devices we want or the detail parts or structural elements needed to create an assembly or a structure that does what we want, parts need to have sizes and shapes that fulfill their intended function. It is thus virtually always necessary to shape the metal into the object, device, part, or structural element. Such shaping can be accomplished in four fundamental ways: (1) by melting and re-solidifying the metal in a mold or die that will yield the shape needed in the newly-created solid part (i.e., by casting); (2) by working the metal while it is either cold (i.e., at room temperature) or hot by hammering, rolling, pressing, stretching, bending, drawing, or other means to create the needed shape (i.e., deformation processing or forming)[6]; (3) at least for small parts, by compacting fine particles of a metal in a mold or die that will yield the desired shape and sintering (i.e., solid-state diffusion welding or transient liquid-phase bonding) the compacted shape (i.e., powder processing); and (4) by bringing together smaller, usually simpler-shaped detail parts or structural elements to produce a larger and often more complex assembly or structure (i.e., joining) (Lindgren, 1990; Swift and Booker, 1997).

Even when detail parts can be produced by casting, deformation processing or forming, or powder processing, it is very often necessary to join these detail parts to one another. Reasons are numerous (Messler, 2004) but include the following especially important ones: first, to produce assemblies that to perform their function(s) must have internal parts move relative to one another, detail parts must be held in proper relative positions, orientations, and alignments. This can only be accomplished using a mechanical joining process, using supplemental fasteners or, less frequently, using integral geometric attachment features. Second, to produce individual parts that are larger or more geometrically complex than can be produced by casting, forging, rolling, forming, or powder processing, smaller parts must be joined to create larger and, possibly, more complex parts, assemblies, or structures. This can be accomplished using mechanical joining, adhesive bonding, or welding (including, possibly, brazing or soldering), depending on the materials involved, the loads/stresses to be tolerated, and the service environment to be survived. Third, to achieve needed functionality (and/or, perhaps, ease of manufacture, including affordability, or aesthetics), different metals or even different materials may need to be combined in a part or from part to part in an assembly or a structure. This can also be accomplished using any of the primary joining processes, depending on the materials, needs, and conditions.

So, hopefully, two things are now clear: (1) metals are particularly attractive materials for producing parts, assemblies, and structures because of their strength, hardness, and ductility, as well as their relative abundance and affordability, and ease of being fabricated into needed shapes, and (2) for a number of reasons, detail parts made from metals need to be joined to create larger, more complex assemblies or structures that have the required functionality and performance. The third thing that should be clear from the discussion in Chapter 1 is that a particularly attractive method for joining parts into assemblies or structures is integral mechanical attachment. But, what is it specifically that suits

---

[6] There is a particular advantage to creating parts by using a deformation processing method. Doing so, cold or hot, leads to the breakdown of the dendritic structure of as-cast metals or alloys through repeated nucleation and growth-based recrystallization. Since recrystallization must be carried out at temperatures exceeding about 40–50% of the absolute melting temperature of the metal or alloy (in degrees Kelvin), any compositional inhomogeneities (which arise naturally in all non-equilibrium solidification!) tend to be reduced by diffusion, and the nucleation and growth of new, equiaxed grains replaces the dendritic grain structure. The result is what is known as a "wrought" structure.

metal parts to being joined by integral mechanical attachment methods? Let's consider this question before looking at how metals can be joined by integral mechanical attachment schemes and attachments.

Recall that there are three fundamental classes of integral mechanical attachments or interlocks: rigid interlocks, elastic interlocks, and plastic interlocks. By looking at the behavior of a typical metal under an applied stress, it becomes clear why metals are unusually well suited to all three approaches.

Figure 7.1 shows the stress–strain curve for a typical (albeit hypothetical) metal. The particular curve shown is a plot of what is known as engineering stress versus engineering strain. It is derived from a test that actually measures the elongation response of a metal to an externally applied tensile force, which tries to pull the test specimen apart in one direction (i.e., uniaxial tension). To help see what is happening to the test specimen during testing, small insets show the test specimen at various points throughout the test.

To remove the effect of the size of the test specimen (or, more practically and importantly, the size of an actual structural member) from the results, the applied load or force, F, is divided by the cross-sectional area of the test specimen (usually in an intentionally-reduced region in the middle half—or so—of the specimen's length), A. The resulting force per unit area (from F/A) is known as the *stress,* in general, and, as long as the initial cross-sectional area A_o before testing is used, is known as the *engineering stress* (F/A_o). The units of stress are

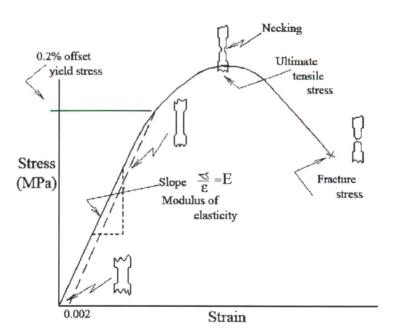

FIGURE 7.1 Plot of the engineering stress versus the engineering strain for a typical metal, showing, as insets, the progression of deformation in the tensile test specimen.

N/m^2 or Pa or MPa (10^3 Pa) or, in English units, lbs/in^2 or psi or ksi (10^3 psi). Similarly, to remove the effect of the length of the test specimen (or, more importantly, the length of an actual structural member) from the results, the resulting elongation, e, or change in length, ΔL, is divided by the length of a so-called "gage length" over which the change in length within the central portion of the reduced section of the test specimen is actually measured, L. The resulting change in length per unit length ($\Delta L/L$) is known as the *strain,* in general, and, as long as the initial length L$_o$ before testing is used, is known as *engineering strain* ($\Delta L/L_o$). Strain is dimensionless, since it is either mm/mm, cm/cm, m/m, or in/in, depending on the preferred length measurement units.

The plot of engineering stress versus engineering strain for a typical metal in Figure 7.1 shows several key points. First, metals initially exhibit an extended region of linear behavior between stress and strain. Throughout this region, all of the strain is fully recoverable (i.e., the metal behaves elastically). Removing the force (stress) allows the specimen or structural member to return to its original length.[7] Furthermore, the relationship between the resulting strain and the applied stress is linear, as given by $\sigma = E\varepsilon$, in which σ is the applied stress, ε is the resulting stain, and E is the modulus of elasticity, known as Young's modulus. The modulus of elasticity gives a measure of the stiffness of the metal, as, for example, how much a cantilevered beam of a given cross-section (and, thus, moment of inertia, I) and length will deflect under a given end-load. The units of E are the same as the units of the applied stress (i.e., Pa, MPa, or even GPa, 10^6 Pa).

Second, the stress level to which metals remain elastic (i.e., their elastic limit as usually measured by the stress required to cause 0.002 or 0.2% strain, known as their "0.2% offset yield strength" or simply yield strength) tends to be high compared to polymers, for example, although it tends to be low compared to ceramic materials and glasses, as well as reinforced composites, for which reinforcement is often specifically intended to increase stiffness.[8]

Given that materials, in general, and metals, specifically for this discussion, do not permanently (i.e., plastically) deform until relatively high stresses are reached means that they are very well suited to joining using rigid integral mechanical attachments. In fact, metals offer a particular advantage because, besides exhibiting high yield strengths, they also exhibit high strains to fracture. This can be seen in Figure 7.1. The strain to failure is given by the vertical dashed line projecting downward from the end of the stress–strain curve.

---

[7] In fact, metals, and other solid materials, when they are caused to lengthen in one direction shorten or contract in the two other orthogonal directions. This arises from the tendency of solids to maintain constant volume, although for some materials (notably, some polymers) the constancy is not perfect. The ratio of strains transverse to the loading direction (say, z) constitutes Poisson's ratio, $\nu = -\varepsilon_x$ or $-\varepsilon_y/\varepsilon_z$. For metals, ν is typically between 0.33 and 0.5.

[8] Actually, the particular advantage of reinforced composites tends to be that their stiffness is particularly high for their weight (i.e., their so-called "specific modulus", E/ρ, is high).

Compared to ceramic materials and glasses, which also tend to have high strengths and, in fact, rarely yield at room temperature, metals are much more ductile. Beyond good ductility, which many, if not most, polymers also exhibit, the combination of high strength and high ductility in metals results in what is known as high "toughness." Toughness is the ability of a material to absorb energy without breaking or fracturing, and a measure of the relative toughness of two materials can be obtained by comparing the areas under their stress–strain curves to the point of fracture. Such a comparison is made between a typical (albeit hypothetical) metal, ceramic, and polymer in Figure 7.2. Good toughness allows rigid interlocks to better tolerate impact loads and makes them less susceptible to problems from stress concentration at geometric notches, such as sharp radii.

The combination of good elasticity *and* high elastic limit or yield strength exhibited by many, if not most, metals and alloys of engineering interest suits metals to joining using elastic integral mechanical attachments. The only possible drawback is that the tendency of metals to be much stiffer than polymers, for example, means that the forces needed to cause an elastic interlock to deflect to allow engagement tends to be high. The good news is that if the forces needed for insertion can be applied, the resulting locking or retention force can be terrific. This behavior has tended to limit the application of snap-fits for assembly of metal parts.

The third feature of the stress–strain curve for a typical metal is that it exhibits a long range of strain over which it can deform past the onset of yielding before it fractures. This relatively high degree of plasticity (especially

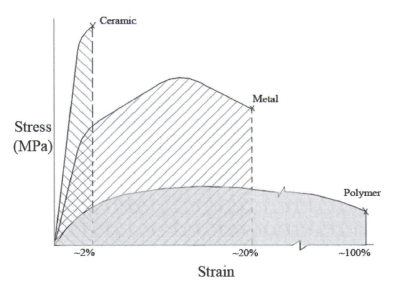

FIGURE 7.2 Plot showing the comparative toughness of a typical metal compared to a typical ceramic and a typical polymer.

compared to ceramics, glasses, and most reinforced composites) allows metals to be easily plastically deformed. At the same time that they are relatively easy to plastically deform, metals also exhibit relatively high strength, and, furthermore, they tend to get progressively stronger as they are deformed progressively more. This can be seen in Figure 7.3, in which the engineering stress–engineering strain curve is shown for a typical (albeit hypothetical) metal that is loaded past its yield strength (Point A) and is then unloaded (from Point B) and allowed to recover, as shown by the dashed line with downward-pointing arrows. Upon the removal of all loading, the material recovers all of its elastic strain (shown by $\varepsilon_{elastic}$) but exhibits some permanent plastic strain (shown by $\varepsilon_{plastic}$). If the metal is immediately reloaded, it will not begin to exhibit further plastic strain until the applied stress exceeds the stress at Point B, beyond which it will continue to plastically deform until it fractures at Point C. The increase in yield strength following plastic deformation is known as "strain hardening" or "work hardening." Strain hardening only occurs for plastic deformation done while the metal is below about its recrystallization temperature (see Footnote 6), in what is known as "cold working."

The combination of the ease with which metals can be plastically deformed without fracturing *and* both the relatively high stress needed to accomplish plastic deformation and the fact that strain hardening occurs suits metals to plastic integral mechanical attachment.

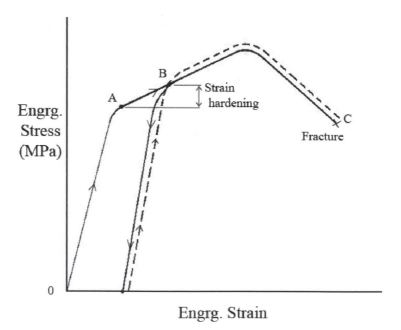

FIGURE 7.3 Plot of engineering stress-engineering strain showing the strain hardening that results from exceeding the yield stress below the temperature at which thermal recovery in the form of recrystallization would occur, typically greater than about 0.35–0.40 $T_{abs, mp}$ in metals.

So, in summary, metals have a combination of high elastic limit, reasonable modulus of elasticity, high yield strength, good hardness, good ductility, tendency to strain harden, and high strain to fracture that suits them to joining by all three classes of integral mechanical attachment: rigid, elastic, and plastic. Without question, metals are the most versatile of all materials when it comes to options for joining by integral mechanical attachment and, also, by all other processes and methods for joining (Messler, 2004).

With this background, it is now appropriate to look at the specific attachment schemes and attachments possible for the various forms in which metals are used.

Appendix B lists important properties of some metals and alloys that are particularly important in engineering.

7.2 SHEET-METAL ATTACHMENT SCHEMES AND ATTACHMENTS

One of the most common forms in which metal is used to produce parts and structures is as rolled sheet. Depending on the area of application (i.e., the industry), precisely what is considered sheet, as opposed to what is considered plate (which is thicker than sheet) or foil (which is thinner than sheet), differs. Builders of large ships and submarines, as well as railcar and locomotive builders, tend to consider flat product that has a width that is a significant fraction (say, $\frac{1}{3}$ to $\frac{1}{2}$) of the length as sheet until it exceeds about $\frac{1}{4}$ to $\frac{3}{8}$ inch (6–10 mm) in thickness. Beyond this, most consider it to be plate. On the other hand, automotive, aerospace, and appliance manufacturers tend to consider flat-rolled product sheet up to about $\frac{1}{8}$ to $\frac{3}{16}$ inch (3.0–4.5 mm) thick, and beyond that as plate. Almost everyone agrees that sheet becomes foil below a thickness of about 0.005 inch (about 0.1 mm), and as thin sheet below about 0.020 inch (0.5 mm).

For this discussion of integral mechanical attachment of sheet metal, we're talking about flat-rolled product that has a thickness of between about 0.010 and 0.150 inch (0.25 to about 4 mm), plus or minus a little bit. Furthermore, such flat-rolled product may be considered sheet, for which the width is a significant fraction of the length, or it may be strip, for which the width-to-length fraction tends to be less than $\frac{1}{6}$ or so, down to relatively much more narrow. Such flat product may have been rolled hot and subsequently cleaned, although most sheet and strip of the thicknesses being considered tends to have been cold-finished. Cold-rolled and cold-finished metals, especially steel, are free of surface scales, have good thickness control (i.e., tolerance on nominal thickness) and uniformity across their width, and have excellent surface finish beyond just being free of scale.

Assembly of sheet-metal parts normally occurs as parts are fabricated by forming the flat sheet into the needed shape and/or to join one simple, flat or formed part to another. Forming can involve simple bending to create an angle, as folding up the sides of a box from a pre-cut flat pattern so that the sides are continuous with the bottom. It can also involve the roll-forming of a flat piece into a curved piece, such as a cylinder. Or, finally, it can involve the more extensive forming or drawing of flat pieces into parts with compound curvature, bowl-like, pot-like, or more complex shapes such as automobile body panels. For any and all of these forming operations, joining many be needed to (1) close open joints (e.g., between the formed sides of the box); (2) add other formed details (e.g., a formed collar or flange); (3) literally join two or more parts that are formed simultaneously (i.e., co-formed) or formed one over the other, into the other, or around the other (i.e., conformed); (4) add detail parts fabricated from other than sheet metal, but from metal; or (5) join one formed sheet metal part to another.

Figure 7.4 schematically illustrates some of the major ways for which sheet-metal parts may require joining.

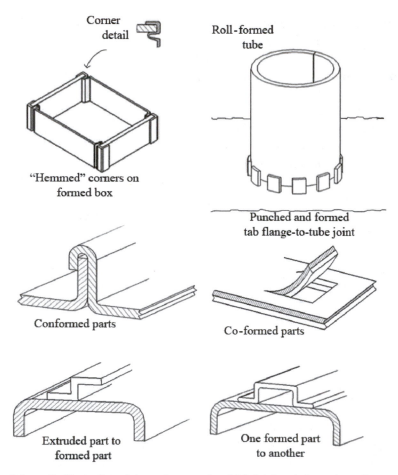

FIGURE 7.4 Schematic illustration of the major ways in which joining, independent of a specific process or method, can be used with sheet-metal part fabrication and assembly.

Without going into all of the processes that could be used to join sheet-metal parts,[9] it is worthwhile considering why integral mechanical attachment might be preferred over mechanical fastening. Using fasteners that impose a concentrated stress on a thin piece of material in its through-the-thickness direction is rarely wise, and it is not particularly different for sheet-gauge metals, although they tend to handle the situation better than polymers and most laminated composites. Without care, fasteners in sheet-gauge metals can cause elongation of fastener holes in bearing and can pull the fastener through the sheet under out-of-plane loads (e.g., bending). Rivets are used extensively and successfully in sheet-gauge metal in the aerospace industry and in certain transportation industries (e.g., buses, RVs, and motor homes), as well as in some appliances. Special sheet-metal self-tapping screws are also used, but, again, care must be exercised, as gauges get thin and as metals used are softer. As a general guideline, as sheet-metal gauges get thinner and/or as the metals involved are less strong and more plastic, integral mechanical attachment methods become more attractive.

Given these various needs, let's consider the options for joining by integral mechanical attachment. The order of presentation of the various options will be from the most integral to the least integral to the parts. In fact, the first several options, namely hemming and crimping, formed tabs, and beaded assembly, and metal clinching, all involve locking features that are fully integral to the parts of the assembly. The latter two options, namely integral fasteners and stitching and stapling, involve the use of supplemental devices that in the first case are obviously fasteners, but in the second case are less obviously fasteners. These are included in this treatment, however, because integral fasteners are made part of the detail parts prior to their assembly, and for stitching and stapling, they become quite permanent in the assembly once they are installed.

7.2.1 Hemming and Crimping

Chapter 5, Section 5.6, describes the processes of hemming and crimping for creating plastic interlocks. Both are ideal for joining sheet-gauge metal using a similar approach, albeit for different applications.

In *hemming,* one sheet of metal is literally wrapped (by press-forming or rolling, as shown in Figure 5.13) around another, at an edge, to trap the inner piece and lock the two together into a sandwiched assembly (see Figure 5.11). It is

---

[9] Obviously, sheet metal can be joined by mechanical fastening, adhesive bonding, various types of welding, as well as brazing or soldering, and even by weld-bonding or weld-brazing. Which of these is most appropriate depends on a number of technical, economic, and practical considerations. For guidance, the interested reader is referred to references on joining in general (Messler, 2004) or on specific processes (Landrock, 1985; *AWS Welding Handbook*, 1990; *ASM Handbook,* Volume 6, 1993; Speck, 1997).

possible, though not usual, to sandwich more than one piece by wrapping all with another piece. The process is virtually identical in intent and practice to the process of hemming used in garment manufacturing. In garments, a hem is defined as a finished edge formed by folding the selvage of the cloth under and affixing it by sewing, for example. In metal, the edge of a single piece could be folded under to form a stiffening edge, for example, but, more often than not with metals, the process is used to wrap an outer piece around an inner piece so that the outer piece has a good appearance, while the inner piece provides structural stiffening.

This is precisely what is done in the automobile industry to attach outer body panels, that will be given a finish for aesthetic appeal, to inner panels that provide structural stiffening and crash protection. Since it is performed primarily on thin-gauge sheet steel (typically 1.2 mm thick outer panels to 2 mm thick inner panels), the process is occasionally called "ironing," which is somewhat descriptive of the way the outer piece is folded around and squeezed tight to the inner piece. Occasionally, hemmed edges of automobile sheet-metal body stock are further secured with either resistance- or laser-spot welds or structural adhesives or sealant-adhesive mixtures, or both in the form of weld-bonding. This is not unlike what is done in garment-making, in which folded (and ironed or steam-pressed) hems are then stitched or basted to stay secure.

The process of hemming is also used in the manufacturing of sheet-metal ducting for heating, ventilating, and air conditioning and of sheet-metal housings or shells for appliances such as washers, dryers, dishwashers, and so on.

Figure 7.5 shows an example of the use of hemming in modern automobile assembly.

Another form of hemming is known as a "roll-and-press joint," an example of which is shown in Figure 7.6.

Crimping also involves the squeezing of one thin sheet-metal part over another to lock the inner piece into the outer piece. The process of crimping is used almost exclusively for making electrical connections, however. In a typical example, a thin-gauge formed sheet-metal terminal connector has its cylindrical-shaped barrel portion squeezed onto a wire, preferably, one consisting of multiple twisted, bundled, or braided strands. This allows the metal of the soft terminal connector to plastically deform down into the crevices between the various strands of wire, locking the connector to the strands by the resulting physical interference and friction that develops. Aluminum or aluminum-alloy or copper or brass terminal connectors are usually employed both for their good electrical conductivity and for their easy plastic forming and deforming/crimping. For harder connectors and/or wires, a soft-metal interlayer can be used to facilitate interlocking after crimping.

Figure 7.7 shows an example of the use of crimping in the connection of electrical leads.

Chapter 12 describes mechanical methods for producing electrical connections in detail, wherein many different designs and approaches for crimping are presented.

FIGURE 7.5 A photograph showing hemming in modern automobile assembly, here the front door of a passenger car. (Photograph courtesy of General Motors Corporation, Detroit, MI; used with permission.)

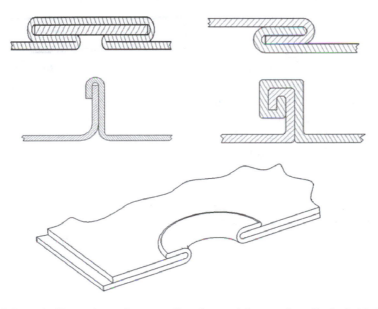

FIGURE 7.6 Schematic illustrations of some roll-and-press joints used to firmly hold sheet-metal parts in place.

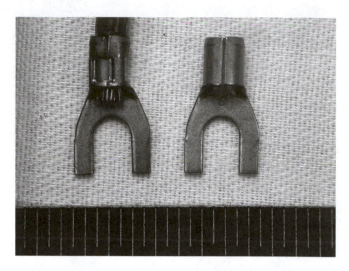

FIGURE 7.7 A photograph showing crimping (left) in electrical terminal connectors. (Photograph taken by Rob Planty for the author, Robert W. Messler, Jr.; used with permission.)

7.2.2 Formed Tabs

Before children's toys were made almost exclusively of plastics, beginning in the late 1950's and blossoming through the 1960's and into the 1970's, many toys were made from metal. Some of the earliest toys, such as coin banks that performed some tantalizing, mechanically-actuated actions, were made from cast iron. But, beginning around World War I, in the 1920's and until the early 1950's, thin-gauge sheet metal was used. Called "tin-type" toys today, most were actually made from inexpensive low-carbon steel. Fabrication of parts was done by stamping and punching flat-pattern details from a very thin (typically, 0.020–0.030 inch or 0.5–0.8 mm) sheet, forming them to the required shapes, color-printing or painting them, and joining formed and printed or painted details and any other parts required by the use of small integral tabs.[10] The concept of formed tabs was briefly described in Chapter 5, Section 5.8.

The integral tabs were one of two general types; they either extended from the periphery of the stamped flat pattern for the part or they were punched into the flat pattern during or immediately following its stamping. Figure 7.8 shows the two general types schematically.

For those tabs located on edges, once the stamped flat pattern was formed to the required shape, the tabs were inserted through punched slots in the mating stamped and punched part. The protruding ends of the tabs were then folded over

---

[10] Occasionally, some toys were soldered and some were riveted, but both methods were actually rare.

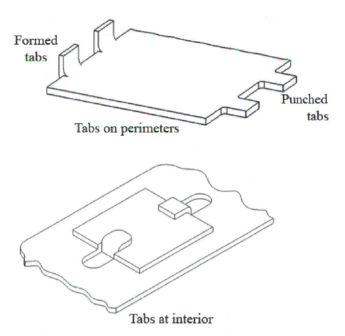

Formed tabs

Punched tabs

Tabs on perimeters

Tabs at interior

FIGURE 7.8 Schematic illustrations of the two general types of integral formed tabs, that is, when tabs are produced around the perimeter of a punched or blanked sheet-metal part to be formed later, or when they are punched at the interior of such parts to be formed later.

the part containing the slots, locking the two parts together. For tabs punched and formed into the interior of the flat pattern, these could be inserted through receiving slots if the tabs were cut free along three edges and were folded around the remaining attached edge. Alternatively, such tabs could be cut free on either two opposite edges and formed out of plane to create an integral loop in the sheet metal part or they could be cut free along three edges and formed to create either an open hook or a nearly-closed loop. More often than not, such interior punched-and-formed tabs were used to attach other parts that were usually not made from sheet metal. One of the best examples most now-grown boys of the tin-type toy era will remember was how such interior punched-and-formed tabs were used to hold the axle of the wheels on a toy car or truck.

In either case, peripheral or interior tabs were formed around mating parts to lock one part to another and, so, are appropriately called "formed tabs." Since they are formed to cause interlocking after parts are assembled, they are properly classified as plastic interlocks. However, in some cases, parts were inserted into already formed tabs by springing them open slightly, making them serve as elastic interlocks.

Such formed tabs were by no means restricted in their use to toys. Rather, they were found used in almost all objects made from thin-gauge sheet metal (especially steel), and they persist to this day.

Figures 1.13 and 5.17 (Chapters 1 and 5, respectively) show examples of the use of formed tabs for integral mechanical assembly of sheet-metal assemblies. Figures 7.9 and 7.10 schematically illustrate a number of different concepts for formed tabs of both types described earlier.

The variety of formed tabs used in sheet-metal parts is actually astounding.

7.2.3 Beaded Assembly

Another way in which sheet-gauge metal parts can be joined into an assembly using integral mechanical attachments is to use what is referred to as "beaded assembly." In *beaded assembly,* there are actually three sub-processes, one of

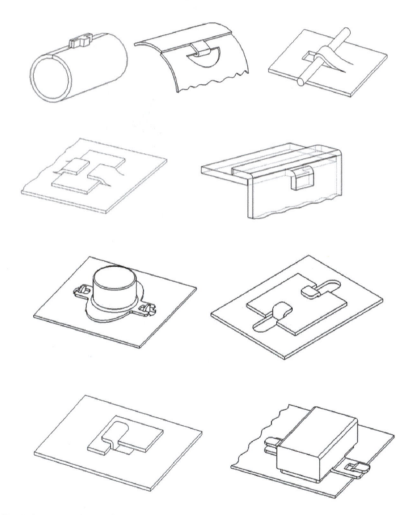

FIGURE 7.9 Schematic illustrations of formed tabs used in sheet-metal part assembly. (Adapted from Laughner and Hargan's long-out-of-print *Handbook of Sheet-metal Joining and Fastening,* formerly owned by McGraw-Hill, but no longer copyrighted.)

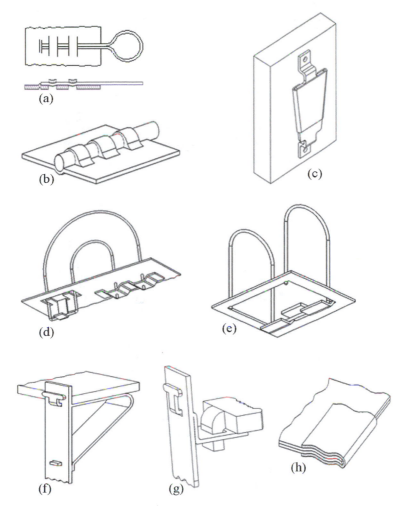

FIGURE 7.10 Schematic illustration of some other examples of formed tabs used in sheet-metal part assembly. (Adapted from Laughner and Hargan's long-out-of-print *Handbook of Sheet-metal Joining and Fastening,* formerly owned by McGraw-Hill, but no longer copyrighted.)

which creates plastic interlocks, one of which could create plastic or elastic interlocks, and one of which creates purely elastic interlocks. All sub-processes, however, are used nearly exclusively for joining tubular structures, although they could certainly be used for joining other closed-form shapes, such as covers on square or rectangular boxes.

In the sub-type known as *beaded joint assembly,* one part, say, a tube, contains a pre-formed circumferential groove into which the other tube is plastically formed after assembly. Variations of the assembly method are shown schematically in Figure 7.11. In Figure 7.11a, the inside bead is pre-formed,

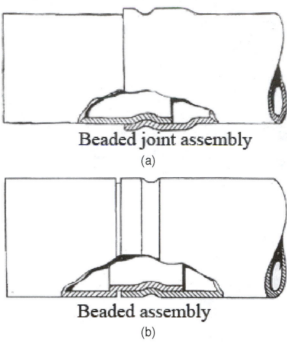

Beaded joint assembly

(a)

Beaded assembly

(b)

FIGURE 7.11 Schematic illustrations of the two ways in which beaded assemblies may be produced, that is, by forming the outer piece into a pre-formed inner piece (a) or vice versa (b).

and the outside bead is made after assembly to lock the two tubes together end to end. Alternatively, the outside bead could have been pre-formed, and the inner tube's bead could have been formed into it after assembly. In Figure 7.11b, a variation is shown in which the outer diameters of the two tubes are simply matched.

In the sub-type known as *indentation-type joints,* the inner part in a nesting pair (of, say, tubes) contains a pre-formed dimple or dimples, as opposed to a continuous circumferential pre-formed groove or recess. Once an outer part is assembled onto the inner part, the wall of the outer part is formed down into the dimple(s). The depth of the dimple(s) determines whether the joint is detachable. As just described, the resulting joint is the result of a plastic interlock. However, if both parts contain matching pre-formed dimples and protrusions, a joint can be formed by an elastic interlock. For this to be possible, the depth of the dimples (and heights of the corresponding bumps) cannot be too great.

A final sub-type, known as a *friction-type joint,* involves a readily detachable elastic interlock. One tube, for example, is designed and fabricated to fit into another with some combination of taper and springiness. The resulting elastic springing of either the inner or the outer part, whichever is less stiff, gives rise to friction that holds the two parts together. A well-known example of this type of joint is found in the extension tubes and end-accessories of vacuum cleaners.

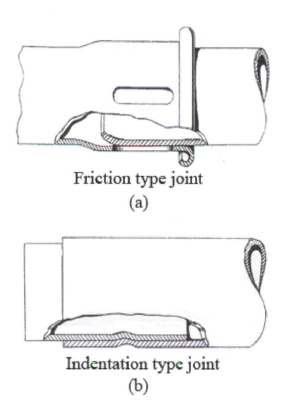

Friction type joint

(a)

Indentation type joint

(b)

FIGURE 7.12 Schematic illustrations of (a) indentation-type and (b) friction-type joints.

Figure 7.12a and b schematically illustrate indentation-type (a) and friction-type (b) joints.

Indentation-type and beaded-assembly–type joints were described earlier in Chapter 5, Section 5.5.

7.2.4 Metal Clinching

Metal clinching was described in Chapter 5, Section 5.4, as a plastic integral mechanical attachment method that creates interlocking features by a combination of forming and drawing of one part into and through another. In older approaches to metal clinching, rather gross deformations were caused at discrete points, usually involving the punching of what are similar to what were referred to earlier as tabs. In newer approaches, less gross interlocking depressions and buttons are created.

Two methods of metal clinching were described in Chapter 5, Section 5.4: BTM Corporation's patented Tog-L-Loc™ and Lance-N-Loc™ sheet metal joining systems and Tempress Technologies' hyper-pressure pulse bonding process™. The former system is shown schematically in Chapter 5, Figure 5.8a and b, and an example of the processes application is shown in Figure 5.9, also in Chapter 5.

In the late 1970's, when the aerospace industry was working with the NASA on then-futuristic concepts for assembling large-array solar collectors in orbit, one approach considered was the use of metal clinching. To construct the huge array (measuring thousands of meters in length and width) standard 1-meter-on-a-side equilateral-triangle cross-sectioned trusses would be fabricated in orbit using rolling mills mounted in the cargo bay of the Space Shuttle to form the very long corner elements from 0.020-inch (0.5 mm) thick Al-alloy strip, all three at once. To these would be attached the truss cross-members. In one concept, these would all be prefabricated to shape and size and stored, nested, in spring-loaded magazines by the thousands. As the corner elements were roll-formed, cross-members would be automatically located into position and then the cross-members would be clinched using any one of a number of different plastic forming methods and designs. Using this approach, problems inherent with fusion and non-fusion welding could be avoided, and sufficiently strong joints could be produced to hold the truss assembly together in the microgravity of Earth orbit.

While the mission didn't proceed at the time, many of the ideas for metal clinching did, albeit for more down-to-earth applications.

7.2.5 Integral Fasteners

There are some situations in which fasteners are used as supplemental devices to accomplish actual attachment, but either the fasteners, though fairly conventional, are used in a special way *or* the fasteners themselves are somehow special. Because of these unusual usages or types, these specialized fastening methods are included within this treatment of integral mechanical attachment, in general, and, because they are used almost exclusively with metals, are included under integral attachment schemes and attachments used for joining metals, in particular.

The specialized usage of fasteners in integral attachment involves what are referred to as "integral fasteners," while the specialized fasteners that facilitate integral attachment are referred to as "self-clinching fasteners" (Messler, 2004).

An *integral fastener* is a device, almost always threaded, that is installed into a component or unit (such as an automobile or truck chassis, aircraft or spacecraft airframe, or even an appliance) to become a permanent part of that component or unit. Such permanently mounted fasteners are used because they facilitate subsequent, often automated, assembly, because they are dependable in service in that they cannot be lost, and that they are generally very cost effective because they facilitate subsequent assembly, disassembly, and re-assembly, whether manual or automated. Permanently mounted integral fasteners may be riveted, mechanically clinched or swaged, welded, brazed, or soldered, or even adhesively-bonded in place. Such integral fasteners are available in many types including sems, captive nuts, and welded on studs or nuts.

Sems are a combination of various standard screws and a captive (as opposed to truly integral) washer. Such screw-washer assemblies are used for their convenience in assembly. They combine two parts into one that is much easier to handle, with no chance of failing to install the washer. This is particularly advantageous for automated assembly. Figure 7.13 schematically illustrates a few examples of sems.

The other common and important example of integral fasteners is the captive nut, also known as "anchor nuts" or "plate nuts." In all cases, these consist of an internally-threaded piece that serves as a nut and is able to "float" within a constraining mounting or cage. With this design, the nut is always there to accept a screw or bolt *and* is able to adjust slightly to account for slight mis-alignments from any source. The cage or body of the captive nut is, in turn, riveted, screwed, adhesively bonded, or soldered, brazed, or welded (usually, spot-welded) to the component on which it is being used. One captive nut is mounted beneath each hole that is to accept a screw or bolt. Figure 7.14 schematically illustrates a typical example of a captive nut that provides "floating position" of the nut to facilitate bolt or screw insertion.

The final type of integral fastener is that which is made integral to a metal component by welding. Threaded studs (to accept a nut), unthreaded studs (to be upset or accept a pin or retaining clip), and nuts (to accept a screw or bolt) can be welded to metal components. The welding process used is almost always either a specialized form of electric arc welding known as "stud arc welding" but can be capacitor-discharge (or percussion) resistance welding or friction welding (Messler, 1999). Figure 7.15 schematically illustrates some welded-on

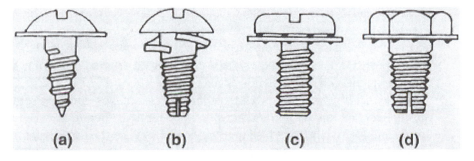

FIGURE 7.13 Schematic illustrations of various types of sems. (Reproduced from *Joining of Materials and Structures,* Robert W. Messler, Jr., Butterworth-Heinemann/Elsevier, Burlington, MA, with permission.)

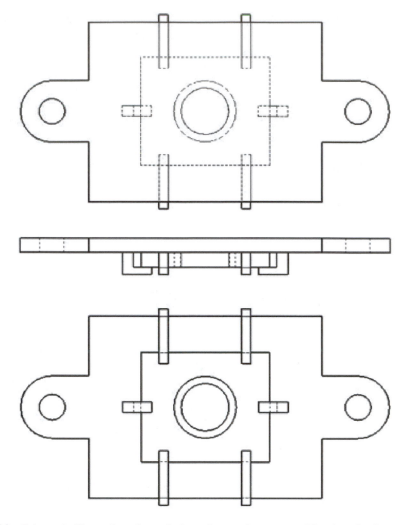

FIGURE 7.14 Schematic illustration of a typical captive nut, known as a "floating plate" nut.

fasteners and Figure 3.26 (see Chapter 3) shows their installation, whereas Figure 7.16 shows their use in service.

Beyond these specialized usages of fairly conventional fasteners, there are specialized fasteners, known as "self-clinching fasteners" that come close to creating an integral attachment system.

Self-clinching fasteners provide economic advantages during assembly and during service, and also provide cosmetic benefits as they tend to be compact and hidden. Self-clinching fasteners squeeze or swage themselves into a previously prepared hole in a sheet metal (or plastic) part. To work, the part must be composed of a metal (or plastic) that is softer than the metal used to make the fastener. In this way, when a portion of the fastener is expanded (often by the

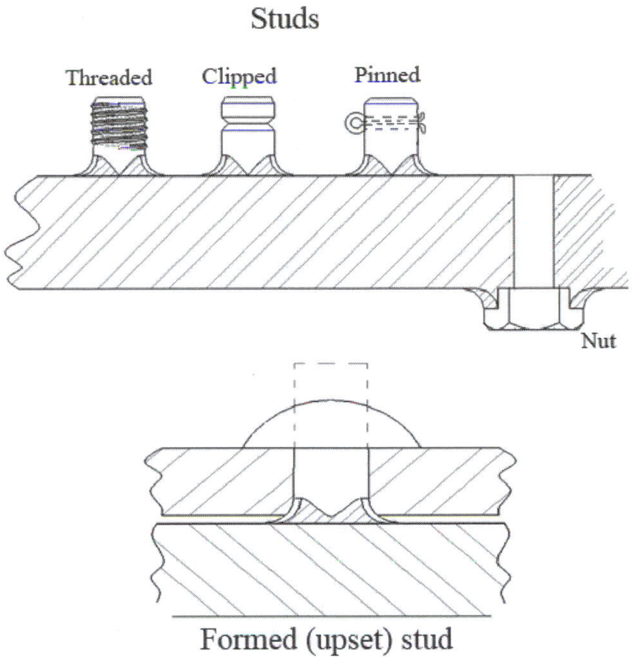

Studs

Threaded   Clipped   Pinned

Nut

Formed (upset) stud

FIGURE 7.15 Schematic illustration showing various welded-on fasteners.

insertion of a threaded screw or bolt), it forces itself into an annular groove around the perimeter of a clinch fastener collar or body under an installation pressure.

A typical installation of a self-clinching fastener is shown schematically in Figure 7.17. A strong permanent fit is ensured by such fasteners even in very thin (e.g., 0.020 inch or 0.5 mm) thick sheet metal (or plastic).

FIGURE 7.16 Photographs showing the welding on of studs to steel plate (left) to allow the subsequent fastening of seats in the Denver Arena using nuts (right). (Photographs courtesy of Nelson Stud Welding, Elyria, OH; used with permission.)

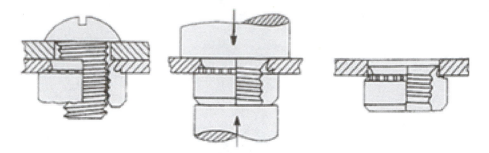

FIGURE 7.17 A schematic illustration of the operation of a self-clinching fastener, including the steps in its installation. (Reproduced from *Joining of Materials and Structures,* Robert W. Messler, Jr., Butterworth-Heinemann/Elsevier, Burlington, MA, with permission.)

Several fairly standard self-clinching fasteners include nuts with free-floating or self-locking threads; self-locking or non–self-locking floating insert nuts; flush-head, heavy-head, and concealed-head studs; through-hole, blind-hole, and concealed-head standoffs; panel fasteners with captive screws; spring-loaded pins; flush-head pins; and electrical grounding solder terminals (Parmley, 1989).

7.2.6 Metal Stitching or Stapling

The *stitching* or *stapling* of thin-gauge sheet metals with wires was, by the late 1950's, developed to the point that machines were available capable of driving steel wire through $1/16$-inch (1.5-mm) sheet metal at speeds up to nearly 300 stitches per minute. The processes could be used to fasten metal to metal or metal with a softer material, including rubber, plastics, leather, wood veneers, plywood, fiberboard, cork sheets, or fire-resistant materials, including, at the time, asbestos.

During stitching, or what really much more closely resembles stapling, the wire is fed through gear-driven feed rolls to a device that grips and holds it in position until a cutoff die cuts it to a predetermined length. The cut wire is carried by a mandrel to forming dies that bend it to an inverted U shape. These U-shaped staples are then driven downward through the sheets to be attached using a ram. The legs of the staple that pierce and project through the material are formed against a clinching die while the top of the staple is pressed tightly against the material to resist the forming force and develop a slight squeezing force through the layers of materials.

Figure 7.18 schematically illustrates various types of stitches or staples, as well as the recommendations for parallel and diagonal stitch placement.

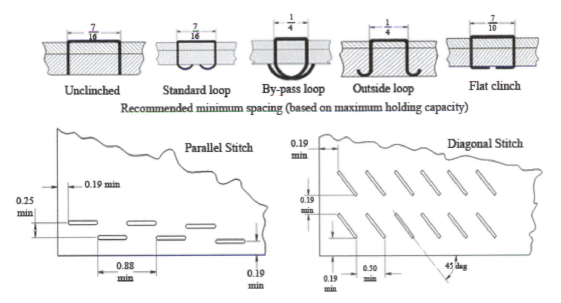

FIGURE 7.18 Schematic illustrations of the various types of stitches or staples used with metals, along with recommendations for parallel and diagonal stitch placement. (Adapted from Laughner and Hargan's long-out-of-print *Handbook of Sheet-metal Joining and Fastening,* formerly owned by McGraw-Hill, but no longer copyrighted.)

Normally, 0.048-inch (1.2 mm) diameter steel wire, cold-drawn to a tensile strength of from 220,000 to 330,000 psi (1,500–2,300 MPa), was used. The lower strength wire was capable of penetrating a single thickness of 0.020-inch (0.5 mm) soft steel, while the highest strength wire was capable of penetrating two sheets of 0.048-inch (1.2 mm) thick cold-rolled AISI 1010 steel.

Metal stitching or stapling is a good replacement for riveting or screwing thin sheet metal to other materials when large areas are involved, and out-of-plane loading would put adhesives under an undesirable peeling force. When attaching softer materials to metal, the stitches or staples should be pushed through from the steel side and clinched on the softer material side.

As important as sheet-gauge material is to the fabrication of components, products, assemblies, and structures from metal, other product forms are also extensively used, and usually for higher-load situations. The major other product forms of metals are (1) castings, (2) extrusions, (3) forgings, and (4) machined parts. For net-shaped castings, extrusions, and net-shaped forgings, integral attachment features can be incorporated into geometry of the part to allow mechanical attachment between parts. For machined parts, such features are machined into the part.

7.3 CASTING ATTACHMENT SCHEMES AND ATTACHMENTS

7.3.1 Cast-in Attachments

For metals,[11] casting is the process of taking metal in its molten state,[12] pouring it into a mold or die containing a cavity the shape of which is the negative of the shape of the part to be produced, allowing the molten metal to cool and solidify, and extracting the resulting solid-metal part. There are many variations of casting including sand casting, shell casting, die casting, investment casting, and others (Lindberg, 1990). For some of these processes, most notably, die casting, the metals and alloys used tend to be more malleable, and, therefore, it is often feasible and economical to design the casting so it provides its own attachments for assembly to other parts. The kinds of attachments incorporated into the design of die casting may be integral rivets, studs, or lugs that can be inserted through corresponding openings in a mating part in an assembly, and then staked, upset-headed, spun, or bent to produce a secure attachment. Figure 7.19 schematically illustrates some cast-in attachment features.

For heavier loads, more substantial integral studs or projections can be incorporated into the design of the casting, over which various nuts and clips used to attach other parts. There are also designs that allow the use of so-called "speed nuts" or "speed clips" that can be simply pushed over unthreaded studs and

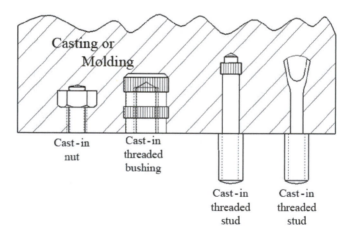

FIGURE 7.19 A schematic illustration of some integral cast-in attachment features.

---

[11] For ceramic materials, casting generally is done by mixing finely crushed or otherwise powdered ceramic material(s) with water (or another liquid binder) to create a slurry that is able to be poured into a mold and allowed to dry (by evaporation and/or absorption of the liquid) to produce the part.

[12] There are, in fact, some modern casting processes, known as "rheo-casting" or "thixotropic casting," in which a metal alloy heated to between its solidus and liquidus temperatures to produce a mixture of liquid and solid which is forced under pressure into a mold or die to produce a casting. The advantage of such casting is lower pouring temperatures and, so, less heat-resistant mold or die materials and/or longer mold or die life.

achieve locking. Special D-shaped cross-sections on studs even allow speed nuts to be removed by giving them a partial turn.

7.3.2 Cast-In or Molded-In Inserts

Many metals and alloys that are cast lack the inherent ductility or malleability and/or strength to allow the creation of cast attachment features to be used for subsequent assembly to other parts.[13] For these, inserts of stronger and more machinable metals are employed. Such inserts can be incorporated into the casting during its creation, to become an integral part of the casting, or can be installed once the casting has been produced.

Casting or molding in inserts can complicate the design (and increase the cost) of the die, virtually always slows down production due to the added handling and placement of the inserts, and cast metal can be contaminated by metal from the insert so that reclamation and reuse (i.e., recycling) is prevented. However, with proper consideration of these potential problems during design of the casting, inserts can be used to advantage and can be among the cheapest and best methods for enabling joining of castings into assemblies.

Specific types of inserts that can be cast-in or molded-in include:

- Threaded bushings (i.e., inserts with tapped holes to accept screws or bolts)
- Threaded studs (i.e., inserts that will accept a nut)
- Bearings
- Contact springs
- Terminal blocks

Besides these, there are a number of patented inserts, some of which are intended to be molded-in, and some of which are intended to be installed in the finished casting. Some of the more common examples of patented inserts include:

- Rosan™ inserts (which consist of both an external and internal threaded member)
- Tap-lok™ inserts (which is a self-locking bushing to be installed after casting)
- Anchor nuts and self-clinching anchor nuts (see Section 7.2.5)
- Heli-Coil™ inserts (which consist of thread inserts for use with metals too hard or too soft to allow durable threads to be produced)

---

[13] Some alloys can only be used to make parts by casting because they cannot be processed into shapes by plastic deformation. Some are even difficult to machine. Many cannot be welded either.

Some metallic alloys can be forced through a die while they are hot enough to flow easily in the plastic state, but without any melting, thereby having the cross-section of a long member take the shape of the die. The process resembles toothpaste being squeezed from a tube. Some alloys of Al, Mg, and Ti are commonly extruded into long parts with shapes that vary from simple L's or T's, for example, to much more complex shapes such as those commonly found used to make the tracks of aluminum windows, sliding-door sills, and so on. It is even possible to produce extrusions that contain holes; from simple hollow squares to elaborate multi-channel heat exchangers.

Extruded shapes[14] can incorporate design features that facilitate joining of the extrusion to other parts, whether those other parts are themselves extrusions or machined parts, castings, and so on. *The Aluminum Extrusion Manual* (Aluminum Extruders Council and The Aluminum Association, 1998) states that "the joining of aluminum extrusions can be accomplished by way of nine distinct methods that can be designed into the profiles themselves," including the following:

1. Nesting
2. Interlocking
3. Snap-fit
4. Three-piece interlock
5. Combination
6. Slip-fit
7. Hinge-point
8. Key-lock
9. Screw slot

Without going into detail, as the interested reader can consult the just-mentioned source, suffice to say here that both rigid and elastic types of interlocks are available. Figure 7.20 schematically illustrates a variety of the more common types of interlocks found in extrusions. *Nesting joints,* for example, include lap joints and tongue-and-groove joints that offer only limited locking action but operate by remaining rigid. *Interlocking joints* involve modified tongue-and-groove designs to create more interlocking by operating a rigid feature. Assembly can be by sliding or, for some, by a slight rotation in what would be classified as a tilt. Full-fledged elastic *snap-fits* or "snap-locks" can also be created. *Three-piece interlocking* joints are designed to use a normally-hidden fastener to interlock two principal extrusions. *Combination joints*

---

[14] Extrusions tend to be produced in very long (typically, 50+ m) lengths, which are virtually always cut into shorter lengths for use. In some instances, such as the frames of aluminum windows and doors, the cut lengths are reasonably long. In other instances, the cut lengths can be very short, for example, only an inch (i.e., 2–3 cm) or less.

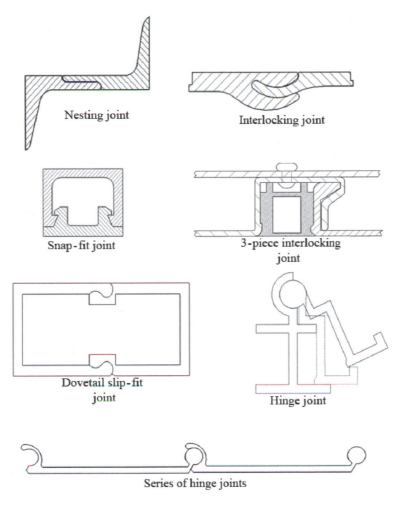

FIGURE 7.20 Schematic illustrations of various ways in which rigid or elastic integral interlocking features can be incorporated into extrusions to allow assembly.

combine nesting, interlocking, and snap-fit designs. *Slip-fit joints* are assembled by sliding rigid members together, while *hinge joints* are assembled using a tilt motion. Both operate by remaining rigid. Finally, *key-locked joints* use a specialized part, called a "key," which is slid into position to lock extruded members that assemble with a simple sliding or tilting motion, and would separate if not locked with the key. The final approach (i.e., *screw slots*) really only involves the incorporation of a geometric feature into the extrusion's profile that allows the easy insertion of a screw to accomplish assembly.

Companies that manufacture extrusions produce a wide variety of standard shapes such as L's, T', single- and multiple-channels, and so on. But, much of what they produce is custom designed to meet the needs of a specific customer for a specific application. For such extrusions, the design of the extrusion belongs to the end-user (not the extruder), and the extrusion dies are usually stored (in inventory) by the extruder until they are needed to fill repeat orders for the same extrusion of the same end-user. As such, it is not possible to speak about standard designs for integral features of extrusions that allow attachment, but such integral attachment features are not at all uncommon.

Figure 7.21 shows a promotional piece by the Aluminum Extrusion Council that shows the possibilities of using integral mechanical attachments of a variety of types to produce an assembly, in this case, a hypothetical C-shaped frame.

Figure 7.22 shows an example of interlocking extrusions produced by Nova Extrusion Company as an "executive toy" and promotional piece.

7.5 FORGING ATTACHMENT SCHEMES AND ATTACHMENTS

Forging is the process of plastically deforming metal, while it is cold or hot, to produce a desired shape. Various types of forging allow a wide range of capability for how close the finished forging is to the desired shape (i.e., net-shape), controlling dimensions, allowing thin sections or thin-to-thick transitions, and so on. Some specific types of forging include hammer forging, drop-forging, press forging, matched-die forging, precision forging, and isothermal and creep forging. The types of metals that can be forged include Al-alloys, Ti-alloys, C- and alloy steels, stainless steels, Ni- and Co-based alloys, and others.

The use of integral design features in forgings is considerably rarer than in castings and extrusions. The reason is that it is usually far more difficult to move metal with precision by forging, so obtaining the precision of shape and dimensions needed to allow such integral geometric features to operate properly is far more difficult. However, the possibility does exist with various methods of precision forging using matched dies and isothermal creep. Obviously, it is also possible to incorporate unfinished (rough) or semi-finished protrusions that must be finish-machined to create the rigid or, possibly, elastic interlock features.

7.6 MACHINED ATTACHMENTS

Any part that is machined can have geometric features that allow rigid or elastic interlocking to another machined part or to a precision (die) casting or extrusion with integral attachment features produced in it. The shapes of such features closely parallel those found in castings and extrusions. Whether the features machined in operate to produce a rigid or an elastic interlock depends primarily on the inherent characteristics of the metal or alloy.

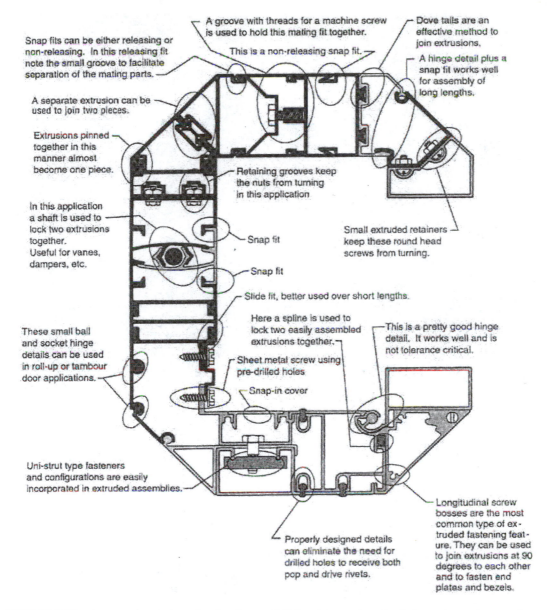

Snap fits can be either releasing or non-releasing. In this releasing fit note the small groove to facilitate separation of the mating parts.

A groove with threads for a machine screw is used to hold this mating fit together.

This is a non-releasing snap fit.

Dove tails are an effective method to join extrusions.

A hinge detail plus a snap fit works well for assembly of long lengths.

A separate extrusion can be used to join two pieces.

Extrusions pinned together in this manner almost become one piece.

Retaining grooves keep the nuts from turning in this application

In this application a shaft is used to lock two extrusions together. Useful for vanes, dampers, etc.

Small extruded retainers keep these round head screws from turning.

Snap fit

Snap fit

Slide fit, better used over short lengths.

Here a spline is used to lock two easily assembled extrusions together.

This is a pretty good hinge detail. It works well and is not tolerance critical.

These small ball and socket hinge details can be used in roll-up or tambour door applications.

Sheet metal screw using pre-drilled holes

Snap-in cover

Uni-strut type fasteners and configurations are easily incorporated in extruded assemblies.

Properly designed details can eliminate the need for drilled holes to receive both pop and drive rivets.

Longitudinal screw bosses are the most common type of extruded fastening feature. They can be used to join extrusions at 90 degrees to each other and to fasten end plates and bezels.

FIGURE 7.21 A promotional piece showing the possibilities of integral mechanical attachment in extrusions. (Reproduced from an original from the *Aluminum Extrusion Manual* of the Aluminum Extrusion Council, Wauconda, IL; used with permission.)

Figures 7.23 and 7.24 show how forgings (such as the large bed of the milling machine shown) and wrought billets (such as the traversing unit) can be machined to create rigid interlocks, here, in the form of T-slots for allowing the rapid installation of hold-down tooling and as rails and ways for allowing precision traversing motions, respectively.

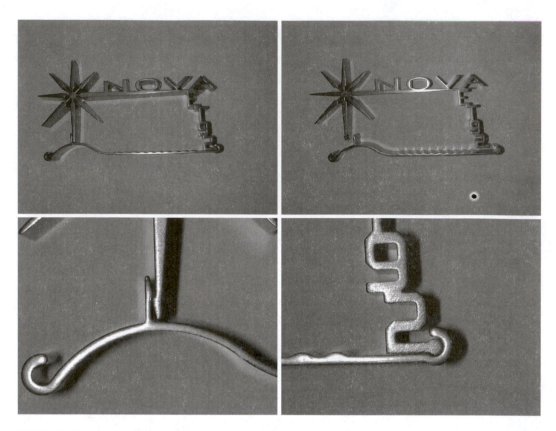

FIGURE 7.22 An "executive toy" and promotion piece by Nova Extrusion Company showing how Al-alloy extrusions can be designed and produced to interlock; assembled (top-left) and pre-assembled (top-right). Details of the rigid interlocks are shown in the close-ups at the bottom of the figure. (Photograph by Donald Van Steele for the author, Robert W. Messler, Jr.; used with permission.)

FIGURE 7.23 The cast and machined working bed of a milling machine showing precision-machined steel shaped rails with traversing units with matched, shaped ways. (Photograph courtesy of Haas Automation, Inc., Oxnard, CA; used with permission.)

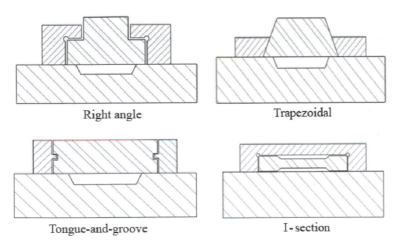

Right angle   Trapezoidal

Tongue-and-groove   I-section

FIGURE 7.24 A traversing unit containing machined ways that fit precisely with a machined rail to allow precision motion. (Photograph courtesy of Haas Automation, Inc., Oxnard, CA; used with permission.)

7.7 SUMMARY

Other than cement, concrete, and stone, metallic materials are the most extensively used and technologically and economically important materials of all. Metallic materials are used in a wide variety of product forms including rolled sheet, castings, extrusions, forgings, and machined parts. For most, if not all, of these product forms, parts made by one process can be joined to parts made by the same or another process using integral mechanical attachments. Because of the inherent elastic–plastic strain behavior of metals, rigid, elastic, and plastic types of interlocks can be employed.

Sheet-metal parts can be joined using: (1) hemming and crimping to form plastic interlocks; (2) formed-tabs to produce rigid, elastic, or plastic interlocks; (3) beaded assemblies to produce plastic or elastic interlocks; (4) metal clinching to produce plastic interlocks; (5) integral fasteners to allow rigid or elastic fastening/attachment; and (6) metal stitching or stapling to produce rigid interlocks.

Castings can be created with integral (1) cast-in or molded-in geometric features that allow joining by integral mechanical attachment or can be created with (2) cast-in or post-casting installation of inserts that allow the use of threaded or unthreaded fastening techniques.

Extrusion allows the production of a variety of shapes into a profile that allows subsequent assembly by either rigid or elastic interlocking, possibly with an assist from a fastener for added security. Specific approaches include: (1) nesting; (2) interlocking; (3) snap-fit; (4) three-piece interlock; (5) combination; (6) slip-fit; (7) hinge-joint; (8) key-lock joint; and (9) screw slot.

Forgings, because of the greater difficulty of creating precise geometric details, tend to use only rather gross features to allow assembly by rigid interlocking.

TABLE 7.1 List of Integral Mechanical Attachment Schemes and Attachments

Used with Metals (with an indication of whether the principal method of creation/operation is as a rigid [R], elastic [E], or plastic [P] interlock)

Sheet-Metal

- Hemming
  - pressed (folded) joints [P/R+E]
  - roll-and-press joints [P/R+E]
- Crimping (normally of wires) [P/R+E]
- Formed tabs
  - straight-through/tab-in-slot interlocks [P/R]
  - bent or folded [P/R+E]
- Beaded assembly
  - friction-type joints [P/E+R]
  - indentation-type joints [P/R]
- Metal clinching [P/R]
- Integral fasteners [NA/R]
- Metal stitching or stapling [P/R]

Castings

- Cast-in features [NA/R, R+E]
- Cast-in inserts
  - threaded bushings or sleeves [NA/R]
  - threaded or unthreaded studs [NA/R]
  - bearings [NA/R, R+E]
  - contact springs [NA/E+R]
  - terminal blocks [NA/P+R]
- Metal stitching (for repairs) [NA/R]

Extrusions

- Nesting [NA/R, R+E]
- Interlocking [NA/R, R+E]
- Snap-fit [NA/E]
- Three-piece interlock [NA/R, R+E]
- Combination [NA/R, R+E]
- Slip-fit [NA/R+E, E]
- Hinge point [NA/R]
- Key-locks [NA/R]
- Screw slot [NA/R]

Forgings

- Forged-in features [P/R]

Machined Parts or Machinings

- Machined-in features [NA/R, R+E]

Machined parts can be designed and fabricated with precision geometric features that allow either rigid or elastic interlocking assembly with other machined parts, or with precision castings, extrusions, or even sheet-metal parts.

Table 7.1 lists all of the integral mechanical attachment schemes and attachments used with metals, with an indication of whether the principal method of creation/operation is rigid (R), elastic (E), or plastic (P).

REFERENCES

Aluminum Extrusion Manual, 3rd edition, Aluminum Extruders Council and The Aluminum Association, Wauconda, IL, 1998.

Ashby, M.F., *Materials Selection in Mechanical Design,* 3rd edition, Butterworth-Heinemann, Oxford, United Kingdom, 2004.

ASM Handbook, Volume 6, *Welding, Brazing, and Soldering,* ASM International, Materials Park, OH, 1993.

AWS Welding Handbook, 8th edition, Volume 2, *Welding Processes,* American Welding Society, Miami, FL, 1990.

Callister, W.D., Jr., *Fundamentals of Material Science and Engineering: An Introduction e-Text,* 5th edition, John Wiley & Sons, Inc., New York, NY, 2000.

Landrock, A.H., *Adhesives Technology Handbook,* Noyes Publications, Park Ridge, NJ, 1985.

Lindberg, R.A., *Processes and Materials of Manufacture,* 4th edition, Allyn and Bacon, Needham Heights, MA, 1990.

Messler, R.W., Jr., *Joining of Materials and Structures: From Pragmatic Process to Enabling Technology,* Butterworth-Heinemann/Elsevier, 2004.

Messler, R.W., Jr., *Principles of Welding: Processes, Physics, Chemistry, and Metallurgy,* John Wiley & Sons, Inc., New York, NY, 1999.

Parmley, R.O., *Standard Handbook of Fastening & Joining,* 2nd edition, McGraw-Hill Publishing Company, 1989.

Speck, J.A., *Mechanical Fastening, Joining, and Assembly,* Marcel Dekker, Inc., New York, NY, 1997.

Swift, K.G., and Booker, J.D., *Process Selection: From Design to Manufacture,* Arnold, London, United Kingdom, 1997.

BIBLIOGRAPHY

Askeland, D.R., and Phuke, P.P., *The Science and Engineering of Materials,* 4th edition, Thomson-Engineering, 2002.

Callister, W.D., Jr., *Fundamentals of Materials Science and Engineering: An Introduction e-Text,* 5th edition, John Wiley & Sons, Inc., New York, NY, 2000.

Schackelford, J.F., *Introduction to Materials Science for Engineers,* 5th edition, Prentice Hall, Upper Saddle River, NJ, 1999.

Schaffer, J.P, Saxena, A., Sanders, T.H., Antolovich, S.D. and Warner, S.B., *Science and Design of Engineering Materials,* McGraw-Hill, New York, NY, 2000.

Smith, W.F., *Fundamentals of Materials Science and Engineering,* 3rd edition, McGraw-Hill, New York, NY, 2003.

8 POLYMER ATTACHMENT SCHEMES AND ATTACHMENTS

8.1 PROPERTIES OF POLYMERS THAT FACILITATE INTEGRAL MECHANICAL ATTACHMENT

Materials can be divided into three fundamental types or classes: metals, ceramics, and polymers, as shown schematically in Figure 8.1.[1] Metals are defined and described in Chapter 7, Section 7.1. Ceramics are defined and described in Chapter 9, Section 9.1. Polymers will be defined and described here, in preparation for consideration of attachment schemes and attachments appropriate to these interesting and important materials.

Polymers are defined as solid, nonmetallic, normally-organic[2] compounds of high molecular weight, the structure of which is composed of small repeat units known as mers (Callister, 2005). Many polymers occur naturally, being derived from plants and animals. Examples of such naturally-occurring polymers are

---

[1] Other well-known, as well as less well known, materials are, in fact, special forms, sub-types, or combinations of the three fundamental types: metals, ceramics, and polymers. Glasses, as known by the average person, are a non-crystalline form of ceramics. However, a non-crystalline, glassy form of metals can also be made to exist. Most polymers can exhibit a glassy form and a crystalline form, and usually exist as a mixture as semi-crystalline materials. Semiconductor materials can be considered sub-types of ceramic or even polymeric materials. Composites are combinations of other materials, possibly of different fundamental types or, possibly, simply of different specific materials within a fundamental type. Wood is a naturally-occurring composite material, and concrete is a composite made from cement, a ceramic material, and naturally-occurring rocks, which are, in turn, composed of ceramics.

[2] In chemistry, something is "organic" if it is a compound of carbon other than a carbide, carbonate, bicarbonate, cyanide, cyanate, or gaseous oxide.

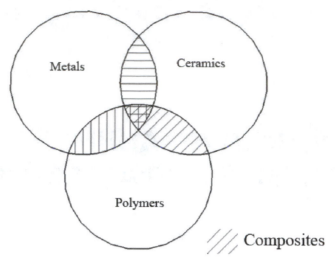

FIGURE 8.1 A schematic illustration, in the form of a Venn diagram, of the three major types of materials: metals, ceramics, and polymers.

wood (which is actually a composite of several polymers), rubber, cotton, wool, leather, and silk. Other natural polymers that are part of plants or animals are cellulose, starches, proteins, and enzymes, all of which are important in the biological and physiological processes that occur in living things.

Modern chemistry has enabled the creation of polymers that do not occur naturally.[3] These are synthesized from small organic molecules to produce much larger, higher molecular weight molecules. All plastics, many types of rubber, and many fiber materials are synthetic polymers. What is particularly exciting about synthetic polymers is that most can be produced relatively inexpensively, and their properties can be engineered to the degree that they are superior to their natural counterparts. This ability to "engineer" synthetic polymers at both the molecular and the aggregate, micro- and macro-structural level in many ways makes polymers the most exciting of all materials!

As stated above, most polymers are organic in origin, and many organic materials are hydrocarbons or derivatives of hydrocarbons. As the name implies, "hydrocarbons" are principally composed of hydrogen and carbon. Within hydrocarbon molecules, the bonding between the hydrogen and carbon is covalent. Within the many derivatives of hydrocarbons that constitute organic chemistry, atoms of other elements can be included, and most of these are also covalently bonded to the carbon and hydrogen atoms. In covalent bonding, the outermost four electrons of carbon (which has six electrons in total) participate

---

[3] In fact, modern chemistry is increasingly even allowing some naturally-occurring polymers to be synthesized.

in the bonding. In hydrogen, its single electron participates in covalent bonding. To do so, atoms of carbon or hydrogen must share their outermost electrons with other carbon atoms, with other hydrogen atoms, with combinations of carbon and hydrogen atoms, or with other atomic species (i.e., elements) also willing to have their outermost electrons be shared. Simple hydrocarbon molecules are called "mers" or "mer units" when they are subsequently involved in the synthesis of a polymer molecule.

Without going into great detail here, a molecule of a particular hydrocarbon mer or monomer can have carbon atoms that share more than one of each one's outermost electrons, creating double and even triple bonds between adjacent carbon atoms within the mer and leading to what are referred to as "unsaturated" molecules. These double or triple bonds can be broken in various ways[4] to cause the previously unsaturated molecule to become "saturated," in which case all carbons will form only a single bond with another carbon. By breaking double or triple bonds in mer units, activated mers with unsatisfied bonds in carbon atoms can bond with one another end-to-end to form a chain-like molecule. Such chain formation occurs in what is called "chain reaction polymerization" or, alternatively, "addition polymerization," producing a molecule made up of many mers (i.e., a polymer). The creation of a polymer by chain reaction polymerization of a simple mer is shown in Figure 8.2a.

An alternative method by which more than one monomer specie can be caused to form a polymer by the process of polymerization is for the reaction to occur by a stepwise intermolecular chemical reaction. In "step reaction polymerization," a small molecular weight by-product, such as water, is usually formed, leading to the alternative name for this type of polymerization being "condensation polymerization." The creation of a polymer by stepwise reaction or condensation polymerization of two monomers is shown in Figure 8.2b.

By combining one or more monomers in various ways, a tremendous number of possibilities exist for creating a vast array of different polymers. Some of the things that lead to the possibility of many, many combinations are:

- The length of the chain of mers comprising a polymer, known as the "degree of polymerization," can differ, with longer chains having the same chemical formula or composition as shorter chains, but exhibiting different properties (e.g., density, melting temperature and/or glass transition temperature, solubility in a particular solvent, etc.).[5]
- The chain of the polymer can consist of one specie of mer or of different species of mers, the latter creating a "co-polymer." Furthermore, in a

---

[4] Bond breaking, and reformation, is what occurs in chemical reactions. Bonds can be broken by supplying energy from heat, light, or electricity, or from chemical activity or reactivity.

[5] In fact, for any particular degree of polymerization, the precise number of mers in the polymer's chains varies around some mean. Chain length can be controlled by chemical engineers by various means, including adding a chemical that terminates the polymerization reaction.

(a)   Addion Polymerization

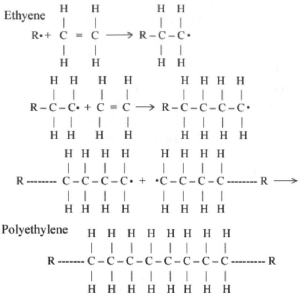

FIGURE 8.2 Schematic illustrations of the two ways in which polymers can be created from "polymerization"; (a) using addition or chain reaction polymerization and (b) using condensation or step reaction polymerization. (Adapted from Callister, W.D., *Fundamentals of Materials Science and Engineering: An Integrated Approach,* 2nd edition, John Wiley & Sons, Inc., New York, NY, 2005; with permission.)

co-polymer, the different mers can link together end-to-end in groups, forming a "block co-polymer," or they can alternate without any pattern, forming a "random co-polymer."

- The chain of a polymer can develop side-branches, producing what is known as a "branched polymer." Furthermore, the branches can be made up of the same mer as the main chain of the polymer or can be made up of a different mer species, producing what is known as a "grafted polymer."

- Various atoms or small molecules, collectively referred to as "radicals," can attach to the main carbon-based chain or "spine" of the polymer changing the chemical formula and properties of the polymer. Examples of radicals are a Cl atom in place of an H atom, an F atom in place of an H atom, an OH radical in place of an H atom, a methyl (CH_3) group, a benzene ring, an amine group, and so on.

- Radicals can arrange themselves on the central chain of a polymer by (1) attaching on both sides[6] of the chain in a regular, alternating pattern, (2) all attaching on only one side of the chain, or (3) attaching on both sides of the chain in a random way. These different arrangements are known as "stereo-isomers," and, while the polymer has the same chemical formula or composition, the properties (melting and boiling points, ease of forming a crystalline form of the polymer, glass transition temperature, etc.) differ from one arrangement to another. The three different arrangements are called "isotactic," "syndiotactic," and "atactic" stereoisomers, respectively.

- The geometric arrangement of the atoms making up certain mers can differ without changing the chemical formula of the monomer, but the different arrangements cause the monomer, and thus the polymer, to have different properties. Two different so-called "geometric isomers" are the "cis-" and the "trans-" isomer forms.

Beyond these many possibilities within polymer molecules, aggregates of long-chain polymer molecules that make up the macroscopic material can sometimes arrange themselves to yield long-range order (i.e., form a crystalline structure) and sometimes cannot (i.e., they remain random and form an amorphous structure). In fact, in most polymers, most of the time, both forms co-exist (i.e., the polymer is semi-crystalline).

The reader interested in more details on polymers is referred to a good reference in basic materials science and engineering such as Callister (2005), Shackelford (2005), or others.

As interesting as the chemistry, molecular structure, and microstructure (i.e., aggregation of molecules) of polymers are, it is the properties that result from these that are of most interest to engineers.

Polymers are best known for the ease with which they can be formed into complex shapes often by being deformed into those shapes. The high level of plasticity found in polymers[7] gives rise to their common name (i.e., "plastics").

---

[6] The term "sides" applied to a carbon chain in a molecule actually refers to one of the covalent bond sites to which another atom can attach, which in carbon occur at conformation angles of 109.5 degrees. When single bonds exist between carbon atoms in the chain, the carbon atoms are free to have their conformation angles rotate about the axis of the chain, so use of "side" relative to the chain becomes less obvious.

[7] There are, in fact, some polymers that are quite rigid at room temperature, at least, and, so, these may not be considered as particularly plastic. Other polymers called elastomers (to be discussed later in Section 8.1) behave elastically almost to the point for rupture, which occurs at very high strains for most.

Many polymers also exhibit a high degree of elasticity, that is, the ability to be deformed or strained under the application of a force and, upon release of the force, recovering to their original shape and dimensions. In some polymers, the elasticity is considerable. Such polymers are called "elastomers." However, the behavior of polymers under an applied force or stress is actually more complex than this, as can be seen in Figure 8.3.

Figure 8.3b shows the response of a typical polymer to a load or stress that is applied at some point in time (shown in Figure 8.3a), held constant for some period, and then is removed. The polymer is seen to immediately respond to the

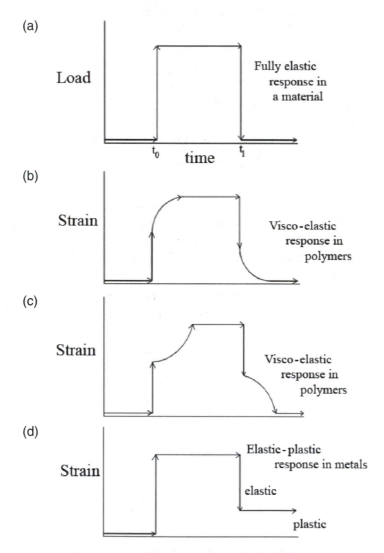

FIGURE 8.3 Schematic plots of the strain behavior of polymers compared to metals and ceramics; (a) load versus time; (b) visco-elastic strain response in a typical polymer; (c) different visco-elastic strain responses for different polymers; and (d) elastic and elastic–plastic strain behavior in metals and ceramics.

applied stress by straining to a level determined by the polymer's modulus of elasticity.[8] However, following this immediate response, the polymer continues to strain with time—for some time—until the total amount of strain reaches some level and then remains constant. There is thus an anelastic component of strain. Similarly, when the applied stress is removed, the polymer immediately recovers at least some of its strain under loading. There is also an additional, time-dependent (i.e., anelastic) component of recovery, however. After some time, the polymer may fully recover to its original shape and dimensions, in which case, the polymer was operating within its elastic range. Contrarily, the polymer may not fully recover to its original shape and dimensions, in which case, it operated into its plastic range and was plastically deformed. Different polymers with different molecular structures and different arrangements of molecules within the macroscopic material exhibit a variety of this so-called "visco-elastic behavior," as shown in Figure 8.3c. Metals and ceramics do not exhibit time-dependent visco-elastic behavior but exhibit elastic or elastic-plastic behavior, as shown in Figure 8.3d.

The fact that polymers exhibit visco-elastic strain behavior has the greatest influence on how they can—or should—be joined. But, first, there is another characteristic of polymers that also affects how they can—or should—be joined. Polymers can be divided into two major types or classes: (1) thermoplastic polymers (or thermoplastics) and (2) thermosetting polymers (or thermosets). *Thermoplastic polymers* are those that reversibly and repeatedly soften upon heating and stiffen upon cooling. As a result, they can easily be caused to permanently change shape (i.e., to plastically deform) by heating them and applying forces to move the material to create the desired shape. Thermoplastic polymers can also be caused to join to one another by heating them while they are in contact and under pressure (i.e., they can be welded).[9] *Thermosetting polymers* are those that do not soften upon heating but remain stiff. In thermosetting polymers, the long-chain molecules of the polymer species comprising the macroscopic material do not slide over one another but tend to bond to one another via linking or, more properly, cross-linking atoms. Thermoplastics easily deform plastically because their long-chain molecules tend not to have radicals or side-groups that are physically bulky, they rarely have branches, and they usually have no cross-linking atoms. They are thus free to slide over one another

---

[8] The modulus of elasticity, E, of a material, is the proportionality constant between the strain that results and the stress that is applied. It gives a measure of the material's stiffness against deflection, as well as a measure of the material's ability to store strain energy. In metals and ceramics, E tends to be fairly independent of composition, so the relationship between strain and stress is linear. In most polymers, the relationship between strain and stress is almost always non-linear, even at low applied stresses. For this reason, the modulus of a polymer is measured using a secant method, rather than the tangent method used for measuring the modulus of a metal or a ceramic (Callister, 2005). The modulus of elasticity of a metal, a ceramic, and especially a polymer varies with temperature, decreasing with increasing temperature.

[9] Welding of thermoplastics is referred to by plastics or polymer engineers as "thermal bonding."

when the weak van der Waals bonding (for example) between chains is overcome by the applied stress. In thermosets, on the other hand, the chain-like molecules are literally cross-linked by atoms that form actual strong, primary ionic or covalent bonds. This is the basis by which rubbers, such as butadiene, are made hard (i.e., by cross-linking between chains using sulfur atoms in the process of vulcanization). As a result of this behavior, thermosetting polymers must be cured from a soft, fluid-like state to a more rigid state by allowing/causing cross-linking to occur. Thermosets cannot be welded or thermally bonded to form joints.

Another consequence of the absence of cross-linking between relatively linear long-chain molecules of thermoplastic polymers is that they tend to be more soluble in organic solvents (acetone, toluene, benzene, etc.). Bulky, branched, and, especially, cross-linked long-chain molecules of thermosetting polymers,[10] on the other hand, tend to be far less soluble in organic solvents. The significance being that thermoplastic polymers tend to be easier to join using adhesives, and only thermoplastics can be joined by softening them with a solvent in what is known as "solvent cementing."

In summary, as was seen for metals in Chapter 7, Section 7.1, and as is true for all materials, the combination of the sub-microscopic atomic-level and microscopic-level structure of a material determines the material's properties; mechanical, as well as physical, electrical, optical, and so on. For polymers, long-chain molecules can vary from polymer specie to polymer specie in length (and, thus, molecular weight), geometric complexity (from type of radicals, stereoisomerism, and branching), presence and degree of cross-linking, and degree of crystallinity. The response of such molecules to applied stresses results in time-dependent visco-elastic behavior in which there are degrees of elasticity, plasticity, rigidity, and time-dependent flow or creep. The ability of a polymer to soften with heating—and even be joined by welding—depends on whether its structure results in thermoplastic behavior. A polymer's ability to resist being softened (i.e., partially dissolved) by organic solvents likewise depends on the degree to which long-chain molecules are cross-linked, to the point that a rigid three-dimensional network results. What all of this means is that: (1) thermoplastic polymers can be welded or thermally bonded, but thermosetting polymers cannot; (2) all polymers can be adhesively bonded using compatible adhesives,[11] which themselves tend to be polymeric; and (3) all

---

[10] Cross-linking in some thermosetting polymers can produce what is known as a "networked polymer," in which virtually all long-chain molecules are cross-linked to other chains, in several directions. The result is a fairly rigid network and, thus, a fairly rigid material.

[11] For polymers, an adhesive tends to be compatible if it is thermoplastic or thermosetting like the polymer(s) to be bonded. As a rule, thermoplastic polymers should be bonded using thermoplastic adhesives, preferably of the same polymer specie. Likewise, thermosetting polymers should be bonded using thermosetting adhesives; preferably of the same specie. Combinations of thermoplastic and thermosetting polymers should be bonded using a mixture of a thermoplastic and a thermosetting polymer, known as an "adhesive alloy" (Messler, 2004).

polymers can be mechanically joined, but only with due consideration of their particular visco-elastic strain behavior.

For integral mechanical attachment, a material must exhibit either reasonable rigidity (i.e., strength to resist tension, shear, or bearing and hardness to resist indenting) or reasonable elasticity (i.e., modulus of elasticity or shear and elastic limit strength to allow and tolerate deflection and recovery) or plasticity (i.e., ductility or malleability to allow plastic deformation). Since polymeric materials exhibit varying degrees of rigidity, elasticity, and plasticity, they can be joined using one or more of rigid, elastic, or plastic interlocks.

With this background, let's look at mechanical joining of polymeric materials, in general, and at integral mechanical attachment methods, in particular.

Appendix B lists important properties for some of the most important polymers (as well as other materials) used in engineering.

8.2 MOLDED-IN INTEGRAL ATTACHMENTS

Next to adhesive bonding, which, by far, is the most common method by which polymeric materials have been and still are joined (*ASM Engineering Materials Handbook,* 1988; Muccio, 1995), joining parts made from polymers is most conveniently done using integral mechanical attachment features that are molded into the parts to be joined.

As described in Chapter 1, Section 1.4, integral mechanical attachment has many advantages in its own right and, for inherently tougher materials, such as polymers (compared to metals and ceramics), is particularly advantageous because specific approaches tend to generate far less point loading than mechanical fasteners. Point loading along the walls of holes used with pins, rivets, and also screws or bolts arises under shear loading as bearing stresses on the joint element. Bearing stresses would tend to cause fastener holes in most polymers to elongate, if not immediately upon loading, over time due to visco-elastic flow. Similar problems can occur at the bearing surfaces under the heads of rivets, screws, and bolts, under the nuts used with machine screws and bolts, and under the formed "heads" of rivets when these fasteners are subjected to out-of-plane loading.

Two additional problems that can arise with fasteners occur with threaded types (Messler, 2004). For bolts and machine screws, which are used with nuts or internally-threaded backing parts, the clamping preload between the head of the screw or bolt and the nut or internally-threaded part that allows these fasteners to work properly to carry applied loads (Shigley et al, 2004) is lost over time in polymers when they flow (due to their visco-elastic nature) and experience stress relaxation. For threaded fasteners, like self-tapping screws (which form their own threads in the part as they are installed) or for machine screws or bolts used with internal threads in the part, such internal threads are prone to irreversible and progressive damage in inherently soft materials like polymers.

It is thus of little surprise that if mechanical fasteners are used to join parts made from polymers, great care must be taken to prevent concentrated loading. Techniques for accomplishing this are presented in Section 8.3, which follows.

Since the principal method by which plastic parts are produced is by molding, it is only logical that molded-in attachment features to allow mechanical joining are increasingly commonplace. All polymers can be molded, although the specific processes used differ between the two major classes. Thermoplastic polymers tend to be molded by taking advantage of the fact that these materials can be softened to the point that they can be made to flow, at least under an applied pressure. Hence, thermoplastic polymers are processed into even very complex shapes, with intricate details, using thermal molding processes that include compression molding, injection molding, blow molding, thermoforming (which involves forming in a die, and not molding, as such), and extrusion (which involves forcing through a die, and not molding) (Muccio, 1995). Thermosetting polymers, on the other hand, can only be molded while they are fluid-like or pliable prior to curing by the creation of cross-links between the polymer's long-chain molecules. But, in their pre-cured state, thermosetting polymers can be molded using reaction transfer molding and reaction injection molding (Muccio, 1995). In both classes of polymers, molding allows the creation of intricate details in each and every part by incorporating those details into the molds or dies. Once the investment is made in the mold or die, no matter how numerous or intricate, the cost per part decreases as the number of parts produced increases.

Because of the inherent nature of most polymers to exhibit a fair degree of elastic behavior or elasticity, elastic integral mechanical attachments or elastic interlocks—known as "snap-fits"—have been increasingly used to allow part assembly as polymers are increasingly used. As described in great detail in Chapter 4, Section 4.3, snap-fits are a simple, economical, rapid way of joining components. Many different designs and types exist, and all share the unique characteristic that they produce an audible and/or tactile "snap" when a protruding feature of one part clears either a different protruding feature or a recess in the opposing part in a mating locking pair.

Figure 8.4 schematically illustrates a wide variety of snap-fits suitable for molding into plastic parts. Figures 8.5 and 8.6 show two applications of molded-in snap-fits to allow and enable assembly of parts.

8.3 FASTENING AND MOLDED-IN AND OTHER INSERTS FOR JOINING PLASTIC PARTS

Parts molded from polymers can be secured together into assemblies using machine screws or bolts, nuts, and washers; however, attention has to be paid to ensuring that bearing pressure from the heads and nuts, in particular, is prevented from causing visco-elastic flow in the polymer(s), damage to the parts,

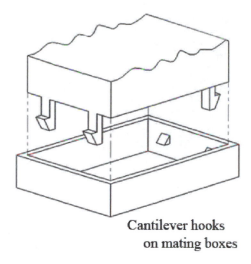

Cantilever hooks
on mating boxes

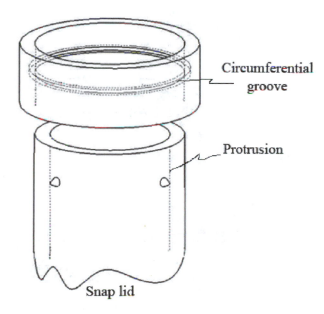

Circumferential
groove

Protrusion

Snap lid

FIGURE 8.4 A schematic illustration showing a variety of molded-in snap-fit integral attachments used in polymers. The use of cantilever hook snaps and a form of annular snaps are shown, top and bottom.

and loss of tightness (and, thus, effectiveness) of the fastener. Flat, as opposed to conical, under-head shapes are preferred because they tend to develop better-tolerated compressive stresses than tensile stresses. Even with such fasteners, however, flat washers are preferred to help distribute the clamping force needed for these types of fasteners to work properly. Figure 8.7 shows the proper use of washers for assembling plastic parts, and some of the preferred "flat-headed" screw and bolt types. Also shown are Heli-coil™ inserts for threaded fasteners.

FIGURE 8.5 Molded-in snap-fits are used in the well-known LEGO blocks (top) that allow the construction of an endless variety of structures and figures (bottom). (Photographs courtesy of LEGO Systems, Inc., Enfield, CT. LEGO and the brick configuration are trademarks of the LEGO Group, copyright 2005 The LEGO Group. Used here with permission, without endorsing this book. All rights reserved.)

Because of the susceptibility of polymers to creep and stress relaxation at room temperature, when fasteners are to transfer loads across joints by operating in shear by applying a bearing stress to the walls of the fastener holes in the parts, it is best to insert metal sleeves or bushings into the holes.

While special self-tapping screws, which either cut or form their own mating threads into a part, have been developed for use with polymer materials, these should only be used if they are to be installed and never removed and

FIGURE 8.6 The use of molded-in snap-fits in hand-held remote control units, here to secure the cover of the battery compartment. (Photograph courtesy of Sam Chiappone for the author; used with permission.)

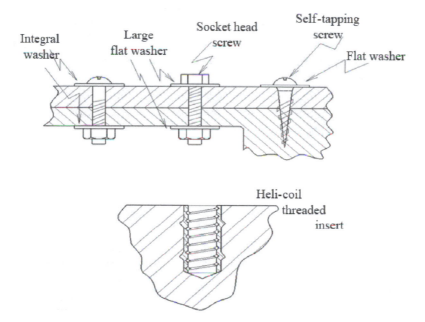

FIGURE 8.7 Schematic illustration of the use of washers and so-called "flat-headed" screws and bolts and Heli-coil™ inserts for assembling plastic parts.

replaced. Removing such screws and attempting to reinstall them causes severe damage to the cut or formed internal threads in the part and prevents the fastener from tightening properly. If thread-cutting type screws are to be used, it is best to use what are known as "Type 23" or "Type 25" (see ANSI standards[12]), as these are designed with relief near the cutting edge that allows material to be

---

[12] ANSI stands for American National Standards Institute (Washington, DC), which, as a private organization, develops and provides standards on mechanical fasteners, among other things, for voluntary adoption.

removed as the threads are cut. This avoids the buildup of high stresses around the fastener.

If a particular assembly involves infrequent disassembly and re-assembly, threads can be molded in. Not surprisingly, coarse threads can be molded in more easily than fine threads (i.e., less than 32 threads/inch, for example).

Special so-called "spring-steel fasteners" are often used in place of conventional nuts and lock washers. *Spring-steel fasteners,* such as the one shown in Figure 8.8a, are designed to be pushed onto a protruding molded post or stud of a plastic part inserted through a hole in a mating part. When properly installed, the spring-steel fastener locks the integral post or stud—and part—with an elastic spring force (see Figure 8.8b).

Other situations where special care needs to be taken in using mechanical fasteners with plastics are:

- If the mating part of an intended assembly is made from metal, care must be taken not to over-torque a threaded fastener into the polymer part, whether the fastener is installed through or into that polymer part into or through the metal part.
- Self-tapping screws that create threads in the part by forming rather than cutting them, induce high stress levels in the polymer part and, so, are not recommended for some types of polymers.

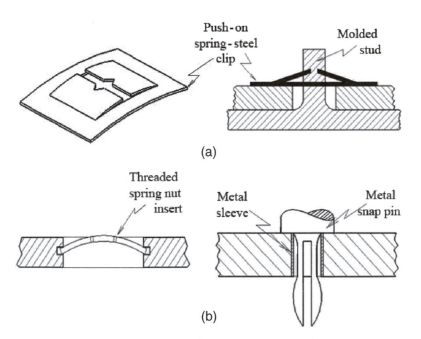

FIGURE 8.8 Schematic illustrations of spring-steel fasteners used with molded polymers.

- Pins can be used to join polymer parts, and tapered types tend to work best, as they tend to wedge themselves in place. Care needs to be taken to ensure that the bearing stress developed where the pin and polymer part contact does not get so high that the polymer cold flows and stress relaxes over time.
- When rivets are used with polymers, large head types, with heads three times the diameter of the shank of the rivet, should be used with washers under the flared (upset) end of the rivet.
- It is inadvisable to use keys and keyways in polymer parts.
- Press-fit metal pins or even other polymer parts can be used with polymers; however, press-fits should not be used if assemblies are to be thermally cycled, which would cause relaxation of the residual stresses needed to maintain the press fit, or subjected to harsh environments, where chemical attack or degradation of the polymer could be accelerated by the resulting residual stress in the part.

If parts molded from polymers are to be assembled, disassembled, and re-assembled repeated times, and/or if the loads on joints between parts are expected to be high, but, for some reasons, joining using mechanical fasteners is still preferred, metal inserts can be employed. Such inserts usually contain internal threads to accept threaded fasteners but can be simple metal sleeves or bushings to line holes for metal pins to prevent problems from bearing stresses or can be soft-metal inserts that can accept thread-cutting screws. There are also various types that operate to hold a fastener by expanding, as a split cylinder, against the plastic part in which they are installed. These metal inserts can be molded into thermoplastics or even thermosets, or other techniques, described below, for installing them following molding can be used.

When metal inserts are to be molded into thermoplastics, care must be taken relative to several factors. First, because the coefficient of thermal expansion (CTE) of most thermoplastics is much higher than that for most metals, molded-in metal inserts should be avoided for applications in which the assembly will be subjected to thermal cycling. The problem is less of a concern with many glass-reinforced thermoplastics, such as polycarbonates, because the CTE of the fiber-reinforced plastic (FRP) is closer to that of the metal. Second, as metal inserts get larger (e.g., weigh more than 1.5 g or 0.05 oz or have diameters greater than about 6.35 mm or 0.250 inch) they should be preheated before they are inserted into the mold. This prevents thermal shock and reduces the differential between thermal contraction during part cooling following molding. Obviously, inserts should be thoroughly cleaned and properly seated to be properly incorporated into the part during molding.

Besides molded-in metal inserts, there are a number of ways in which metal inserts can be installed into plastic parts after they have been molded. Examples of several other methods include:

- *Expansion inserts.* These are placed in pre-formed or drilled holes and are expanded against the inside walls of the holes when a screw or bolt is installed. Recognize, however, that the allowable interference between the expanded insert and the hole is very small, as it is for press-fits, in order to prevent buildup of detrimental stresses.
- *Ultrasonically-installed inserts.* These are internally-threaded inserts that also have a series of concentric rings located along the length of their external body. They are installed into thermoplastics using ultrasonic vibrational energy to soften the plastic around the pre-molded or drilled hole and accept the insert. Once the insert is fully installed and the plastic cools, the external concentric rings lock the insert in place.
- *Heat-installed inserts.* These inserts are preheated prior to installation into pre-prepared holes in thermoplastic parts. The preheat, rather than the heat generated from friction by ultrasonic vibration, allows the metal insert to be installed.

Metal inserts, whether molded in or installed using the expansion of the insert once a threaded fastener is installed, can be used with either thermoplastic polymers or thermosetting polymers. Ultrasonically-installed and heat-inserted inserts can only be used with thermoplastic polymers, as only thermoplastics soften with heating.

Obviously, although it is rarely done, metal inserts can be installed by adhesively bonding or "potting" them into the molded part, whether the part is made from a thermoplastic or thermosetting polymer.

Examples of some inserts used in molded polymer parts and installed using ultrasonics are shown schematically in Figure 8.9.

8.4 PROCESSED-IN ATTACHMENTS

Thermoplastic polymers allow some other means to be used to either add attachment features to molded parts or to create interlocks from molded-in features. Basically, should provision not have been made in the molded part to provide a feature to allow and enable attachment to another part, a thermoplastic feature could be welded or, in the language used by polymer engineers, "thermally bonded" to the part. The heat for such welding could come from friction, as it does in ultrasonic welding, vibration welding, spin welding, hot-plate welding, or, for thermoplastics containing some micro-sized ferromagnetic particles dispersed within the thermoplastic matrix, electromagnetic or induction welding. These so-called "plastic welding processes" can be used to join thermoplastic parts without integral attachment features.

The final and most common method for creating processed-in attachments is "thermal staking." *Thermal staking* creates a plastic interlock by deforming the free end of a molded-in post or stud in a thermoplastic polymer to produce a

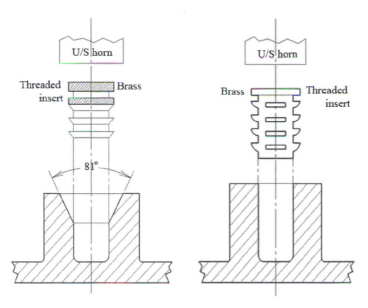

FIGURE 8.9 Schematic illustrations of some metal inserts used in molded polymers and installed using ultrasonic vibration-induced friction heating.

locking head over another part. The heat to cause such deformation can come from an ultrasonic vibration (in "ultrasonic staking") or from a hot-plate. Some examples of the types of thermal stakes that can be produced using ultrasonic vibration, specifically, are shown in Figure 8.10.

Figure 8.11 shows an RF seal, which, like a thermal stake, is formed into a thermoplastic polymer part.

8.5 RUBBER CONNECTIONS

So-called "rubbers" are special types of polymers known as elastomers. *Elastomers* are not a fundamentally different class of polymers, however. That is, they are either thermoplastic polymers or thermosetting polymers, as either sub-class can exhibit the exceptional degree of recoverable elasticity that typifies all elastomers. Hence, elastomers are really just highly elastic or elastomeric polymers. What gives elastomers this characteristic of strain behavior are very long molecular chains that tangle, kink, and coil in order to increase their entropy and, thereby, lower their energy state, in accordance with the laws of thermodynamics. Rubbers are almost always amorphous and seldom, if ever, exhibit any degree of crystallinity, despite that their molecules tend not to be too bulky or branched or cross-linked. Thermoplastic elastomers soften upon heating and stiffen upon cooling and tend to have very little if any cross-linking

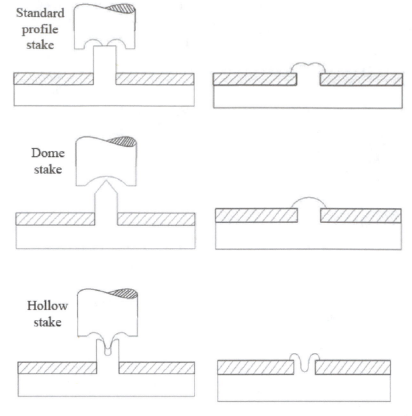

FIGURE 8.10 Schematic illustrations of some examples of thermal stakes that can be produced using ultrasonic vibration.

between molecular chains. To exhibit elastomeric behavior, thermoplastic polymers must consist of segments of different polymers within their central, long molecular chain or spine, with the different segments of one polymer specie being more inherently flexible than another. Such thermoplastic blocky co-polymers are relatively new developments, first appearing in the late 1970's or early 1980's. Thermosetting elastomers do not soften and stiffen with heating and cooling but tend to maintain a fairly constant stiffness over a wide range of temperature once they are in their cured condition. There is, in all thermoset polymers, some degree of cross-linking between molecular chains. In thermosetting or thermoset elastomers, however, the degree of cross-linking is less and less for greater and greater elasticity.

Table 8.1 lists some common elastomers by thermoplastic and thermosetting types. It can be seen in this listing that there are natural rubbers (that occur in Nature, as the secretions of certain types of trees or other plants) and there are synthetic rubbers. Other than the fact that synthetic rubbers may not occur in Nature, they are still elastomers. Unlike natural rubbers, synthetic rubbers can be engineered to have a set of functionally-specific properties, including

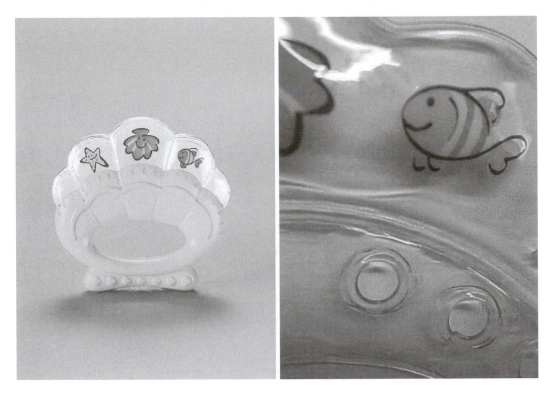

FIGURE 8.11 A photograph showing a special type of thermal stake, or spot and seam welds, employed in molded thermoplastic parts of a child's toy. Here a soft plastic teether is shown (left) in which spots and seams (right) were created by the heat from an RF system. (Photograph courtesy of Fisher-Price, East Aurora, NY; used with permission.)

strength, stiffness, hardness (measured on a so-called "durometer scale"), solvent-resistance, and so on. Like all semi-crystalline or amorphous polymers, rubbers exhibit a glass transition temperature, T_{glass} or T_g. Above its T_g, a polymer tends to behave in a more fluid-like or viscous state. Below its T_g, a polymer tends to behave more rigidly or "glass-like." Thus, at lower temperatures, rubbers tend to be harder, stiffer, and stronger, but, also, less elastic.

What is meant by the title of this section (i.e., "rubber connections") actually refers to two fundamentally-different aspects of joining associated with rubber. The first way that rubber can be used to aid in joining is to help produce joints with certain properties between other materials (e.g., metals). Such properties imparted by inserting rubber into a joint between two metal parts, for example, include: (1) softening the joint feel; (2) reducing susceptibility of joint elements to damage from impact (i.e., "cushioning" the joint); (3) making up for mis-fit or mis-alignment; (4) making up for changes in joint tightness caused by thermal expansion/contraction in the joint elements; (5) damping vibrations; and (6) sealing against fluid intrusion or leakage. Figure 8.12 shows some of the different shapes used for the retaining grooves for rubber seals, such as O-rings.

TABLE 8.1 Some Elastomers by Thermosetting and Thermoplastic Types

Thermoset Elastomers (cross-linked rubbers)

- Butadiene (butadiene rubber) (BR)
- Butyl rubber (isobutylene/isoprene) (IIR)
- Chlorobutyl rubber (CBR)
- Chlorosulfonated polyethylene (CSM)
- Ethylene-Polypropylene-Diene Terpolymer (EPDM)
- Fluorocarbon elastomers
- Fluorosilicone elastomers
- Natural rubber
- Neoprene (polychloroprene) (CR)
- Nitrile rubber (butadiene-acrylonitrile) (NBR)
- Polyacrylate (polyacrylate rubber) (AMC or ANM)
- Polypropylene oxide (polypropylene oxide rubber) (PO)
- Polysulfide rubber (PSR)
- Polyurethane elastomers
- Silicone rubber (polydimethylsiloxane)
- Styrene-Butabiene rubber (SBR) (Buna S)
- Synthetic rubber (polyisoprene) (IR)

Thermoplastic Elastomers (block copolymers)

- Polyester (Hytrel–DuPont)
- Polyolefin (TPR thermoplastic rubber)
- Polystryene-Butadiene-Polystyrene (S-P-S) (Krator–Shell Development)
- Polystryene-Isoprene-Polystryrene (Solprene–Phillips Chemical)
- Urethane

Compiled from several sources, must notably Arthur H. Landrock's *Adhesives Technology Handbook,* Noyes Publications, Park Ridge, NY, 1985.

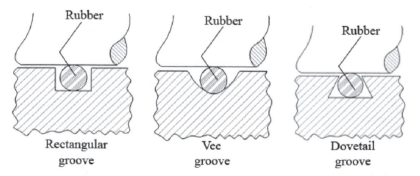

FIGURE 8.12 Schematic illustration of various rubber gasket or O-ring groove shapes.

In other applications, the rubber part is fit between joint elements to take up mechanical play or "slop," ensure a degree of compliance in an otherwise rigid joint, or provide damping of unwanted vibrations or shock loads. In many applications, the rubber serves as an integral element in the joint.

Figure 8.13 shows rubber inserts in the rigid coupling between the driving and driven units of a machine tool. Such inserts are used to reduce unwanted mechanical movement, vibration, and noise.

The second major way that rubber can be used in joining is to help produce an integral mechanical attachment and interlock. One example is "rubber crimping." In *rubber crimping,* a piece of rubber (often in the form of a ring) is placed between metals parts (for example) and is subjected to a compressive stress that causes the rubber to transfer the stress to one of the metal parts, forming that part into a pre-prepared recess or groove in the mating part. The result is a beaded-joint assembly (see Chapter 5, Section 5.5). This is shown in Figure 8.14 for the two situations where rubber is used to force a metal part to form an interlock with an inner part and where it is used to form an interlock with an outer part. In this same figure, the types of crimps used for various groove or recess designs are shown.

Finally, rubber can be used to hold other parts together without fasteners or without adhesives, thereby allowing easy intentional disassembly and/or providing joint flexibility. Figure 8.15 shows examples of (a) rubber sleeves (simply stretched to fit, or thermally shunk-fit), (b) rubber grommets, and (c) rubber couplings.

FIGURE 8.13 A photograph showing rubber inserts between the teeth of matching faces in a rigid coupling for a machine tool. Such inserts are used to reduce unwanted mechanical movement, vibration, and noise. (Photograph courtesy of Haas Automation, Inc., Oxnard, CA; used with permission.)

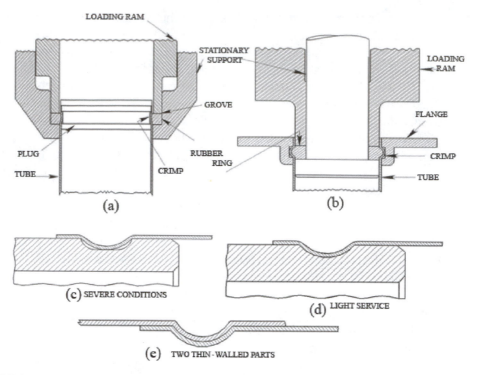

FIGURE 8.14 Schematic illustrations of rubber crimping to form an outer part into an inner part to produce a beaded assembly (a), and vice versa (b). Also shown, [in (c), (d), and (e)] are a number of crimps used with different groove designs.

8.6 SUMMARY

Polymers are, at once, the most complex-structured fundamental material type and the most diverse in compositions and properties. All polymers are produced from simple molecular building blocks known as "mers," linked together into long chains. The chemical reactions that cause polymerization of monomers are either (1) addition or chain reaction polymerization or (2) condensation or stepwise reaction polymerization.

Polymers share the properties of low thermal conductivity (i.e., thermal insulation), low electrical conductivity (i.e., electrical insulation), low strength, low stiffness, and low hardness compared to metals and ceramics, and, often, an ability to deform elastically to high degrees. Unlike metals and ceramics, which both exhibit elastic–plastic strain behavior, polymers all exhibit visco-elastic strain behavior. In visco-elastic strain behavior, an instantaneous response to the application or release of a load is followed by a time-dependent component of strain. The fact that different polymers have interesting combinations of rigidity, elasticity, and plasticity ideally suits them

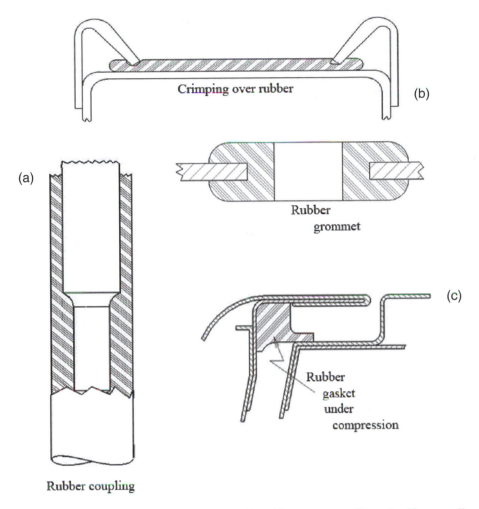

Crimping over rubber (b)

Rubber grommet

(a)

(c)

Rubber gasket under compression

Rubber coupling

FIGURE 8.15 Schematic illustrations of rubber sleeves (a), rubber grommets (b), and rubber couplings (c) used in joining.

for joining using integral attachments that produce rigid, elastic, or plastic interlocks, respectively.

All polymers can be formed into complex shapes by molding. Thermoplastic types are molded using heat to soften them, after which, once they have taken the shape of the mold, they cool and stiffen. Thermosetting types are molded before they are caused to cure, during which time they form cross-links between their long-chain molecules and become irreversibly stiff.

Polymers can be joined using mechanical fasteners, but only if care is taken to ensure that the fasteners do not create point loading that causes the polymer part(s) to deform visco-elastically or to allow stress in the fastener (being reacted to by the polymer part) to be lost by stress relaxation in the polymer part. If fasteners are used, parts should not be assembled, disassembled, and

TABLE 8.2 List of Methods by Which Polymers Can Be Joined Using Integral Mechanical Attachment Schemes and Attachments

Molded-in Integral Attachments

- Snap-fits
- Hook-and-loop attachments or fasteners

Molded-in Inserts

- Adhesively-bonded inserts/fasteners
- Expansion inserts/fasteners
- Heat-installed inserts/fasteners
- Ultrasonically-installed inserts/fasteners

Processed-in Attachments

- Thermally-bonded/welded attachments
  – friction welding (e.g., spin welding, linear vibration welding)
  – hot-gas welding
  – hot-plate welding
  – induction welding
  – radio-frequency (RF) welding
  – ultrasonic welding
- Thermal staking
  – heat
  – friction

Rubber Connections

- Inserts (for softening, damping, etc.)
- Gaskets or O-rings
- Crimping
- Sleeves
  – elastic
  – shrink fit
- Grommets
- Couplings

re-assembled. If multiple assembly operations are to be expected, polymer parts should contain metal inserts to take the point-loading of the fasteners. Metal inserts can be molded-in when the part is created or can be added later using any of a variety of methods.

Thermoplastic polymers can be thermally staked using heat from friction or from a hot source.

Rubbers are elastomeric polymers that are either thermoplastic or thermosetting. Rubber connections can be of two major types: (1) those that help provide a joint between materials other than rubber (e.g., metal, wood, concrete) with special properties or characteristics or (2) those that actually create joints using rubber. Rubber pieces can be caused to create beaded assemblies. Rubber sleeves and grommets can be used to actually lock parts together.

Table 8.2 summarizes the methods by which polymer materials can be joined using integral mechanical attachment schemes and attachments.

REFERENCES

Engineering Materials Handbook, Volume 2, *Engineering Plastics,* ASM International, Metals Park, OH, 1988.

Avallone, E.A., and Baumeister, T., III, *Mark's Standard Handbook for Mechanical Engineers,* 10 edition, McGraw-Hill, New York, NY, 1996.

Callister, W.D., Jr., *Fundamentals of Materials Science and Engineering: An Integrated Approach,* 2nd edition, John Wiley & Sons, Inc, New York, NY, 2005.

Messler, R.W., Jr., *Joining of Materials and Structures: From Pragmatic Process to Enabling Technology,* Butterworth-Heinemann/Elsevier, Burlington, MA, 2004.

Muccio, E.A., *Plastic Part Technology,* 3rd printing, ASM International, Materials Park, OH, 1995.

Shackelford, J.F., *Introduction to Materials Science for Engineers,* 6th edition, Pearson/Prentice-Hall, Upper Saddle River, NJ, 2005.

Shigley, J.E., Mischke, C.R., and Brown. T.H., *Standard Handbook of Machine Design,* McGraw-Hill, New York, NY, 2004.

9   CERAMIC AND GLASS ATTACHMENT SCHEMES AND ATTACHMENTS

9.1   INTRODUCTION TO THE JOINING OF CERAMICS, GLASSES, AND GLASS CERAMICS

Ceramics, inorganic glasses, and glass-ceramics comprise what are known collectively as "ceramic materials." Ceramics are crystalline and are generally harder, stronger (especially in compression), stiffer, and higher melting than metals, but are generally electrically and thermally insulating and are far less ductile and tough; in fact, they are generally brittle. Most are also generally much more resistant to chemical attack. Inorganic glasses derived from or related to ceramics exhibit generally similar properties to ceramics but usually exhibit quite different optical properties, being transparent to light, in most cases. Glass-ceramics, being crystalline materials created from amorphous inorganic glasses by thermal treatments known as "devitrification," are more like ceramics than glasses.

Many of the properties that make ceramics and glasses attractive for certain functionally-specific properties and for harsh operating environments render them difficult to process into large and/or complex shapes (see Sections 9.2 and 9.5). To produce parts with sizes or shape complexity that exceeds the limits of available processing methods, joining of smaller, less complex parts becomes especially important. Furthermore, while attractive for their special physical, electrical, thermal, optical, and chemical properties compared to metals, ceramics and glasses often need to be combined with metals in devices or structures where structural integrity is important. Thus, joining of dissimilar combinations of ceramics or glasses with metals is particularly important.

The requirements imposed on joints in ceramic materials vary with the application but include one or more of (1) hermeticity[1]; (2) strain accommodation (or ductile strain relief); (3) high-temperature strength and stability; and (4) corrosion resistance. As will be clear from detailed descriptions of the properties of ceramics (in Section 9.2) and glasses (in Section 9.5), more than for most materials, the ability to tailor the physical and mechanical properties of the joints in ceramics and glasses, and, especially, between these and other materials, is essential. Nowhere among materials is there a greater need to engineer the joint than for ceramic joining.

The joining technologies appropriate to ceramic materials can be classified by the types of materials involved in the joint. Specific ceramic joining technologies include:

- Ceramic/ceramic (or ceramic-to-ceramic) joining (normally referring to oxides)
- Advanced joining of non-oxide ceramics
- Ceramic/metal (or ceramic-to-metal) joining
- Glass/metal (or glass-to-metal) sealing
- Glass-ceramic/metal joining

These technologies can be further sub-classified by the specific bonding mechanism or process. For each technology of joining mentioned earlier, mechanical methods, chemical/adhesive methods,[2] and welding methods exist. Welding methods can, of course, include brazing or soldering.

While only mechanical joining, in general, and integral mechanical attachment, in particular, is of interest in the treatment of this book, a very brief summary of the various adhesive joining, chemical bonding, and welding (including brazing and soldering) methods available for joining ceramic materials will be presented in the remainder of this section, more for background and context than for completeness. Readers interested in these other methods should consult dedicated references on ceramics and ceramic joining (*Engineering Materials Handbook,* 1991; Nicholas, 1990; Schwartz, 1990).

Before considering the various specific adhesive joining, chemical bonding, and welding methods, it is important to point out that specialists in ceramic materials (i.e., ceramists and glass technologists) tend to divide these methods into two categories: (1) direct joining methods and (2) indirect joining methods. With *direct joining methods,* one ceramic material is joined to another without anything in between. No intermediate material is required. For this to be

---

[1] "Hermeticity" means leak tightness to fluids, whether gases or liquids, from vacuum to high pressure.

[2] In the context of ceramic joining, there is a subtle difference between "adhesive joining" and "chemical bonding." Adhesive joining employs an intermediate material between the ceramic or glass and an opposing substrate material, whatever it might be. That intermediate material, known as an "adhesive," is a chemical agent that develops adhesion to the joint materials using some combination of surface adsorption, diffusion (or inter-diffusion), electrostatic attraction, and mechanical interlocking of surface asperities on opposing surfaces (Messler, 2004). Chemical bonding involves an actual chemical reaction between the joint materials or the joint materials and some specially-chosen intermediate material.

possible, one material in the joint must be able to form chemical bonds with the other; which requires that the materials be fundamentally similar (e.g., oxide ceramics to oxide ceramics, glass to glass). With *indirect joining methods,* inherent differences in the atomic-level structure and bonding of the materials to be joined hinders them from forming bonds with one another. Hence, an intermediate material that is compatible with each of the materials comprising the joint members must be used.

The list of adhesive joining, chemical bonding, and welding methods includes:

- Adhesive joining
- Diffusion bonding
- Sinter bonding
- Friction welding
- Fusion welding
- Refractory metal brazing
- Precious metal or noble metal brazing
- Active metal brazing
- Ceramic brazing
- Combustion synthesis joining

As each of these methods is described below, whether the method is a direct or indirect joining method will be identified.

9.1.1 Adhesive Joining

Adhesive joining of ceramic materials is almost always an indirect joining method, as an intermediate material that is chemically compatible with all of the different materials comprising joint members is used. The intermediate material is called the "adhesive." It is a chemical substance that develops an adhesive force with each joint member material by some combination of surface adsorption (atomic-level, often secondary van der Waals or hydrogen, bonding), diffusion or inter-diffusion (involving actual atomic-level mass transfer), electrostatic attraction (due to polarization), and, usually, some mechanical interlocking of the adhesive into nooks and crannies (i.e., asperities) on the surfaces of the substrates, known as "adherends" in the adhesive joining parlance. Adhesives used to join ceramics can be organic polymers or can be inorganic, non-metallic ceramics or ceramic-glass mixtures.[3] Polymer adhesives tend to be called "adhesives," while ceramic adhesives tend to be called "cements." Ceramic-glass mixtures used as adhesives tend to also be called "cements" or "frits."

---

[3] One should not confuse "ceramic-glass mixtures," which are mixtures of actual crystalline ceramics and amorphous glasses, with "glass ceramics," which are crystalline ceramics derived from an amorphous glass by heat treatment.

Details of how inorganic cements develop adhesion vary with specific types but often involve the formation of fairly strong hydrogen bonds from waters of hydration in what are known as "hydraulic cements," typified by well-known Portland cements used in modern-day construction. Some ceramic cements are applied as slurries, are allowed to harden by setting or curing (i.e., forming hydrogen bonds), and are used as-is. Other cements, particularly those containing some glass, are fired following application, with the firing causing sintering (by solid-phase diffusion) or wetting by the glass.

When the adhesive used is made from the same ceramic as the ceramics being joined, the joining can approach a direct joining method. Two examples are the formation of adhesive joints within a structure made from Portland cement and the formation of sintered joints within structures made from oxide ceramics, like alumina. Once one pre-cast cement unit is joined to another using an interlayer of the same cement, it can be difficult to tell where the original interface was. The same is true for one piece of mono-lithic[4] alumina joined to another piece using an alumina cement, which is sintered.

It is possible to produce both ceramic-ceramic and ceramic-metal joints using adhesive joining.

9.1.2 Diffusion Bonding

Diffusion bonding of ceramic materials is usually a direct joining method that accomplishes joining between two often like materials through atomic-level mass transport and exchange. The mechanism is solid-state diffusion or, more properly, inter-diffusion. Atoms (or, for many ceramics, ions) from one crystalline material held in intimate contact with another move across points of contact by thermally-activated jumps to take up positions on the crystal lattice of the other, and vice versa. The resulting exchange, usually driven by a difference in the chemical potential of the diffusing species in the two joint elements due to compositional differences, leads, in time, to the elimination of the original gap between the two pieces. Another driving force, of course, is the thermodynamic force trying to reduce the area of the surfaces between the two materials, as any interface has positive energy. This latter driving force is what predominates when diffusion occurs between two pieces of precisely the same material during sintering (see Section 9.1.3).

For ceramic joining, the process tends to be called "diffusion bonding," while for metals (where it may also occur), the process tends to be called

---

[4] The term "monolithic" in materials is often used to distinguish just the material from a reinforced, composite form.

"diffusion welding." One factor that tends to distinguish one from the other is whether the diffusion occurs entirely in the solid state or whether any liquid (even if only present as a transient phase) is involved. Diffusion welding involves no liquid phase but occurs entirely in the solid state.

Diffusion in and between ceramics is more complicated than it is for metals, since (as is discussed in Section 9.2) ceramics are non-metallic, inorganic compounds of one or more metallic elements with one or more non-metallic elements. In ionic compounds, the metal ions (cations) are much smaller than the non-metal ions (anions) and, so, move much more easily from lattice site to lattice site by jumping into adjacent vacant sites. While it is easier for them to do so, cations alone cannot move because doing so would disrupt the electroneutrality that must exist in ionic compounds. Hence, diffusion in ceramics tends to be limited by the diffusivity of the anions.

Sometimes to speed inter-diffusion between two ceramics (or metals, for that matter), an intermediate material is used. The higher diffusion rate may arise from the intermediate material being softer and, so, complying more with each substrate material to increase the number of point contacts across which mass transport via atom jumps can occur. Alternatively, the greater diffusion may occur because the actual rate with which one atomic or ionic species can move is faster in the intermediate than it is in the substrate materials.

It is possible to produce both ceramic-ceramic and ceramic-metal joints by diffusion bonding.

9.1.3 Sinter Bonding

Sinter bonding is actually diffusion bonding between particles of ceramic within a part or between particles of ceramic in parts in contact with one another across an interface. It can take place entirely in the solid state or it can be accelerated (rather dramatically, in most cases) by employing some liquid phase. Sinter bonding is unique to ceramic materials (i.e., to produce only ceramic-ceramic joints) and has no counterpart in metals, except, perhaps, when they are powder processed (German, 1996; Rahaman, 1995) via transient liquid phase bonding (TLPB). Sinter bonding, when the term is used correctly, is always a direct joining method. It tends to be used far more often to join small, simple details during the initial fabrication of a larger, more complex part. It is rarely used for joining pre-existing parts later on during a pure assembly or construction process.[5]

---

[5] Joining done at the time a part is initially produced is called "primary joining." Joining done between pre-fabricated parts later on to create an assembly or construction is called "secondary joining" (Messler, 1993, 2004)

9.1.4 Friction Welding

It is possible to join some ceramic materials to similar or even dissimilar ceramic materials, or even metals, using friction welding. In *friction welding,* parts are held in contact with one another and are caused to move relative to one another under pressure to generate friction. The combination of intimate contact between atoms on each side of the joint, the scrubbing action to level asperities and bring more atoms into intimate contact, and the heat that results from the work by both friction and deformation that is being done all contribute to the formation of a weld, in which a "weld" is a physical union between materials at the atomic level involving primary bond formation. Since no intermediate material tends to be used, friction welding is a direct joining method. The process can be used to produce either ceramic-ceramic or ceramic-metal joints.

9.1.5 Fusion Welding

It is much more difficult to produce welds in or between ceramics than within or between metals by causing some melting to occur in each part involved in the joint, followed by intermixing of the melts, followed by solidification of the intermixed materials. Processes that do this are called *fusion welding* processes. The reasons for fusion welding being difficult with ceramics include the facts that: (1) ceramics tend to melt at much higher temperatures than metals (placing much greater demands on the heat sources needed); (2) ceramics tend to have poor resistance to rapid temperature changes (i.e., poor thermal shock resistance) and, so, tend to crack upon heating or cooling; and (3) ceramics are inherently brittle, so tend to crack upon transformation of the liquid phase to the almost always lower specific-volume solid phase. But, despite these difficulties, some ceramics can be fusion welded by some processes. Many of the oxide ceramics (e.g., alumina, magnesia, zirconia, titania, and others) have been welded in similar and dissimilar combinations using electron-beam, laser-beam, or plasma-arc welding (Schwartz, 1990).

Fusion welding of ceramics to form ceramic-ceramic joints is almost always a direct joining method, as fillers of any kind are rarely used. In other words, welding is done autogenously.

9.1.6 Metal Brazing

There are three general approaches for joining ceramic materials to one another (i.e., ceramic-ceramic joints) or to metals (i.e., ceramic-metal joints) using metallic fillers. All three are *brazing* processes, wherein a metallic filler is caused to melt, wet the surfaces of the substrate materials (which must remain entirely solid throughout the process), fill a controlled gap between joint elements by capillary action, and solidify to produce a joint. The strength of the joint comes from a fair degree of inter-diffusion between at least some components of the braze filler and the substrates and from the formation of primary bonds between the filler metal and the substrate. The three general metal brazing approaches are: (1) refractory metal brazing; (2) precious (or noble) metal brazing; and (3) active metal brazing. All three approaches, since they require an intermediate material to work, are indirect joining methods.

Refractory metal brazing employs a filler alloy based on a refractory metal, such as W, Mo, Ta, or Hf, usually. As such, refractory metal braze joints tend to have excellent temperature resistance, which is often a necessary asset for ceramic substrates, which were likely chosen by a designer for their refractoriness. To allow the molten refractory metal/alloy filler to wet the surface of the ceramic substrates, the substrates are often "metallized" with a similar refractory metal. Means for metallizing the ceramic substrate include: (1) incorporating particles of refractory metal during the powder-processing of the ceramic part from particulate ceramic; (2) coating or plating a thin refractory layer by one of several chemical, electrochemical, thermal, or mechanical means (Messler, 2004); or (3) implanting ions of the refractory metal by high-energy bombardment.

One specific, common, and old method of metallizing ceramics so they accept a refractory metal braze is known as the "molybdenum-manganese process" or "moly-manganese process," for short. In this process, a slurry composed of a mixture of powdered manganese and molybdenum metal and oxides and a volatile binder is painted onto the surface of the ceramic parts whereupon brazing is to take place. The parts are then fired in a furnace with a hydrogen atmosphere with a controlled dew point.[6] A chemical reduction reaction takes place on the surface of ceramics for which this process has been successful (e.g., Al_2O_3, BeO, and some others) and leads to the formation of Mo metal that tenaciously adheres to the ceramic through MnO that dissolves into the surface of the ceramic. The Mo metal tends to form on the surface as particles sinter together to form a porous layer. Much of the adhesion of this layer arises from a glassy phase that forms when the oxide ceramic reacts with the MnO wicking up into the pores, locking the layer to the ceramic. Beyond this mechanical contribution to adhesion, however, is a chemical contribution from the reaction of this glass phase with the Mo to form MoO, which tenaciously adheres to the Mo.

In *precious metal* or *noble metal brazing,* Au, Pt, Pd, or Ag filler metal is caused to adhere to oxide ceramics (such as Al_2O_3, MgO, ZrO_2, UO_2, BeO,

---

[6] A "dew point" is a measure of the amount of water vapor in a gaseous atmosphere. The more water vapor present, the higher the dew point, which is the temperature at which liquid water would condense from the vapor at some temperature. Higher dew point atmospheres are more oxidizing than lower dew point atmospheres.

SiO_2), glasses, and graphite by causing these normally very inert or noble metals to oxidize slightly in an oxygen-rich atmosphere. The noble metal's oxide then bonds to the substrate oxide to create a joint between joint elements by solid state diffusion.

Active metal brazing uses an active metal like Ti, Zr, or Nb to react with the non-metallic component(s) of ceramic substrate materials to form atomic-level bonds and, thus, joints. Essentially all active-metal braze fillers are alloys based on these inherently reactive metals. The particular advantage of reactive metal brazing is that both oxide-type and non-oxide–type (e.g., carbide, nitride, boride, and so on) ceramics can be joined in similar and dissimilar combinations, which is normally quite difficult to accomplish.

9.1.7 Ceramic Brazing

It is possible to join ceramic materials to one another, in ceramic-ceramic joints, using a filler that is, itself, a ceramic or ceramic-glass mixture. When a ceramic filler or ceramic-glass filler mixture is used to braze ceramics, the process is called *ceramic brazing.* A major difference between metal brazing and ceramic brazing is that in ceramic brazing the filler rarely flows into the joint—to form the joint—by capillary action. The reason is that molten ceramics tend to be too viscous to flow. Hence, ceramic braze fillers are normally pre-placed at the joint faying surfaces as powders or slurries and are caused to simply melt in place and wet the substrates to form bonds. There are a variety of ceramic mixtures or ceramic-glass mixtures used for ceramic brazing (Messler, 2004; Schwartz, 1990).

9.1.8 Combustion Synthesis Joining

An interesting process for joining ceramic materials is combustion synthesis. Combustion synthesis[7] is a highly exothermic reaction that occurs between metals and certain non-metals, for example, to create carbides, nitrides, borides, silicides, and other ceramic materials in the first place. The tremendous heat released during such reactions often produces a liquid phase, and, so, the process has been used to accomplish joining. *Combustion synthesis joining* is, in fact, classified by the American Welding Society as an "exothermic brazing process." As such, it is an indirect joining method that can be used with ceramic

---

[7] Combustion synthesis (CS) is also known as "self-propagating high-temperature synthesis" (SHS). The former term most aptly refers to when the reaction occurs throughout the reactants, all at once, once the triggering temperature for the reaction is reached. The latter term most aptly refers to when the reaction front moves through the reactants, raising the temperature to cause further reaction as it moves.

materials to produce ceramic-ceramic or ceramic-metal joints. Details of the process can be found in the literature (e.g., Messler, 1995).

9.1.9 Soldering

Finally, it is possible to join ceramic materials to themselves or to metals using soldering. *Soldering* involves the use of a filler material that melts at low temperatures, typically below 450°C (840°F). In soldering, as in brazing, the molten filler must wet the substrates, flow to fill the joint by capillary action, and solidify in place to form atomic-level bonds with the substrates. The problem with soldering ceramics is trying to find fillers that wet the inherently chemically-inert ceramics. Metallic solders that work best are those based on indium (In) which has the unique characteristic that it wets virtually all materials. There are also low-melting glasses that can be used to solder glasses or ceramic-glass mixtures. These are called "solder glasses" (Messler, 2004). Joining is always indirect, as an intermediate material is used.

9.2 PROPERTIES OF CERAMICS THAT FAVOR INTEGRAL MECHANICAL ATTACHMENT

With Section 9.1 as a background on adhesive joining, chemical bonding, and welding (including brazing and soldering) methods by which ceramic materials can be joined, it is time to turn to the mechanical methods that can be used to join ceramic materials. But, first, it is necessary to fully define and characterize ceramics.

Ceramics are surely the most abundant materials on Earth. They occur naturally,[8] usually as mechanical mixtures of more than one type. They are defined as non-metallic, inorganic compounds of one or more of the metallic elements with one or more of the non-metallic elements. In the crust of the earth, they are solids, while well below the crust, in the mantle and, perhaps, in the core, they are molten. The reason they form with abundance is that metals tend, by their nature, to oxidize, that is, give up some of their outermost electrons to establish a stable electron configuration. Once the metals undergo this chemical oxidation process (which does not require oxygen to occur, only highly electronegative elements), they exist as positive ions, which, in turn, seek out negative ions. Negative ions are created when non-metallic elements (i.e., whether the various gaseous elements other than the inert gases, or non-gaseous elements, such as S, P, C, and so on) take on additional electrons to complete their outermost

---

[8] In Nature, ceramics are found in the form of minerals, rocks (which are mixtures of minerals), and ores (which contain desirable metals in either their nascent or oxidized state).

electron shells to become stable. This occurs by a chemical reduction process. The negative ions of the non-metallic elements tend to seek out, find, and bond with the positive ions of the metallic elements to form non-metallic inorganic compounds we call ceramics.

While this process of oxidation and reduction to produce ceramics can and does occur in Nature, it can also be made to occur to produce ceramic materials synthetically. In either case, the chemical compositions, atomic-level structures, and resulting properties tend to be the same. The major difference is that synthesized ceramics tend to be more pure.

Positive metal ions (called "cations") are smaller than the neutral metal atoms from which they form, while negative non-metal ions (called "anions") are larger than the neutral atoms from which they form. To form a solid aggregate, the small positively-charged cations try to surround themselves by as many negatively-charged, but larger, anions as they can, based on their relative sizes (as measured by cation-to-anion ratio). This occurs to keep like-charged ions separated from one another, as otherwise they would repel one another. This tendency of small cations and larger anions to pack together as closely as they can must be done in a way that the net positive and net negative charges exactly balance through the solid aggregate. This balance is achieved through the establishment of the proper ratio of positively-charged and negatively-charged ions, that is, the achievement of the proper "stoichiometric ratio" in the compound. The force that holds such ionic compounds together is Coulombic electrostatic attraction.

Ceramic compounds can be conveniently and neatly divided into two groups: oxides and non-oxides. The oxides are compounds of metallic or metalloid elements (e.g., Si, Ge, Sr, B) with oxygen, and include alumina (Al_3O_2), beryllia (BeO), magnesia (MgO), titania (TiO_2), urania (UrO_2), silicon dioxide (SiO_2), zirconia (ZrO_2), and many others, including types with more than one metallic element with oxygen (like $BaTiO_3$, $PbTiO_3$, and complex high-temperature "1-2-3 superconducting ceramics" like $YBa_2Cu_3O_7$). The non-oxides are oxidized compounds of one or more metallic elements and one or more non-metallic elements, one of which could be oxygen. Examples are carbides (e.g., SiC, B_4C, TiC, WC), borides (e.g., TiB_2, MoB), nitrides (e.g., BN, AlN, Si_3N_4, TiN), sulfides (e.g., ZnS), silicides (e.g., $MoSi_2$), and beryllides (e.g., $ZrBe_{13}$), as well as carbonates (e.g., $CaCO_3$, $MgCO_3$), sulfates (e.g., $CaSO_4$, $Mg,Al[SO_4]$), and others.

Because the metallic and non-metallic atoms in a ceramic are bonded either ionically (as has been described) or covalently (in some other ceramics comprised of metallic and non-metallic elements that are located closer to one another in the periodic table, that is, that have more similar electronegativities), their outermost electrons are totally tied up in the bonding. None are free to move around under the influence of an applied electric field. As a result, most ceramics are electric insulators, as opposed to being conductors. Many are also thermal insulators, as a significant contribution to thermal conductivity also comes from outermost electrons that are free to move under the influence of a thermal field.[9]

Because they are already in an oxidized state, ceramic materials tend to be very chemically stable. There is little that tends to attack them. They exhibit virtually no corrosion, in the conventional sense, although some do degrade in certain service environments. Unfortunately, when it comes to joining, they also tend to be difficult to wet with molten metals.

The properties of high melting temperature (arising from the very strong ionic or covalent bonding), electrical and thermal insulating quality (arising from localized versus delocalized bonding such as is found in metals), and chemical stability cause ceramics to be chosen for service in harsh environments. Unfortunately, all of these desirable properties tend to come at a cost. While ceramic materials also tend to be mechanically strong (also as a result of the strong bonding found in them), they lack ductility (i.e., tolerance of strain) and toughness (i.e., tolerance of absorbed strain energy). In fact, also as a result of their alternating arrangement of metallic and non-metallic atoms or ions, they do not tend to exhibit any significant degree of plastic deformation, since the slipping of one crystalline plane in a ceramic over another would disrupt the alternating arrangement of anions and cations. As a consequence of all of this, ceramic materials (1) tend to be hard and difficult to cut by machining, drilling, and so on; (2) are generally impossible to shape by plastic deformation processes (forging, drawing, forming, etc.); and (3) are difficult to cast because they are so high melting or because they do not melt without usually decomposing or subliming to vapor before forming a liquid. This greatly limits both the size and the geometric complexity that can be processed into parts made of ceramics. To produce parts that are large and/or geometrically complex requires that smaller, simpler parts must be joined. Furthermore, such joining is often best accomplished by using strictly mechanical means.

Figure 9.1 shows the strain behavior of a typical ceramic material to a load that is instantaneously applied, held constant, and then instantaneously removed. The ceramic almost always exhibits purely elastic strain behavior, as little or no plastic strain occurs in most ceramics. This behavior, combined with the inherent mechanical properties of high strength (especially in compression[10]), high stiffness, and high hardness suggests the following relative to mechanical joining:

---

[9] The other contribution to thermal conductivity in a material is the stiffness of inter-atomic bonds (i.e., a phonon contribution). There are ceramics that conduct heat but do not conduct electricity. The most notable example is carbon in the form of crystalline diamond. However, other compounds, like AlN, exhibit the same behavior.

[10] The reason ceramic materials exhibit higher strengths under compressive loading than under tensile loading is that they almost always contain randomly oriented micro-flaws, usually in the form of micro-cracks. As a result, tensile loading tends to cause any such flaws oriented such that they tend to open under the loading to extend rapidly, without any yielding at the growing crack tip to slow or arrest propagation. Under compressive loading, there is no force that tries to open such flaws, so the material is not weakened by their presence.

- Because of high chemical stability and difficulty to cause melting, joining relying on chemical and physical forces is generally difficult with ceramic materials
- Use of mechanical fasteners can be problematic, as preparing holes for fasteners that require them is difficult due to the hardness (and wear resistance) of most ceramics, and point-loading under bearing is not well tolerated by inherently brittle materials
- Ceramic materials tend to be most amenable to mechanical joining using rigid interlocks because they are hard, although care must be taken to avoid severe stress concentration from sharp geometric features
- While ceramics exhibit high elastic limit strengths, their high modulus of elasticity makes deflection difficult, and their sensitivity to fracture in tension makes deflection risky. Hence, the use of elastic interlocks is unknown
- The absence of plasticity in almost all ceramics precludes the use of plastic interlocks

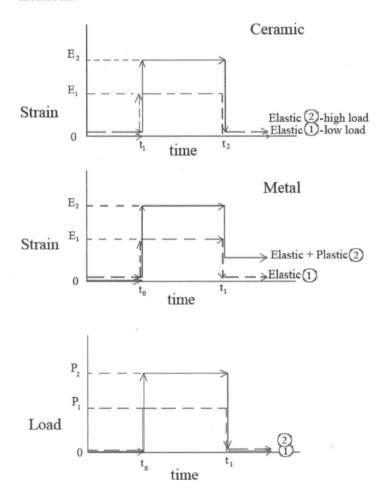

FIGURE 9.1 Plot of the strain behavior of a typical ceramic material.

More than simply facilitating the use of integral mechanical attachments for assembly, ceramics and glass-ceramics, as well, tend to favor this scheme for joining.

With this as a background, let's look at how ceramic materials can be joined using integral mechanical attachments. Basically, there are two possibilities: (1) using geometric features that can be cast or molded into the ceramic part during its creation as rigid interlocks or (2) using attachments (such as inserts) that can be embedded in ceramics made by casting or other means as rigid interlocks.

Appendix B lists important properties for the most important ceramics and glasses used in engineering.

9.3 CAST, MOLDED, AND COMPACTED ATTACHMENTS IN CERAMICS

Crystalline ceramics can be processed into shapes by casting or by powder processing. While casting of metals always means taking the metal or alloy from its fully or partially molten state[11] and pouring it into a mold or die containing a cavity shape that will yield the desired part in positive, this is not always so for ceramics.

Some ceramics do not melt when heated. Rather, some sublime directly to the vapor phase under normal pressures and never produce a liquid. Others decompose into another form or forms rather than form a liquid phase of the same composition as the solid being heated. For those that do melt, melting temperatures tend to be very high, fluidity tends to be poor (i.e., they are viscous), and susceptibility to cracking upon solidification and subsequent cooling is problematic. As a result, another method is often used to cast ceramics into shapes. It is known as "cold casting" or "slip casting." It is accomplished by mixing finely crushed ceramic particles with a suitable liquid binder, often, but not only, water. The resulting slurry is poured into a mold, the water or binder is allowed to evaporate or is caused to be absorbed into the mold, and the slurry becomes progressively thicker as it consists of more solid and less liquid. When all of the liquid binder or carrier is gone or has been taken up in binding waters of hydration, the casting becomes solid and able to be handled. Some cold cast parts are used "as-is," while others are subsequently fired to bind particles together by sintering.

---

[11] Most casting of metals is done by heating them to above the point at which they are 100% liquid. For pure metals, this is above the melting temperature. For an alloy, it is above the liquidus temperature for the alloy; the liquidus temperature is the temperature above which the alloy is completely liquid. Some modern casting processes only heat the alloy to between its solidus and its liquidus temperatures, so that it consists of a mixture of solid and liquid and is in a mushy state. Such casting is known as "rheocasting" or "thixotropic casting" and offers the advantage of requiring less high-temperature mold or die materials. However, to get the mushy mixture to fill the cavity, pressure is needed.

In lieu of casting, ceramics can be produced into shapes by compacting or molding mixtures of the ceramic(s) to comprise the final part and, possibly, a volatile binder into a die under pressure. This is the basis for powder or particulate processing, including ceramic powder injection molding. Once again, compacted parts, which are said to be "green," can be used "as-is," but most are exposed to a sintering or firing heat treatment to increase particle-to-particle binding.

Regardless of which of these two approaches is used to produce a ceramic part, geometric features can be designed into the part to allow interlocking. As was described in Chapter 3, Section 3.2, combinations of protruding features and recessed features are used to allow mating and interlocking using simple push, slide, or tilt motions. Such features tend to be used on large, flat faces as well as on smaller surface-area edges Use in pre-cast cement on concrete units is also quite common (see Chapter 10, Section 10.2).

Figure 9.2 schematically illustrates some possible interlocking features that can be cast or molded into cast or powder processed ceramic parts. Figure 9.3 shows an example of the use of such cast in rigid interlocks.

Great care must be taken to ensure that features to be used to accomplish rigid interlocking do not contain severe stress raisers, that is, notches, radii, corners, and so on.

9.4 EMBEDDED OR INSERTED ATTACHMENTS IN CERAMICS

It is sometimes necessary or desirable to make attachments to ceramic shapes and parts, not just between them. For such applications, various attachment features can be cast in, molded, or embedded as inserts during part production or, alternatively, can be installed or inserted later, during a subsequent operation, perhaps at the site of assembly or structure erection. In fact, many of the possible inserts are completely analogous in form, although they may differ in size, geometric details, and material of construction, from inserts described for use with polymer parts (see Chapter 8, Section 8.3). Threaded and unthreaded studs are very common, as are threaded or threadable or clamping inserts.

Threaded studs can be simple threaded rods or conventional bolts or even specially designed devices intended for use by being cast or compact-molded into ceramics. These are used to attach another part, often made from a material other than ceramic, such as metal, by tightening a nut, without or with a conventional or locking washer. Unthreaded studs tend to be upset at their free end once a mating part is positioned over the stud. Inserts containing threads are used to receive machine screws or bolts, for example, soft-metal (e.g., Pb) anchor inserts and various split-cylinder inserts are used to accept self-tapping screws and so-called "lag bolts."

Many of these attachments, particularly, threaded, soft-metal, or split-cylinder anchoring inserts can be installed into an already cast or pressure-compacted ceramic part.

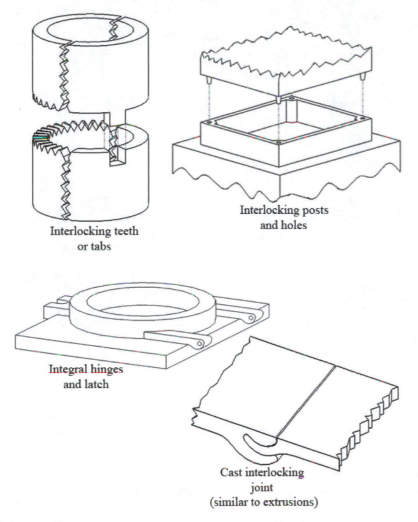

Interlocking teeth or tabs

Interlocking posts and holes

Integral hinges and latch

Cast interlocking joint (similar to extrusions)

FIGURE 9.2 Schematic illustrations of some possible rigid interlocking features in cast or powder processed ceramic parts.

Figure 9.4 schematically illustrates some inserts that can be cast in, pressure-compacted in, or installed into ceramic parts.

9.5 PROPERTIES OF GLASSES THAT FACILITATE MECHANICAL ATTACHMENT

Glasses, as they are known by the general public, are non-metallic, inorganic materials composed of combinations of metallic and non-metallic elements, just as ceramics are. The difference is that glasses are amorphous, that is, they

(a)

(b)

FIGURE 9.3 Photographs showing examples of cast-in rigid interlocks in fired refractory bricks or "fire bricks." Refractory blocks lining the ceiling of a furnace are typically hung from "dog-bone"–shaped ceramic or metal hangers, which mate with cast-in recesses in the blocks (a), while refractory blocks making up the walls are typically shaped at their cool side with interlocks to prevent heat leakage (b). (Photographs courtesy of Morgan Thermal Ceramics, Augusta, GA; used with permission.)

are non-crystalline. Ceramics are crystalline. This difference is a big one in terms of the behavior of glasses compared to ceramics.

Glasses are produced by firing crystalline inorganic, non-metallic materials (e.g., oxides, fluorides, borides, nitrides, and silicates) to allow them to join together to form a three-dimensional network structure often built up of silicate

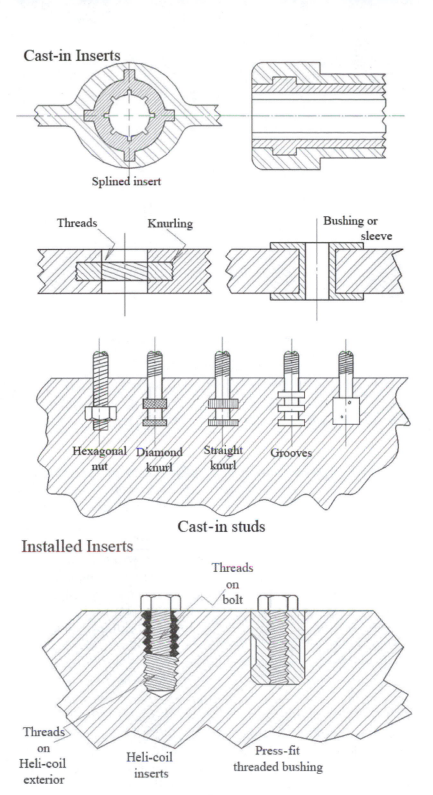

FIGURE 9.4 Schematic illustrations of some inserts that can be cast in, pressure-compacted in, or installed into ceramic parts.

(*Continued*)

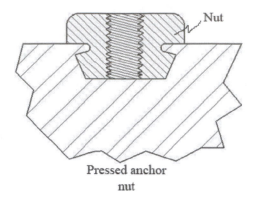

Nut

Pressed anchor
nut

FIGURE 9.4—Cont'd

units consisting of four oxygen atoms forming a tetrahedron around a silicon atom, with each oxygen atom covalently bonded to the silicon atom. Between these tetrahedral building blocks, atoms from the other ingredients modify the network in various ways, sometimes making it stronger, sometimes weaker. However, the end-result is a very complex three-dimensional structure with a particular chemical makeup and arrangement that turns out to be fairly immobile. Hence, by cooling the molten glass down, the atoms are unable to rearrange themselves into a perfectly orderly, crystal structure. Instead, they retain essentially the same structure they had while the glass was fully molten and fluid (albeit viscous). Solidification, as such, does not occur, and there is no distinct melting point. Rather, the molten mass simply becomes less and less fluid, more and more viscous, to the point that it eventually appears solid-like. In other words, glasses are solids with the structure of a liquid.

In fact, glasses are never truly solids, although they behave mechanically as typical solids around room temperature by being rigid. They exhibit good strength in the short-term until a so-called glass-transition temperature (T_{glass}) is reached. Above the T_{glass}, the glass becomes increasing soft and workable as the temperature is increased more and more. Thus, at room temperature and below, and up to near the T_{glass} or T_g, glasses exhibit elastic behavior. They can act rigidly to resist loads without changing shape and can resist penetration under bearing stresses, or they can be caused to deflect under modest loads, from which they can fully recover. At these temperatures, glasses do not exhibit any plastic behavior. As temperature is increased, this elastic behavior becomes increasingly visco-elastic (much like that of a polymer, see Chapter 8, Section 8.1), until finally they behave like a viscous fluid, exhibiting viscous behavior.

The strain behavior of glass is shown schematically in Figure 9.5 at a number of different temperatures.

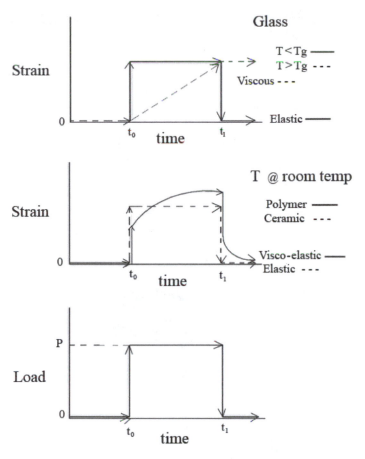

FIGURE 9.5 A schematic illustration of the strain behavior of glass at different temperatures.

The most common glasses (e.g., window glass, known as "soda-lime glass", Pyrex,™ which is a borosilicate glass, and fused silica, which is simply molten SiO_2), like many other glasses, are optically transparent, electrical, and thermal insulators, with quite good resistance to most chemicals. They are hard and strong, and tend to have relatively high stiffness at room temperature, but they are notoriously brittle.

Because of their many unusual, if not unique, properties, glasses are important engineering materials. However, because of their lack of toughness and susceptibility to severe fracture-inducing stress concentration from scratches, nicks, chips, or notches, glasses are often joined to other materials, including metals, crystalline ceramics, polymers, and even wood and cement or concrete.

Like their crystalline cousins, ceramics, glasses can be joined by organic adhesives, by welding, and by solders. However, they can also be joined mechanically using techniques that have been adapted to suit their idiosyncrasies.

Glass-to-glass joining is, by far, the easiest kind of joining to accomplish within the general category of ceramic materials. The options include welding, adhesive bonding, soldering, and even mechanical joining using either fasteners or integral mechanical attachment features. In welding, a fusion approach can be used in which two pieces of glass are heated in a furnace or by an oxy-fuel gas flame while they are held in contact. When they are made suitably soft, they begin to flow together to form a joint in an effort to minimize their combined surface energy. Figure 9.6 shows what is known in the glass industry as "fusing" or, in the case shown, "flame fusing" to join smaller, simpler-shaped pieces of glass into a larger, more complex assembly.

FIGURE 9.6 Glass parts can be fused, or welded, to form a continuous, hermetically-sealed assembly, here, using an oxy-fuel gas flame. (Photograph courtesy of Corning Inc., Corning, NY; used with permission.)

More recently, glass has been welded to glass (and to ceramics and even some specially surface-conditioned metals) using friction welding techniques. In this case, the heat of friction softens the glass over the area of contact and material flows back and forth to form a weld.

While welding or fusing can be done in its own right, heat is often used to form a cap or head on a post to lock one glass part to another by the process of *thermal staking,* producing an integral processed-in plastic interlock.

Various polymeric adhesives can be used to join glass to glass. When this is done, the adhesive is almost always chosen to be optically transparent. A comfmon, albeit perhaps not easily recognized example, is the adhesive bonding of multiple layers of glass in the windshields of automobiles using a transparent thermoplastic interlayer. The tough interlayer helps prevent glass fragments from causing damage to occupants of the vehicles in the event of an accident that shatters the glass.

For other applications, particularly those where the glass is to be used at elevated temperatures, special inorganic adhesives made from powdered glass are used for adhesive joining. Such adhesives are called glass "frits."

Glass-to-glass joints can also be made by soldering, using either of two approaches. While most metallic solders do not wet glass sufficiently to form a good bond for structural integrity, metal solders can be used to produce reasonably good hermetic seals. Specific techniques will be described in Section 9.9.

9.7 MECHANICAL METHODS FOR GLASS-TO-GLASS JOINING

Glass parts can be joined to one another using mechanical methods. In fact, parts made from glass and from a different material can be, and often are, joined using strictly mechanical means. So important is the joining of glass and metal parts to produce hermetic seals that the method of producing glass-metal mechanical seals will be dealt with separately in Section 9.9. In this section, the use of designed-in geometric features for attaching glass parts will be described. Strictly for completeness, the use of mechanical fasteners, though obviously not integral attachment, will be briefly discussed as well.

Integral attachment features that are designed and processed into glass parts to allow their assembly to other glass parts with mating features include integral threads and tab-and-groove interlocks as shown schematically in Figure 9.7, and examples of both integral threads and posts are shown in Chapter 3, Figures 3.10 and 3.23. Because glass is inherently brittle and especially susceptible to fracture due to stress raisers, threads have to be coarse and rounded at both their crowns and their roots, tabs or other protrusions have to be blunt or rounded and receiving grooves need to also be round bottomed. Since most complex-shaped glass objects are molded (using compression or blowing), production of such relatively geometrically complicated shapes as threads, protrusions, and grooves is not difficult. Again, because of the sensitivity of glass to fracture under point of concentrated loading, the fit between external and internal threads and between mating protruding features and grooves needs to be loose to keep friction forces low. In fact, in many designs for integral threads and tabs-and-grooves, insertion and locking are designed to be accomplished with very few turns, and often only a partial turn. Furthermore, at least with tabs-and-grooves, rotation is usually limited with a designed and processed in stop feature. Such designs are analogous to so-called "bayonet fittings" described in Chapter 3, Section 3.2.

Integral threads and even tab-and-groove mechanical attachments are capable of producing quite tight, even leak-tight, joints, if not of their own accord, by using greases or rubber gaskets.

Not surprisingly, integral threads and, to a lesser extent, tabs-and-grooves can be used to join glass parts to parts made from other materials, usually metal, but occasionally ceramic or polymer, or vice versa, parts made from other materials to glass parts.

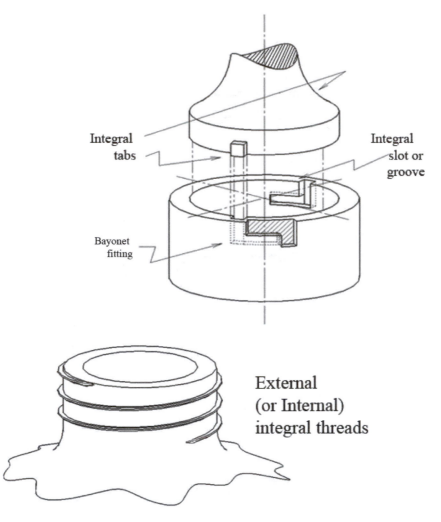

FIGURE 9.7 Schematic illustrations of integral threads and tab-and-groove attachment features used in glass parts.

Mechanical fasteners are rarely used with glass to accomplish joining to any material, especially glass. The reasons are: (1) producing holes in inherently-hard glass for fasteners that require them (e.g., pins, rivets, bolts) is non-trivial and (2) such holes, being small, are severe stress raisers in inherently-brittle materials; (3) the high elastic strength and modulus of glass makes the development of clamping forces needed for bolts to operate properly for tension-loading virtually impossible; and (4) the extreme brittleness of glass, and the resulting sensitivity to point loading, means that bearing stresses on fastener hole walls (e.g., from pins, rivets, or bolts) would be poorly tolerated. For these reasons, pins, rivets, and bolts (including machine screws) are seldom used. When they are, however, fastener holes are—or should be—lined with a strain-accommodating, load-spreading sleeve or grommet. Rubber grommets are, in fact, generally used. For similar reasons, load-spreading, compliant washers made from rubber, soft plastics, cork, or felt are also used when nuts and bolts are used.

As for brittle ceramics, special mechanical fastener devices have been developed for use with glass. Examples include wire clips, metal clamp rings, wire brackets, and T-pins for use in molded slots.

9.8 GLASS-METAL SEALS

Glass-to-metal joints or *glass-metal seals,* as they tend to be called, are important for a host of electrical and electronic, vacuum, and chemical applications. The combination of glass and metals is important because of some of the functionally-specific properties each possesses but neither completely offers. For glasses, these functionally-specific properties include: (1) visual optical transparency (to allow direct observation of some function or operation occurring within a closed system); (2) transparency of some glasses to infrared or visible laser of a select wavelength (to allow such light to be used for sensing or for control, as in laser-guided weapons); (3) excellent chemical inertness and stability to many normally corrosive environments; (4) excellent stability in high vacuum (e.g., 10^{-8} Torr and greater) and ease with which any contamination can be removed; and (6), for some, unique electrical, optoelectronic, and photonic properties. For the metal portion of glass-metal systems, the properties of interest are generally (1) superior structural/mechanical integrity, including strength, ductility, and toughness, and (2) superior thermal conductivity to aid in the dissipation of unwanted heat.

Glass-metal seals can be classified a number of ways, some of which provide an indication of the terminology used in the glass community and all of which, together, reflect the wide variety of approaches that can be used (Schott & Gen., Inc.). These ways include:

- Classification based on the geometry of the metal parts, including bead, tubular, disk, ribbon, and feather-edge

- Classification based on the type of metal or alloy used in the metal parts, including Mo, Pt, Cu, Kovar,[12] Fe-Ni alloys, as well as Pb and liquid Hg
- Classification based on the type of glass used, including soft glasses (e.g., soda-lime-silica) and hard glasses (e.g., borosilicate or fused silica), which offer different working temperatures
- Classification based on the type of joining, including mechanical, adhesive, fused glass, solder glass, and metallization
- Classification based on the technique used to make the seal, including flame, induction, fritting, pressure-diffusion, tape, and high-frequency resistance welding
- Classification based on the level of stress on the glass in two categories: matched or balanced and unmatched or unbalanced

The last classification is the one used most frequently, and is generally considered the most important, the reason being that the coefficients of thermal expansion (CTEs) for metals and for glasses (and ceramics) tend to be quite different, so thermally-induced stresses can develop between these dissimilar materials that can destroy joints.

As a rule, glasses and ceramics have CTEs that are lower than most metals. The CTE of a typical glass might be $3–5 \times 10^{-6}$ per degree C ($1.7–2.8 \times 10^{-6}$ per degree F), while that for a typical metal might be $12–15 \times 10^{-6}$ per degree C ($6.7–8.3 \times 10^{-6}$ per degree F). While it is rarely if ever stated how great the difference in the CTE can be before thermally-induced stresses cannot be tolerated, years of asking designers seems to suggest the difference in CTE between two adjoined materials must be not more than 10–25%, depending on the relative brittleness of one or more of the joint materials. For inherently brittle materials, the difference in CTEs across joints must generally be less than about 10–15%. For inherently ductile, tough materials, the difference in CTEs across joints can be greater, often 25% or more, in fact. Hence, if fused silica glass (0.4×10^{-6} per degree C) is to be intimately bonded to copper (16.8×10^{-6} per degree C), there's going to be a very serious problem with thermally-induced stress.

There are actually two problems, both due to the difference in CTE. In making the seal—at elevated temperature—the bonded materials must cool back to room temperature without developing stresses so high that the more brittle material (in this example, the fused silica) doesn't shatter. Presuming the seal could be made and the assembly could be cooled back to room temperature without ruining the joint, comparable stresses, but of the opposite sign in each joint element, would develop if the assembly was caused to heat up again. It is, after all,

---

[12] Kovar is an alloy of 29% Ni, 17% Co, and 54% Fe used for its very low coefficient of thermal expansion, typically 5.1×10^{-6} per degree C or 2.8×10^{-6} per degree F. This compares to metals like stainless steel ($15.0–17.5 \times 10^{-6}$ per degree C or $8.3–9.7 \times 10^{-6}$ per degree F), copper and Cu alloys ($16–22 \times 10^{-6}$ per degree C or $8.9–11.9 \times 10^{-6}$ per degree F), and to glasses ($0.4–9.0 \times 10^{-6}$ per degree C or $0.22–5.0 \times 10^{-6}$ per degree F).

the change in temperature from an excursion, up or down, that causes thermal stresses. Or, alternatively, severe stresses can develop in the presence of a temperature gradient in a joint between materials with fairly different CTEs.

Table 9.1 lists the CTEs for a variety of important materials.

TABLE 9.1   Coefficients of Thermal Expansion for a Variety of Important Engineering Materials ($\times 10^6$ degrees C^{-1})

Ceramics and Glasses		Polymers	
Alumina (99%)	6.5–7.6	Epoxy	81–117
Alumina-zirconia	~8	Kevlar (aramid)	60
Borosilicate glass (Pyrex)	3.3	Melamin	15
Brick (clay) materials	~6	Nylon 6,6	144
Diamond	0.8–1.2	Phenolic	122
Graphite	1.2–8.2	Polycarbonate	122
Graphite (pyrolitic)	2.7	PET	117
Mullite	5.7	Polyethylene HD	106–198
Pyroceram (glass-ceramic)	5.7–6.5	Polypropylene	145–180
Silica (99%, fused)	0.05–.04	Polystyrene (not foamed)	90–150
Silicon carbide	4.7	PTFE (Teflon)	126–216
Silicon nitride	2.7–3.7	PVC (polyvinylchloride)	90–180
Soda lime glass	7–9	Rubbers (elastomers)	500–800
Zirconia	10.1		
Metals and Alloys		**Composites**	
Al alloys	18–24	Concrete	10.0–13.5
Ag	19.7	CFRP (graphite-epoxy)	
Cast iron (gray)	10.1–12.6	– transverse	32
Co alloys (high-temp.)	11.4–16.0	– longitudinal	–3.6
Cu alloys	16–22.0	E-glass/epoxy (60 v/o)	
Invar (64Fe–36Ni)	1.6	– transverse	30
Mg alloys	25.0–27.0	– longitudinal	6.6
Mo	4.9	Wood (parallel to grain)	3.0–4.5
Ni alloys (high temp.)	12.9–14.6		
Solder (60Sn–40Pb)	24.0		
Si (semiconductor)	2.5		
Steels			
– C steels	11.0–12.8		
– low–alloy steels	~14		
– stainless, austenitic	15.0–17.5		
– stainless, ferritic	10.0–11.7		
– stainless, martensitic	9.9–11.2		
– stainless, PH	10.6–11.0		
Ti alloys	7.6–9.8		
Zr alloys	5.9–6.3		

It is interesting that the problem of different or differential CTEs can be dealt with in two diametrically opposed approaches. CTEs of the different materials (here, a glass and a metal) comprising the joint can be matched with one another, or not! The former is the approach taken with matched or balanced seals. The latter is the approach taken with unmatched or unbalanced seals. But how can both approaches work?

In *matched seals,* alloys and glasses of similar coefficient of thermal expansion are joined to one another so that the resulting joint is substantially free from induced strain and potentially destructive stresses. To "match," the CTEs need to be within about 10–25% of one another, depending on the degree of brittleness of one or more of the joint materials. For alloys and glasses with CTEs that differ by *much* more than 25%, it will almost surely be necessary to create the joint by making a "sandwich" of materials with CTEs that differ by less than 25% between adjoining layers. This result is a stepped joint that could consist of many layers of various metals, alloys, and glasses with CTEs that differ by no more than about 25% between adjoining steps. To further minimize problems, the metal parts are typically rings, tubular forms, or disks to avoid sharp stress-raising features.

In *unmatched seals,* strains and stresses that would normally arise within adjoining materials with CTEs that differ by more than about 15–25% are dealt with without attempting to match the CTEs in steps. Four techniques are used. First, the higher-CTE metal parts are kept to very small dimensions (e.g., wire diameters typically less than 0.8 mm or 0.035 inch). This works because the total strain induced by mismatched CTEs depends on the dimension over which it acts (i.e., the units of CTE are mm/mm/degree C or in./in./degree F). Small dimensions result in small strains, regardless of the difference in CTEs. Second, ductile metal parts can be used because they yield under the induced stresses form mismatched CTEs, thereby shielding the brittle glass from stress buildup. Metals commonly used to take advantage of this technique include Cu, Ni, Pb, and Mo. One option here is to create so-called "soldered seals" in which the glass element of the intended joint is metallized and then soldered to the metal element. The inherently soft, ductile solder alloy accommodates any strain mismatch by yielding. In fact, there are even examples of seals where a liquid metal, specifically Hg, is used (often as an intermediate material between the metal and glass parts) so that no stresses develop. Third, metal parts can be designed to deflect or distort under thermally-induced stress, much like rubber expansion strips operate in concrete pavement. An excellent and familiar example is the use of pleated bellows. Fourth, glasses of intermediate CTEs can be used to bridge the difference between the metal and the glass joint elements. In this technique, glass-to-glass bonds are made everywhere except between the final intermediate glass and the metal joint element. This technique is known as "graded seals" and is less common now that alloys and glasses with a wider range of CTEs are available.

Unless special effort is taken, joints or seals between a glass and a metal will be mechanical. To achieve actual atomic-level bond formation is difficult because these two material types are so fundamentally different. Glasses are amorphous, while metals are crystalline. Metals tend to be reactive to many of the chemical agents to which glasses are non-reactive, including oxygen. Physical properties (e.g., melting point,[13] coefficient of thermal expansion, specific volume versus temperature) and mechanical properties (e.g., strength, elastic limit, modulus of elasticity, ductility, toughness) differ greatly. But, despite these differences, some methods for creating actual bonds between glasses and metals have been described, including adhesive joining and chemical bonding using reactions between oxidized metals and glasses via diffusion or metallized glasses and metals via soldering, as examples. Now it's time to consider purely mechanical methods, in keeping with the theme of this book.

Extensive review of decades of literature on mechanical glass-metal seals reveals that techniques fall into one of four basic types. These four types are:

1. Precision fits or ground joints
2. Elastic compression fits
3. Deformation fits
4. Liquid metal seals

The first three of these are, in reality, integral mechanical attachment schemes that rely solely on designed-in or processed-in mating geometric features. Even more interestingly, these first three create rigid, elastic, and plastic interlocks, respectively. The fourth type of joint is really only a seal, in that electrical and/or thermal interconnection often occurs, along with hermeticity, but mechanical/structural integrity must come about from something other than the liquid metal seal.

> *Precision Fits or Ground Joints.* In *precision fits* or so-called *"ground joints,"* a metal pin (often made from a low-CTE alloy, such as Invar[14]) is precisely ground to fit tightly into a slightly tapered hole in the glass (typically low-CTE fused silica). The preciseness of the fit between the pin and the hole in the glass creates a surprisingly tight seal. The joint interlock occurs even though both parts involved remain rigid.

---

[13] Recall that glasses, being amorphous, do not exhibit a melting point!
[14] Invar is an alloy of Fe—36% Ni by weight and has a CTE of 1.3 per degree C between 0 and 100°C, the lowest of all metals and alloys.

Elastic Compression Fits. Compression seals achieve both mechanical integrity and hermeticity by having a soft, ductile metal part be forced tightly against, into, or around a rigid glass part. Elastic recovery (or spring-back) in the metal part is what produces tightness in the joint. This type of glass-metal seal results from the creating of an elastic interlock, and, unlike most elastic interlocks described elsewhere (e.g., Chapter 4), compression joints rely on residual stresses to maintain joint tightness. Alternatively, the glass member can be formed down around the soft part to produce what is known as a "pinched seal."

Deformation Fits. An example of a *deformation fit* glass-metal seal is when a metal annulus of a soft metal, such as Pb, is pre-placed, and the glass is heated to collapse tightly around the metal component. If the glass is forced against the soft Pb with enough force, the Pb deforms to create a perfectly conforming seal via plastic deformation. Since the seal is created by plastic deformation after parts are assembled, the joint is fully analogous to more conventional plastic interlocks (see Chapter 5).

Liquid Metal Seals. Liquid metal seals use a liquid metal, almost always Hg, to create a hermetic seal between glass and metal parts of an assembly.

Figure 9.8 schematically illustrates a number of different designs for mechanical glass-metal seals gathered from various sources, while Figure 9.9 illustrates some additional glass-metal sealing methods.

In order for mechanical glass-metal seals to work, the materials involved, not just the design of components and the method of actually producing the joint, must be correct. A metal that is to create a mechanical seal with a glass should have the following general attributes: (1) its melting point must be higher than the working temperature of the glass[15]; (2) it should be commercially available in the quantities needed, and with a high metallurgical quality (i.e., freedom from non-metallic inclusions); (3) it must exhibit sufficient ductility and malleability that it can be processed into wire or strip without mechanical defects (e.g., cracks); (4) its CTE as a function of temperature should, for matched seals, closely follow that of the glass with which its to be used; (5) it should not exhibit an allotropic phase transformation (with attendant discontinuous volume changes) over the planned range of temperatures to which the joint will be exposed; (6) any layer of oxide formed in making the seal should adhere firmly to both the metal and the glass; and (7) its ease of joining to other metals by welding, brazing, or soldering should generally be good.

Figure 9.10 shows examples of glass-metal seals used in vacuum tubes and incandescent light bulbs, while Figure 9.11 shows a typical "pinched" seal

---

[15] The *working temperature* of a glass is defined as the temperature at which the glass reaches a viscosity in the range of 10^3–10^5 Pa-sec (10^4–10^6 poise). At this viscosity, the glass can be easily shaped, yet will retain its newly created shape.

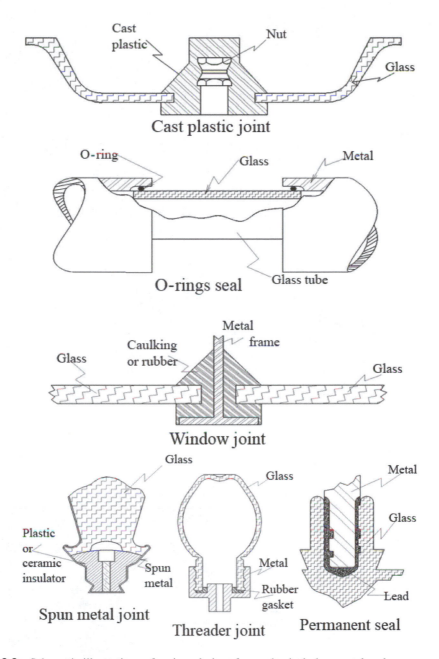

FIGURE 9.8 Schematic illustrations of various designs for mechanical glass-metal seals.

between a glass envelope and electrically-conductive metal leads. Figure 9.12 shows some glass-to-metal seals in various other electronic devices. A more modern example is shown in Figure 9.13 in which several glass-to-metal seals, as well as some ceramic-to-metal seals are found in a micro-electro-mechanical system or MEMS device.

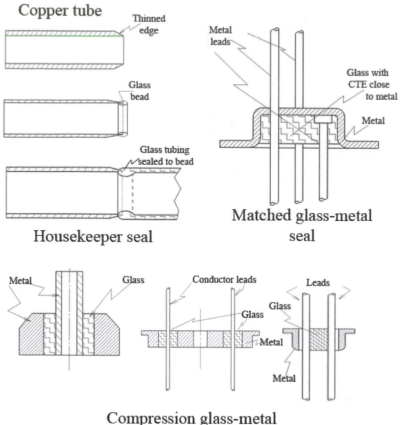

Copper tube

Thinned edge

Glass bead

Glass tubing sealed to bead

Housekeeper seal

Metal leads

Glass with CTE close to metal

Metal

Matched glass-metal seal

Metal   Glass   Conductor leads   Leads

Glass

Glass

Metal

Metal

Compression glass-metal seals

FIGURE 9.9 Schematic illustrations of additional types of glass-metal seals.

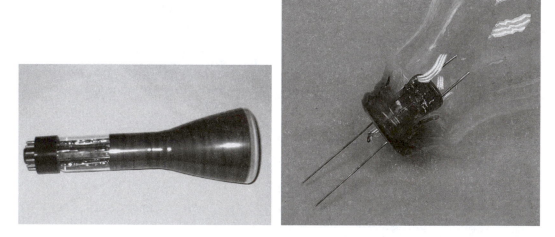

FIGURE 9.10 A photograph showing a cathode ray tube (left) and a typical incandescent light bulb (right) in which the glass envelopes for containing the vacuum or inert gas atmospheres for the electrons are joined to produce a hermetic seal with the metal tube bases. (Photographs courtesy of Corning, Inc., Corning, NY, and GE Consumer Products/GE Lighting, Cleveland, OH, respectively; used with permission.)

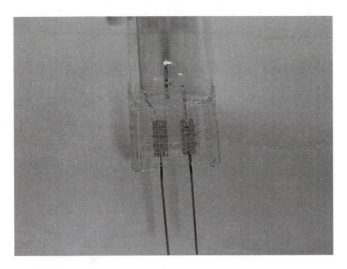

FIGURE 9.11 A photograph showing a "pinched seal" between glass and the electrical leads running into the inert or vacuum atmosphere of a glass bulb. (Photographs courtesy of Corning, Inc., Corning, NY, and GE Consumer Products/GE Lighting, Cleveland, OH, respectively; used with permission.)

Table 9.2 lists suitable metals and alloys for use with low-CTE ("hard") and high-CTE ("soft") glasses.

Glasses to be used to make glass-metal seals must be able to wet and fuse to metals, if the seals are to be created by chemical bonding, and/or must have CTEs as a function of temperature that closely follow the CTE-as-a-function-of-temperature of the mating metal.

Table 9.3 lists some of the important physical properties of glasses and metals used successfully in seals.

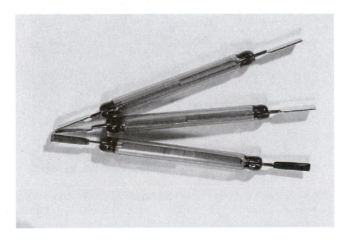

FIGURE 9.12 Photographs showing various glass-metal seals used to fabricate various electronic devices. (Photographs courtesy of Schott Technologies, Schott North America, Elmsford, NY; used with permission.)

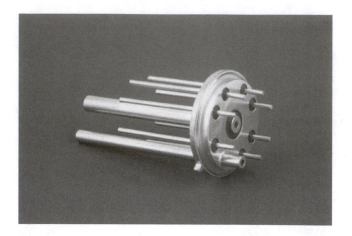

FIGURE 9.12—Cont'd

FIGURE 9.13 A photograph showing glass-metal and ceramic-metal seals in a MEMS device, here, an accelerometer. Silver-filled glass was used to bond the die to the ceramic package base seen through the cutout. Ultrasonic aluminum wire bonds were made between aluminum bond pads on the die and an Alloy 42 lead frame, also seen through the cutout. A glass frit was used to seal the package lid to the package base. (Photograph courtesy of Analog Devices, Inc., Cambridge, MA; used with permission.)

TABLE 9.2 Suitable Metals and Alloys for Use with Low-CTE ("Hard") and High-CTE ("Soft") Glasses

Combination	Metal	$a \times 10^6$ (Metal)*	Glass	$a \times 10^6$ (Glass)*	Annealing Range (Glass)** (°C)	Color of Seal	State of Strain (annealed seals viewed at right angles to longitudinal axis)	Diameter of Wire (2a) (mm)	Diameter of Sheetbed Single Wire Seal (2b) (mm)	Ratio b/a	Maximum Tensile Stress (after normal annealing) kg/cm²
1 a	Tungsten	4.4	Corning 720MX	3.3	553° 510°	Straw to light brown	Severe compression	2.5	7	2.8	480 rl†
1b	"	"	Pyrex	3.2	– –	" "		1.0	4.1	4.1	520rl†
2a	Molybdenum	5.5	Corning 705 AJ	4.6	496 461	Light brown	Severe Compression	2.5	7	2.8	215 rl†
2b	"	"	Corning G71	5.0	513 479	" "	Slight tension	2.5	7	2.8	02 cl†
4a	Platinum	9.4	G.E.O. X4	9.6	520 450	Bright metallic	Severe tension	0.8	4.1	4.1	000 cl
5a	26% Cr-Fe	10.2	Corning G5	8.9	429 404	Greenish grey	Compression	2.5	7	2.8	128 rl†
6a	Fernico I	–	Corning 705AJ	4.6	496 461	Grey	Compression	2.5	7	2.8	118 rl†
6b	(64% Fe, 28% Ni, 18% Co)		Corning 705AO	5.0	495 463	"	Slight tension	2.5	7	2.8	59 cl†
7a	Fernico II (54% Fe, 31% Ni, 15% Co)	–	Corning 705 AO	5.0	495 463	Grey	Strain free	2.5	7.5	3.0	About 10 ‡
8	British Kovar type alloys, e.g. "Nicosel," and "Darwin's F" alloys	4.5	B.T.H. C40	4.8	497 –	"	Slight compression	1.0	3.0	3.0	0–100 rl (according to metal) 12 rl
			G.E.C. FON	5.1	500 440	"	Strain free	2.5	7	2.8	
9a	Fernichrome (37% Fe, 30% Ni, 25% Co, 8% Cr)	9.95	Corning G5	8.9	429 404		Very slight tension	2.5	7	2.8	14 cl†

(Continued)

TABLE 9.2 (*Continued*)

Combination	Metal	a × 10⁶ (Metal)*	Glass	a × 10⁶ (Glass)*	Annealing Range (Glass)** (°C)		Color of Seal	State of Strain (annealed seals viewed at right angles to longitudinal axis)	Diameter of Wire (2a) (mm)	Diameter of Sheetbed Single Wire Seal (2b) (mm)	Ratio b/a	Maximum Tensile Stress (after normal annealing) kg/cm²
9b	–	–	Corning G8	9.2	510	475	"	Compression	–	–	2.8	84 rl†
10	50/50 Ni-Fe Alloy	9.5	G.E.C. L1¶	9.1	410	350	"	Slight tension	1.04	3.75	3.6	34 cl
15	Copper	17.8	Many glasses if suitably shaped	3.5–10.2	–	–	Red to gold	Strain free a fraction of a millimeter from joint	–	–	–	–

rl. Signifies in radial direction.

cl. signifies in circumferential direction.

* a = mean coefficient of linear thermal expansion between 20° and 350°, except for Corning glasses, where range is 0–310°.

† Values taken from Hull and Burger's paper.

‡ Values taken from Hull, Burger, and Navias' paper.

Data taken from T. Takamori, "Solder Glasses," *Treatise on Materials Science and Technology*, Vol. 17, Glass II, M. Tomozawa and R.H. Doremus (Eds.), Academic Press, New York, 1979, page 187.

Reproduced by kind permission of Society of Glass Technology.

Reprinted from *Joining of Advanced Materials*, by Robery W. Messler, Jr., Butterworth-Heinemann, 1993, Table 15.5, page 523, with permission of Elseiver Science, Burlington, MA.

TABLE 9.3 Some Important Physical Properties of Glasses and Metals Used Successfully in Seals

Metal	Melting Point (°C)	Maximum Operating Temp. (°C)		$a \times 10^4$ (20° C– 350°)[a]	Ultimate Strength (tons/in²)	Yield Stress (tons/sq. in)	Elongation (% on 100 mm)	Specific Electric Resistance (ohm/cm)	Thermal Conductivity (cals/aq cm/ cm /° C/Sec)
		In Vacuo	In Air						
Tungsten	3350	3000	300	4.4	99	85	4	5.6	0.38
Molybdenum	2450	2000	200	5.5	47	41	15–20	4.8	0.35
50% W, 50% Mo alloy	ca. 2800	2000	200	5.0	80	74	25–30	8.6	–
84% W, 12% Ni, 4% Co	–	–	–	6.8	–	–	–	–	–
Fernico I (54% Fe, 28% Ni, 18% Co)[b]	ca. 1450	ca. 1000	ca. 600	4.5	40	28–30	24	46	–
Kovar (54% Fe, 29% Ni, 17% Co)	ca. 1450	ca. 1000	ca. 600	4.7 (39)[c]	38–40 (27)[c]	25 (26c)[d]	32	44	0.04
Tantalum	2800	2500	–	6.5	–	–	–	15.5	0.13

[a] The usual symbol a is used throughout in referring to the linear coefficient of thermal expansion.

[b] J. App. Physics, 1941, 12, 698.

[c] The figures in brackets were determined on specimens of iron–nickel–cobalt alloys made by pressing and sintering pure metal powders and subsequently fabricating into wire.

All temperature, unless otherwise stated, are expressed in degrees Celsius.

Data taken from T. Takamori, "Solder Glasses" *Treatise on Materials Science and Technology*, Vol. 17, Glass II, M. Tomozawa and R. H. Doremus (Eds.), Academic Press, New York, 1979, page 186.

Reproduced by kind permission of the Society of Glass Technology.

Reprinted from *Joining of Advanced Materials*, by Robert W. Messler, Jr., Butterworth-Heinemann, 1993, Table 15.3, page 521, with permission of the Elsevier Science, Burlington, MA.

TABLE 9.3B

Metal	Melting Point (°C)	Maximum Operating Temp. (°C) In Vacuo	In Air	$a \times 10^4$ (20° C–350°)[a]	Ultimate Strength (tons/in²)	Yield Stress (tons/sq. in)	Elongation (% on 100 mm)	Specific Electric Resistance (ohm/cm)	Thermal Conductivity (cals/aq cm/ cm /° C/Sec)
Platinum	1750	1600	1400	9.25	8–9	ca 2.	30–40	10.6	0.166
Copper	1083	400	150	17.8	16–17	9–10 (50–60)	30	1.75	0.920
Nickel	1452	900	400	14.5	34	16 (35)	25	7.5–10.0	0.14
Iron	1530	500	200	13.2	15–17	8 (40–50)	30	9.6	0.17
50% Ni, 50% Fe alloy	–	1000	–	9.5	35–36	22–25	25–28	49	0.025
26% Cr–Fe alloy	–	1000	1000	10.2	39–41	28–30 (35)	18–20	68	0.03

[a]The elongation figures in brackets were determined on specimens of larger diameter over a gauge length = 4 area.

Data taken from T. Takamori, "Solder Glasses," *Treatise on Materials Science and Technology*, vol. 17, Glass II, M. Tomozawa and R.H. Doremus (Eds.), Academic Press, New York, 1979, p. 187. Reproduced by kind permission of the Society of Glass Technology.

Reprinted from *Joining of Advanced Materials*, by Robert W. Messler, Jr., Butterworth-Heinemann. 1993, Table 15.4, page 522, with permission of Elsevier Science, Burlington, MA.

Figure 9.14 schematically illustrates some methods by which seals can be made in glass ceramics.

9.10 SUMMARY

Ceramics and glasses are radically different materials than metals but are close cousins to each other. Both typically exhibit high strength, high hardness, high elastic modulus, unusually high chemical inertness, and are electrical and thermal insulators. Ceramics are crystalline, while glasses are amorphous. Hence, glasses progressively soften upon heating and never melt, as such. Ceramics almost always exhibit high melting temperatures and/or thermal stability.

Both ceramics and glasses are difficult to join to themselves, to each other, and to other materials (e.g., metals) because of their relative intolerance of point loading and/or stress concentration, their chemical inertness and inability to wet or be wet by other materials, and because of the sensitivity to thermal shock. A particular challenge is to ensure that any difference in CTE between a ceramic or glass and another material never exceeds about 15–25%, or the more inherently brittle material in the joint will fracture.

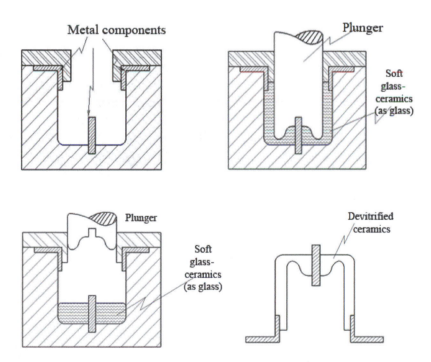

FIGURE 9.14 Schematic illustrations of various ways in which seals can be made in glass ceramics.

TABLE 9.4 List of Various Methods for Joining Ceramics (C), Glasses (G), and Glass-Ceramics (G-C), with Emphasis on Mechanical Methods (in general) and Integral Mechanical Attachment (in particular)

Chemical Methods

- Adhesive bonding (C, G, G-C)

Thermal Methods

- Diffusion bonding/Sinter bonding (C, G-C)
- Flame fusion (G)
- Fusion welding by laser-beam, electron-beam or plasma arc (C)
- Friction welding (C, G-C)
- Combustion synthesis joining (C)
- Fused frits (C, G-C)

Mechanical Methods

- Cast- or molded-in integral attachment features (C, G, G-C)
- Embedded or inserted attachments (C, G, G-C)
- Glass-metal seals (G, G-C)
  - pinch seals/deformation fits
  - precision fit/ground joints
  - elastic compression joints
  - liquid-metal joints

Adhesive joining, chemical bonding, and welding (including brazing and soldering) are often viable options for joining ceramics and glasses to themselves, to each other, and even to other materials. Mechanical joining is also possible, but great care must be taken to avoid stress concentrations at the points of fastening or attachment. Integral mechanical attachments are used for both ceramics and glasses; with rigid types and plastic types predominating.

Glass-metal seals, including ones produced using only mechanical forces, are technologically important, and four general techniques are (1) precision fits or ground joints, (2) elastic compression fits, (3) deformation joints, and (4) liquid metal seals. To be successful, both the metal and the glass to be used to create glass-metal seals must have not only compatible but often complimentary physical and mechanical properties.

Table 9.4 lists the various methods that can be used to join ceramics, glasses, and glass ceramics, with emphasis on mechanical methods, in general, and integral mechanical attachment methods, in particular.

REFERENCES

Engineering Materials Handbook, ASM International, Metals Park, OH, 1991.
German, R.M., *Sintering Theory and Practice,* Wiley-Interscience, New York, NY, 1996.
Messler, R.W., Jr., *Joining of Advanced Materials,* Butterworth-Heinemann, Stoneham, MA, 1993.

Messler, R.W., Jr., *Joining of Materials and Structures,* Butterworth-Heinemann/Elsevier, 2004.

Nicholas, M.G, Editor, *Joining of Ceramics,* Chapman and Hall, London, United Kingdom, 1990.

Rahaman, M.N., *Ceramic Processing and Sintering,* Marcel Dekker, New York, NY, 1995.

Schott & Gen., Inc., "Glass-to-Metal Seals" product information, No. 4830e, Schott, Jena Glasswerk, Schott & Gen., Inc.

Schwartz, M.M., *Ceramic Joining,* ASM International, Metals Park, OH, 1990.

Zurbuchen, M.A., Messler, R.W., Jr., and Orling, T.T., "Welding with self-propagation high-temperature synthesis"; *Welding Journal,* 74(10), 1995, 37–42.

10 CONCRETE AND MASONRY-UNIT ATTACHMENT SCHEMES AND ATTACHMENTS

10.1 INTRODUCTION TO CEMENT, CONCRETE, AND MASONRY UNITS

Materials have evolved tremendously over the millennia, and the rate of evolution seems to be ever accelerating. And, yet, as exciting—and life-improving—as the developments of steel, plastics, synthetic reinforced composites, and nanomaterials are, the most widely used engineering material in the world is concrete. In fact, no other material comes close by any measure: volume or value. It is used everywhere for constructing roads, dams, bridges, buildings, and other civil structures, in huge quantities. And, while not officially recognized as part of ceramics by the American Ceramic Society, it is a ceramic by its nature and a masterpiece of materials engineering by its structure and properties. Together with cement and so-called masonry units, concrete is so important that it not only deserves to be included in a book like this, it merits its own chapter.

As used in this treatment, and generally elsewhere, "cement" refers to a mixture of lime, CaO (obtained by roasting or "calcining" limestone, $CaCO_3$), and various other naturally-occurring oxides including alumina (Al_2O_3), silica (SiO_2), iron oxide (Fe_2O_3), and magnesia (MgO).[1] Known as "hydraulic cement," construction cements or masonry cements achieve their strength in cast and cured solid entities from the formation of hydrogen bonds by water

---

[1] So-called "high alumina" cements used for higher-temperature service contain much more alumina (Al_2O_3) than lime cements but still contain CaO and SiO_2.

molecules between fine particles of these oxides or more complex silicates and aluminates with calcium. "Masonry units" refer to any ceramic or glass-based building block or shape intended to be bedded in mortar or cement, including stone, clay bricks, fired clay and ceramic tiles, pre-cast cement and concrete units, and even some glass blocks. "Stone" refers to naturally-occurring rocks of sedimentary, igneous, or metamorphic types. The most common rocks or stone used in construction or masonry are granite, marble, slate, or, to a far lesser degree today, because of its lack of durability, limestone. "Concrete" refers to a mixture of cement, sand, and aggregate, which is usually sized stones or crushed and sized rock.

10.2 PROPERTIES OF CONCRETE AND MASONRY UNITS THAT FACILITATE MECHANICAL ATTACHMENT

Concrete is actually a composite that consists of a continuous matrix composed of sand and cement and a reinforcing phase, known as an "aggregate," usually consisting of whole stones or crushed stones or rock screened to obtain a mean size. One important reason for the widespread and extensive use of concrete is that the sand and the aggregate needed to create it are provided by Nature—rather inexpensively. The cement, too, while the technological marvel in the system, comes from widely available, abundant, and inexpensive ingredients found in Nature, albeit processed by humans. The cement literally bonds (as opposed to simply surrounds and locks together) each grain of sand and each fragment of stone in a strong and durable matrix, and it does so with water!

Ancient civilizations, including the Greeks and the Egyptians, discovered mortars thousands of years ago.[2] The first mortars were made by heating lime-stone or chalk ($CaCO_3$) to convert it to anhydrous CaO, which could then be rehydrated by adding water to harden into a strong solid material. The CaO (lime) reacts with water to form CaOH, which, in turn, reacts with SiO_2 (silica) particles from sand and particles of other naturally-occurring ceramic minerals, and hardens within itself as well as bonds to stones, bricks, and already hardened or set or cured cement. The reason cement bonds to these other materials is also due to the formation of strong hydrogen bonds from water molecules with similar-composition compounds in these masonry units.

While impressive at the time of its development, lime masonry cement is not very strong. A great advancement occurred in the early 19th century when Portland cement was invented. Portland cement allowed concrete to be made and made the inclusion of building stones or bricks in structures unnecessary,

---

[2] Before discovering mortars, but surely a driving force for seeking such a material, ancient people used mud that they found or made from clay and water either to join naturally-shaped or cut sticks or stones to create strong structures. Later, as an advancement, ancient people even discovered the benefits of adding straw to reinforce the mud against fragmenting when it dried; as in adobe (mud).

as very strong, large structures could be made from just the concrete mixture, which was poured into forms. As in all ceramic-based construction cements, the bonding that occurs in Portland cement involves a chemical reaction that is triggered by the addition of water. While this reaction, which occurs in what are known as "hydraulic cements," begins within minutes, it continues for more than 100 days, with near-final strength being achieved in about 28–30 days. Throughout the period of time when curing or setting is occurring, the concrete must be kept wet to obtain the highest strength. The reaction by which all hydraulic cements are formed involves the solution, recrystallization, and precipitation of layered sheet structures consisting of tehrahedral silicate units with calcium and oxygen atoms in the interstices and polar water molecules holding the sheets together.

Portland cement is, in actuality, a complex compound of CaO, SiO_2, Al_2O_3, Fe_2O_3, MgO, and small quantities of other ceramic compounds. It is made by heating (roasting) chalk ($CaCO_3$) and a clay (which, by definition, always contain silicates bonded together in layers by water molecules) together at high temperatures. This heating drives off all of the water always found in clays to produce a dehydrated mixture of minerals that essentially re-form as "rock" when water is added. Thus, concrete is sand and stones embedded in an artificial rock.[3]

While there are five grades of Portland cement used in the production of concrete, they all contain 19–25% SiO_2, 5–9% Al_2O_3, 60–64% CaO, and 2–4% FeO or Fe_2O_3. As can be seen from the ternary phase diagram for the CaO-Al_2O_3-SiO_2 system shown schematically in Figure 10.1, Portland cements all contain a mixture of several minerals that result from the roasting process, including: tricalcium silicate, C_3S ($3CaO.SiO_2$); dicalcium silicate, C_2S ($2Cao.SiO_2$); and tricalcium aluminate, C_3A ($3CaO.Al_2O_3$); as well as some tetracalcium aluminate ferrite ($4CaO.Al_2O_3.FeO$). The reaction can be accelerated or retarded or caused to generate less exothermic heat of reaction during the adsorption of water, and/or the final properties of the cement can be varied by changing the proportions of these four minerals, the proportions of sand and water, and the addition of certain additives (e.g., gypsum, $CaSO_4$). Table 10.1 lists the five grades of Portland cement given by the ASTM.

The term "mortar," used earlier, properly refers to a mixture of materials to enable the "bedding" and bonding of masonry units. In its most common form in modern usage, it is referred to as "Portland cement-lime mortar," and it contains equal parts of Portland cement and lime (CaO), with sand volume six times the Portland cement volume (or three times the volume of total cementitious materials) and enough water to produce the desired consistency. The hardening process actually forms an amorphous solid network of atoms with

---

[3] Much rock is made in Nature by a similar process of grinding (under geological pressures), heating, and re-hydration.

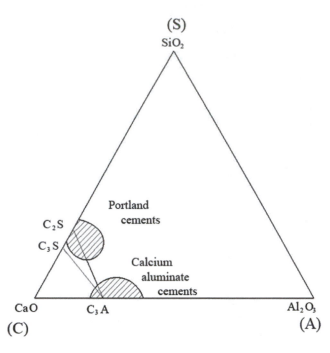

FIGURE 10.1   A schematic ternary phase diagram for the CaO-Al$_2$O$_3$-SiO$_2$ system, showing the location of Portland versus high-alumina cements.

TABLE 10.1   Typical Compositions and Compressive Strengths of the ASTM Five Grades of Portland Cement (Adapted from data of Portland Cement Association, Skokie, IL)

ASTM		Typical Percentage Composition			
Grade	Use	C$_3$S	C$_2$S	C$_3$A	C$_4$AF
I	General purpose or normal	50	24	11	8
II	Some sulfate protection; less heat generation than Grade I	42	33	5	13
III	High early strength	60	13	9	8
IV	Low heat generation	26	50	5	12
V	High sulfate-soil resisting	40	40	4	9

Fraction of Compressive Strength Developed at Indicated Time, Based on Unity Value for Type I				
	1 day	7 days	28 days	3 months
I	1.0	1.0	1.0	1.0
II	0.75	0.85	0.90	1.0
III	1.90	1.20	1.10	1.0
IV	0.55	0.55	0.75	1.0
V	0.65	0.75	0.85	1.0

trapped water (known as a "gel") and crystals of the various ceramic ingredients.

While concretes can have innumerable compositions (depending on the relative amounts, sizes, and specific types of reinforcing aggregate, as well as the specific grade and mix of Portland cement), typical properties given by ASTM Standard C 150 are:

Density	2,453 kg/m^3 (147–150 lb/ft^3)
28-day Compressive Strength	20–35 MPa (3,000–5,000 psi)
Elastic Modulus	30–150 GPa (4–22 × 10^6 psi)
Approx. Tensile Strength	4 MPa (580 psi)
Approx. Fracture Toughness	0.5 MPa m$^{1/2}$ (0.4 ksi in$^{1/2}$)

It is noteworthy that, like all ceramic materials, concrete is considerably stronger in compression than in tension, the reason being that ceramics (and glasses) are inherently prone to manufacturing-induced flaws, often microcracks, that open under tensile loading but do not open under compressive loading.

The combination of good compressive strength, high hardness and modulus of elasticity, but poor fracture toughness and non-existent plasticity in concrete correctly indicates that this material would only permit the creation of rigid integral mechanical interlocks.

Masonry units include all naturally-occurring, cut-to-shape, and processed materials, as well as product forms used to erect structures using mortar or cement. Naturally-occurring and cut-to-shape rock or stone, including granite, marble, limestone, slate, and other types, are used. Bricks and tiles made from fired clays are two processed materials that come in a wide variety of product forms. The difference between bricks and tiles is that the latter, once fired, tend to contain a considerable proportion of glassy phase surrounding crystalline ceramic phases. Finally, pre-cast cement and concrete blocks, as well as various more complex and larger shapes, are widely used in masonry. The latter materials, of course, are also poured as slurries directly into forms to create more monolithic structures.

In summary, cement, concrete, and masonry units are extremely important materials in engineering and, therefore, should never be excluded from any serious treatment of joining. For cement and concrete, which can be cast or pre-cast into desired shapes, as well as for clay units, such as bricks and tiles, which are molded and fired, geometric features capable of allowing rigid mechanical interlocking can be produced (see Chapter 12, Section 12.2). For poured concrete construction, as well as for some construction requiring cementing or mortaring together masonry units, metal devices can be embedded in the cement or concrete during its pouring or application or can be installed later as inserts (see Chapter 12, Section 12.3). For stone masonry units, natural shapes can be caused to interlock, or shapes can be altered by cutting to allow interlocking. Sometimes minor alteration of the shape of a cast or molded masonry unit (e.g., cement or concrete block, or brick or tile) is done to allow interlocking. Various approaches are presented in Section 10.4.

10.3 PRE-CAST INTEGRAL ATTACHMENT FEATURES

As stated earlier, what makes concrete such a marvelous material is its versatility combined with its low cost from the ready availability of its raw ingredients.[4] It can be cast into useful shapes to produce masonry units such as concrete blocks as well as larger and generally more complex-shaped concrete construction units. Well-known examples of such larger pre-cast construction units are the concrete inverted-Y–shaped barrier units used to divide opposing lanes of vehicular traffic on highways, various square or rectangular box or cylindrical shapes for constructing sewers, and cylindrical pipes. Figures 10.2 and 10.3 show two examples of pre-cast concrete products.

For very large structures, such as sections of elevated roadway, bridge supports, piers, coffer dam sections, and so on, much larger custom-designed pre-cast modules are made, as shown in Figure 10.4.

Pre-cast masonry units are available in a wide variety of shapes, whether for use in pre-cast concrete or molded and fired clay brick. Figure 10.5 schematically illustrates a variety of masonry unit shapes. Some of these, and some others, come with cast-in features to allow one unit to be rigidly interlocked with another, either top-to-bottom face or end-to-end, or both, as shown schemati-

FIGURE 10.2 Pre-cast concrete product forms are used for rapid on-site construction. Here, pre-cast, interlocking panels are used to construct the floor/ceiling of a building; left and right. (Courtesy of the Portland Cement Association, Skokie, IL; used with permission.)

---

[4] Even a homeowner can buy small quantities (e.g., 40- and 80-lb/20- or 40-kg bags) of fully pre-mixed (e.g., "ready-mix") cement with sand and/or aggregate for less than 5 cents (U.S.) per pound. In larger quantities, the cost per pound (or kilo) drops dramatically to just a few cents!

FIGURE 10.3 Pre-cast concrete pipes being assembled into an aqueduct. Note the use of cast-in rigid mating interlocks at the opposite ends of each pipe section. (Courtesy of the Portland Cement Association, Skokie, IL; used with permission.)

cally in Figure 10.6. Almost always, masonry units are bedded in mortar to provide additional strength, especially out of plane. But, even when cemented into a wall, for example, integral rigid interlocking protrusions and grooves enhance strength. In addition, such interlocks facilitate assembly (by providing self-keying joints), help establish needed spacing for proper mortar joints, and force any cracks that form in the mortar to move out of a single plane, thereby providing an added degree of structural damage tolerance.

In larger pre-cast concrete units, such as pipes (see Figure 10.3), large retaining-wall blocks (see Figure 2.6), sewer sections, highway-dividing barriers, and so on, similar protruding features and mating receiving grooves are often

FIGURE 10.4 Photograph showing very large pre-cast modules being used to construct an elevated roadway. Such modules speed on-site construction by reducing it to assembly. (Photograph courtesy of the Portland Cement Association, Skokie, IL; used with permission.)

included. So effective are some of these interlocks, because of a taper fit and resulting friction, that mortar may not even be needed.

In poured concrete construction, rigid interlocking joints are also used, as shown in Figure 10.7. In all cases (except Figure 10.7a), protruding and recessed features are designed to mate and provide interlocking against movement in certain directions (albeit never preventing separation when mating blocks are subjected to a tensile force). In Figure 10.7b, a pre-cast block is used to key the two notched construction units, with the block functioning as a fastener.

A special type of pre-cast concrete block is known as a "pilaster." Pilaster blocks are used to structurally reinforce a load-bearing concrete block wall. They do so through the obvious use of cast-in rigid interlocking features, as shown in Figure 10.8.

While not intended to perform as rigid interlocks, some geometric features are cast-in or added to a cast-in feature to provide some special function. Two examples are contraction joints and waterstop joints. *Contraction joints* are placed in or between large pre-cast or poured concrete construction units to allow the concrete to expand during rises in temperature or contract during temperature drops and/or to allow shrinkage during curing without producing damage from random cracking. There are specific guidelines on the placement of

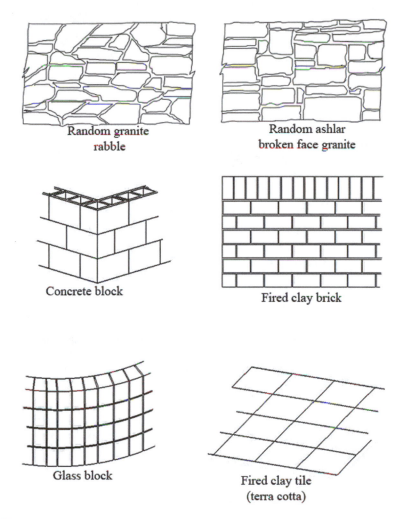

Random granite rabble

Random ashlar broken face granite

Concrete block

Fired clay brick

Glass block

Fired clay tile (terra cotta)

FIGURE 10.5 Schematic illustrations of typical masonry unit shapes.

and spacing between contraction joints, also known as "control joints" in the concrete construction industry. Two types of contraction joints, one employing a cast-in or saw-cut groove, the other employing a similar groove with a pre-placed or inserted strain-compliant material such as rubber, compacted fiber, mastic, or caulking, are shown in Figure 10.9.

Waterstop joints are used at concrete joints to prevent unwanted liquid transfer, whether above or below grade. Waterstops are usually made from polyvinylchloride (PVC) polymer, as this material is resistant to acids, alkalis, and water-borne chemicals. Shapes vary and include serrated, splits, dumbbells, center bulb, end bulb, tear web, and base seals. Figure 10.10 shows a typical example.

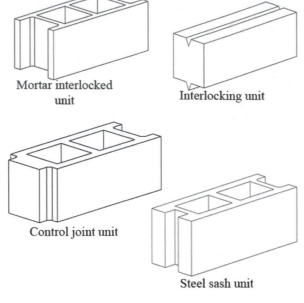

Mortar interlocked unit

Interlocking unit

Control joint unit

Steel sash unit

FIGURE 10.6 A schematic illustration showing some interlocking features used on pre-cast masonry units.

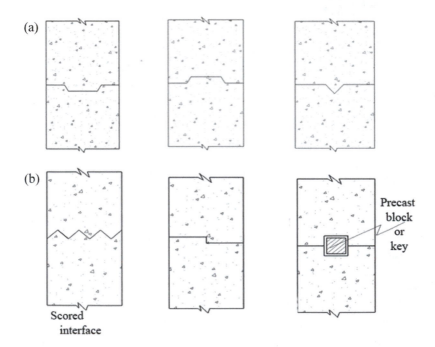

(a)

(b)

Scored interface

Precast block or key

FIGURE 10.7 Schematic illustrations of rigid interlocking features found in poured concrete construction.

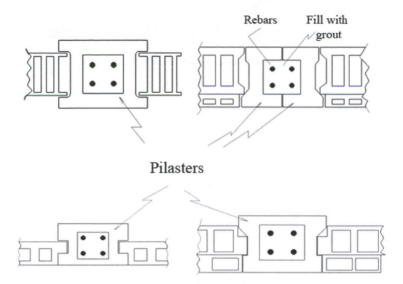

FIGURE 10.8 A schematic illustration of pilaster blocks, showing their obvious use of cast-in rigid inter-locking features to provide structural reinforcement.

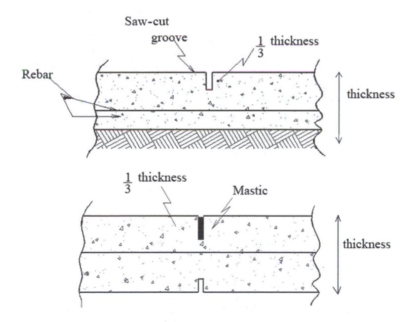

FIGURE 10.9 Schematic illustrations of two types of contraction joints used in concrete.

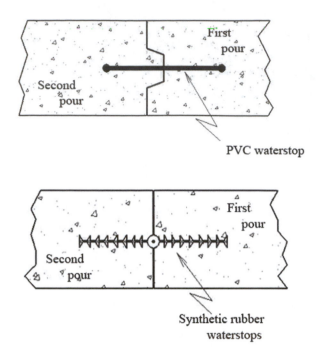

PVC waterstop

Synthetic rubber
waterstops

FIGURE 10.10 A schematic illustration of a typical waterstop joint.

10.4 EMBEDDED AND INSERTED FASTENING ATTACHMENTS

Besides the enhancement of the strength of Portland cement by the aggregate in concrete, large pre-cast concrete units such as those shown in Figures 10.2, 10.3, and 10.4 are often further strengthened by the incorporation of steel reinforcing bars (known as "rebars") or heavy-wire mesh. Similar steel reinforcement is often used in poured concrete as well, including poured roadways, bridge supports, dams, and so on. Such reinforcements give the concrete greater strength in the direction (for bar) or directions (for mesh) of reinforcement. In very large civil structures that have to be erected section by section, such as skyscrapers, bridge supports and towers, elevated roadways, and dams, the steel reinforcing bars extend out of the concrete so that sections can be joined to one another more soundly by first joining cast-in-place reinforcing bars and then pouring additional concrete in the joint to completely envelop the rebars.

Reinforcing bars operate like integral mechanical attachment features by being part of the structural units to be joined, and not added during actual joining, as fasteners are. Admittedly, when reinforcing bars are to be joined end-to-end, to make multiple pre-cast modules act more like an integral structural entity, some form of fastening or welding needs to be done. A variety of devices for connecting rebars end-to-end (often with some overlap) are commercially available. Figure 10.11 schematically illustrates so-called "ladder-type" and

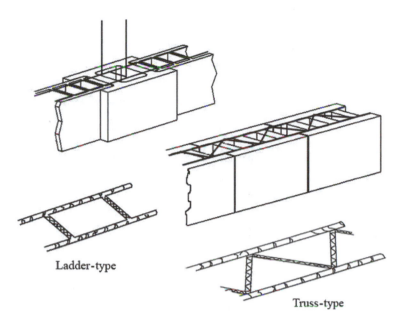

FIGURE 10.11 Schematic illustrations of ladder-type and truss-type joint reinforcement in concrete walls.

Ladder-type

Truss-type

"truss-type" joint reinforcing in concrete block walls. Figure 10.12 schematically illustrates how reinforcing bars are used to structurally reinforce corner and tee intersections of concrete block walls. Figures 10.13 and 10.14 show some actual examples.

Another breakthrough in concrete engineering occurred in the early 1950's with the introduction of pre-stressed concrete. *Pre-stressed concrete* pre-cast modules (e.g., cantilevered beams) are produced by applying a tensile load to the pre-placed steel reinforcing bars, stretching them slightly within their elastic range, pouring the concrete into the form to completely envelop the bars, and allowing the concrete to cure, if not fully (after about 30 days), at least until it sets (in about 72 hours). Once the concrete has set sufficiently, the tension on the rebars is released and the bars try to elastically recover. However, being unable to contract because they are interlocked to the concrete with raised ribbing on their surfaces, they introduce a compressive residual stress into the concrete. In service, this compressive residual stress must be overcome by applied tensile loads before any net tensile stress develops in the concrete. In this way, the inherent relative weakness of concrete (like all ceramics and glasses) in tension compared to compression is offset.

To take advantage of the structural enhancement afforded by pre-stressing, pre-cast pre-stressed concrete modules need to be attached to one another by joints that also offer high integrity. Two examples are shown in Figures 10.15 and 10.16. In Figure 10.15, a single-tee pre-cast pre-stressed roof member

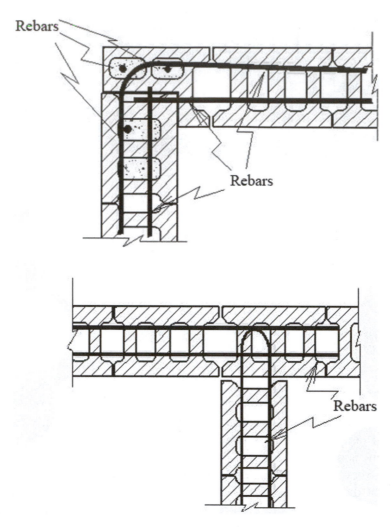

FIGURE 10.12 Schematic illustrations showing how steel reinforcing bars can be cast into concrete walls to produce structurally more sound corner and tee joints.

frames into a pre-cast slot or opening in poured or pre-cast wall panel. This is an excellent example of a rigid integral mechanical interlock. Additional structural integrity can be added to prevent movement of the beam along its axis by employing a steel dowel either pre-cast or inserted into a drilled hole in the bottom of the pre-cast opening at the top of the pre-cast wall panel. In Figure 10.16, the use of steel rebar connections is shown for joining two large pre-cast pre-stressed double-tee at their flanges. In this case, the rebar, being added after the two double-tee sections are positioned for joining, acts more like a fastener than an integral interlock.

Figure 10.17 shows the use of steel reinforcing bars and provisions to lock together abutting modules.

FIGURE 10.13 Example of the use of steel reinforcing steel bars in poured concrete roadways. (Photographs courtesy of the Concrete Steel Reinforcing Institute, Schaumburg, IL; used with permission.)

FIGURE 10.14 Example of the use of steel reinforcing bars in building construction using poured concrete. (Photograph courtesy of the Concrete Steel Reinforcing Institute, Schaumburg, IL; used with permission.)

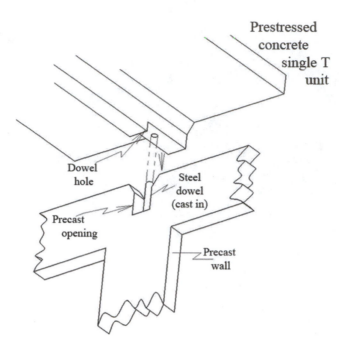

FIGURE 10.15 Schematic illustration of the use of a cast-in rigid interlock to join a single-tee roof member into a pre-cast opening in a pre-cast wall panel.

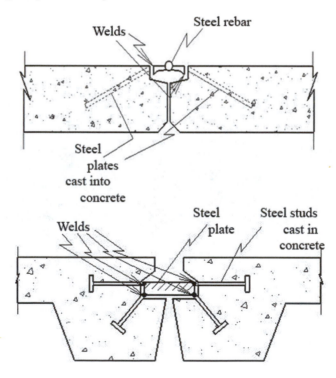

FIGURE 10.16 Schematic illustration of the use of rebar for joining two double-tee flanges on pre-cast pre-stressed concrete modules.

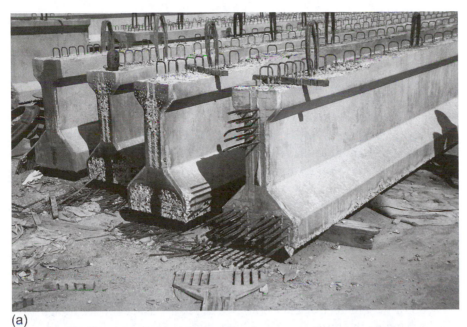

(a)

(b)

FIGURE 10.17 Example of the use of steel reinforcing bars and provisions for locking together large, precast concrete modules (a); here for constructing elevated roadways or overpasses (b). (Photograph courtesy of the Portland Cement Association, Skokie, IL; used with permission.)

From the preceding, it can be seen that steel reinforcing bars can be used in either of two modes for joining concrete: (1) they can be embedded in the pre-cast unit or poured structure, or (2) in rarer instances (such as shown in Figure 10.16), they can be inserted into cast-in or drilled holes. In either case, the rebars themselves can be joined end-to-end to provide connection between modules.

Two very important and popular types of inserts used in the mechanical joining of concrete or other masonry units are "anchor bolts" and "anchors." Metal *anchor bolts* are a basic element in most modern concrete structures. Anchor bolts are usually cast-in or embedded in concrete walls, footings, piers, or bases with length projecting from the concrete to permit the assembly of major structural components. Such components are positioned over the projecting threaded end of the anchor bolt so that the threaded portion extends through a hole in the mating component. A nut is then applied to the threaded end of the anchor bolt, locking the component to the concrete structure.

For anchor bolts subjected to tension forces, they should be designed to engage a mass of concrete that will resist uplifting to 150% of the estimated force. Resistance to pullout can be achieved by placing a nut on the embedded end of the threaded anchor bolt, by adding a washer for additional strength, or by bending the threaded anchor bolt shaft to form an ell or bend leg.

Figure 10.18 schematically illustrates the use of anchor bolts with a nut and with a bend leg.

Anchors are inserts that are installed into holes drilled into pre-cast or poured concrete. They provide an integral point for fastening another component to the concrete structure using a threaded bolt, screw, or lag bolt. Anchors are usually made from metal but can, for lighter loads, be made from plastic. There are five general categories of anchors, as follows:

- Self-setting cone anchors secured by torque
- Hammer-set anchors
- Self-drilling anchors
- Plastic anchors
- Grouted anchors (held in place with a ceramic cement known as "grout")

Figure 10.19 schematically illustrates a variety of common anchors.

10.5 NATURAL SHAPES OR CUT-IN (HEWN) SHAPES FOR INTEGRAL MECHANICAL INTERLOCKING STONE MASONRY UNITS

After describing various integral mechanical attachment schemes and attachments for metals, polymers, ceramics, glasses, and even cement and concrete, we have come full circle, back to our very earliest methods for joining (see Chapter 1, Sections 1.1 and 1.2). Rocks and stones can be assembled into quite strong and durable structures simply by stacking them together; taking advantage

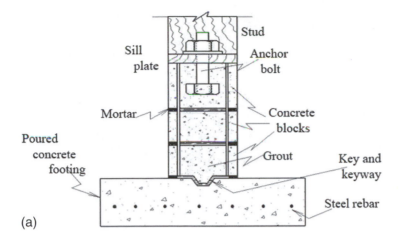

(a)

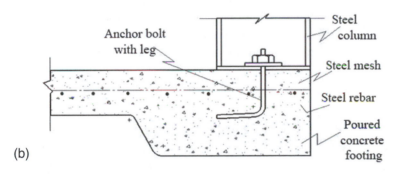

(b)

FIGURE 10.18 Schematic illustrations of (a) an anchor bolt with a nut to help anchor it into the concrete and (b) a bend leg to do so.

of their natural shapes to cause mechanical interlocking as protruding features nest into recessed features.

Some of the best examples of using the natural shapes of rocks and stones for erecting structures are the stone walls of New England made famous by the poet Robert Frost in "Mending Walls." A typical example is shown in Figure 10.20.

The impetus for building such walls throughout Maine, Vermont, New Hampshire, Massachusetts, Rhode Island, and even northern Connecticut and New York was that rocks, stones, and boulders of all sizes and shapes where left scattered over the entire landscape as the great glacier of the last Ice Age receded ten thousand years ago. For more than a hundred years after the first Europeans settled in Plymouth, Massachusetts, and spread from there, people were faced with the daunting task of clearing away the stones to permit them to till and plant their fields. The only way to move the stones was by hand or by horse, and good sense—along with a measure of frugality—suggested they

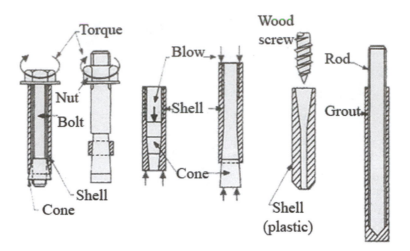

FIGURE 10.19 Schematic illustrations of a variety of common anchors designed to be installed or inserted into holes in pre-cast or poured concrete. (Adapted from and courtesy of the American Cement Institute.)

should only move them as far as necessary. So they moved them to the edges of the field they wanted to clear and stacked them into walls to keep in their grazing animals and keep out their unwanted neighbors. But, they seemed to build the walls, not just stack the stones! These walls typically showed great care in placing the largest stones at the base of the wall (which would make sense,

FIGURE 10.20 A typical stone wall like those found throughout New England; this one near Camden, ME, along U.S. Route 1. Note how flat stones were used to top the wall. (Photograph by Robert W. Messler, Jr., used with permission.)

because they'd be too heavy to pick up, as opposed to roll or drag!) and arranging smaller stones in tiers, using the natural shapes of the stones to get them to fit and nest into a stable structure. The smallest stones were often wedged into open crevices to provide tighter interlocking.

A similar technique was used to construct the foundations of their homes. Old, colonial-period homes throughout New England still rest securely on foundations whose walls are made from stacked uncut stones. The same type of construction (i.e., using naturally-shaped and/or hewn stones) was used to construct dams, spillways, breakwaters, and so on. Figures 10.21 and 10.22 show two examples.

As in our ancient ancestors, it was a natural progression to move from using naturally-shaped stones in simple stacked arrangements to using such stones as

FIGURE 10.21   Naturally-shaped and hewn stones simply interlocked, without mortar, were used in the construction of dams, spillways, seawalls, and retaining walls in New England for two hundred years. Here, in a spillway at a reservoir (left) and in a retaining wall along a stream emptying into the harbor (right) in Camden, Maine, large stones can be seen simply set in place to interlock, without the use of mortar. (Photograph by Robert W. Messler, Jr.; used with permission.)

FIGURE 10.22   A particularly impressive example of the use of naturally-shaped and hewn stones set to interlock without mortar is found in the "breakwater" that protects the harbor of Rockland, Maine. This 7/8-mile long breakwater is trapezoidal in cross-section, with a base of over 120 feet and a top of nearly 40 feet in width. A side view (left) and top view (right) are shown. (Photograph by Robert W. Messler, Jr.; used with permission.)

masonry units bedded in mortar for added strength and added protection from wind and weather. It was also a natural step to cut some stones to the shape needed or wanted, as opposed to searching for just the right-shaped stone. Hence, the use of cut-stone in construction began in the New World.

It began in New England, where there were both the people and the natural resources. Large stones and boulders were cut into smaller shaped masonry units, and quarries were harvested for endless quantities of granite and marble and limestone and slate. To this day, the New Hampshire is known as "the Granite State," and Vermont is renowned for its marble. Cut stone masonry units from both places enabled the construction of large buildings throughout America, but most apparently in America's new capitol, Washington, D.C. (see Figure 10.23).

While cut stone is almost always bedded in mortar today, there are many examples in old roads, bridges, buildings, dams, and dikes around the world where cut-in shape alone was used to achieve mechanical interlocking.

Construction with cut-to-shape stones, as well as cast or molded-and-fired masonry units, still uses stacking arrangements that either help nested units interlock and/or moves the joints in and out of an "opening mode" for cracking under an applied force, thereby increasing structural damage tolerance. Figure 10.24 schematically illustrates various examples.

FIGURE 10.23 A photograph showing buildings in Washington, D.C., made from cut-to-shape granite from New Hampshire and marble from Vermont; here, the U.S. Capitol. (Photograph obtained from a public website, for which there was no acknowledgment of the photographer.)

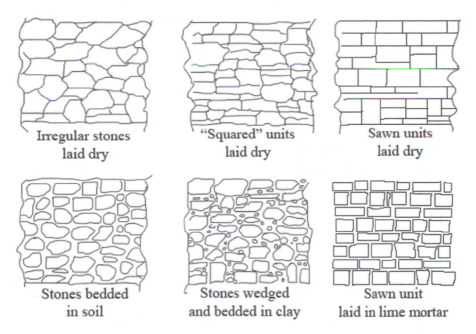

FIGURE 10.24 Schematic illustrations of various ways in which cut-to-shape or cast- or molded-to-shape masonry units are arranged in walls.

10.6 SUMMARY

Concrete is a marvel of ceramic engineering. It consists of a mixture of an hydraulic cement (e.g., today, Portland cement), sand, and aggregate (e.g., sized stone or crushed and sized rock). The addition of the aggregate to the cement eliminates the need for using stones or bricks for obtaining structural strength.

The cement from which concrete is created is itself a marvelous material. Small particles of naturally-occurring, abundant, widespread, and inexpensive minerals, including limestone ($CaCO_3$), SiO_2, Al_2O_3, Fe_2O_3, and MgO, as well as some others, are roasted together to drive off all water and begin to form some more complex minerals such as various calcium silicates and calcium-aluminate silicates. When water is added to a finely crushed mixture of these complex minerals and sand (SiO_2), in the right proportions, the complex minerals hydrogen bond together through waters of hydration to form a strong gel that surrounds even stronger crystalline particles. This is Portland cement. By adding aggregate, the mixture becomes concrete that is made even stronger and bulkier and less expensive than the cement alone, all of which help account for concrete being far and away the most extensively used material in the world.

The inherent high strength, virtually non-existent plasticity, and high hardness of concrete, as well as other cast- or molded-and-fired masonry units made

TABLE 10.2 List of the Various Methods by Which Pre-Cast Cement or Concrete and Various Masonry Units Can be Joined Using Integral Mechanical Attachment

Pre-Cast Integral Attachment Features

- Cast-in shape interlocks
- Cast-in slip/friction fits
- Contraction joints[*]
- Waterstop joints[*]

[*]Not intended to perform as rigid interlocks.

Embedded and Inserted Fastening Attachments for Cement, Concrete, or Masonry Units

- Steel reinforcing bars or mesh
- Anchor bolts
- Anchors
  - self-setting cone anchors
  - hammer-set anchors
  - plastic anchors
  - grouted anchors

Natural Shapes or Cut-In (Hewn) Shapes for Stone Masonry Units

from plain cement or clay (e.g., bricks and tiles), suggest that joining using fasteners could be difficult, but joining using integral rigid mechanical interlocks would be natural.

A variety of methods exist by which concrete and other masonry units can be joined mechanically by taking advantage of integral attachment features *or* embedded or inserted integral fasteners. Examples relying on the natural shape of stones, or the cut-in, cast-in, or molded-and-fired-in shapes of stones, pre-cast cement or concrete, or bricks or tiles exist and are used with great success. Examples wherein metal anchor bolts or metal or plastic anchors are cast into or inserted into drilled holes in pre-cast concrete, respectively, are also commonly used with great success.

Table 10.2 summarizes the various ways in which poured cement or concrete and various masonry units (including pre-cast concrete, bricks, tiles, and stone) can be joined mechanically using integral designed-and-processed-in features or designed-in and added integral fasteners.

REFERENCES

Drysdale, R.G., Hamid, A.A., and Baker, L.R., *Masonry Structures: Behavior and Design,* Prentice Hall, Englewood Cliffs, NJ, 1994.

Nawy, E.G., Editor-in-Chief, *Concrete Construction Engineering Handbook,* CRC Press, Boca Raton, FL/New York, NY, 1997.

Parmley, R.O., *Standard Handbook of Fastening & Joining,* 2nd edition, McGraw-Hill, New York, NY, 1989.

11   WOOD ATTACHMENT SCHEMES AND ATTACHMENTS

11.1   INTRODUCTION TO WOOD AS A MATERIAL

Wood is unique among all engineering materials used for structural applications in that it is the only one that is renewable. Wood, after all, comes from trees, and trees grow naturally. This is the "good news" and the "bad news," however. The good news is that most of the work of "manufacturing" wood is done for us by Nature through the processes of life and growth in the tree from which the wood comes. Hence, wood is a particularly practical and economical material. Even the lumber we extract from the wood of trees is relatively inexpensive. All that's needed is to harvest trees, cut the usable parts of the trees to size to produce the lumber desired,[1] and dry (or "season") the lumber before it is ready for use. The bad news is that because wood is produced by trees and trees are a product of Nature, we have little control over the product. Unlike for other structural materials such as metals, polymers, ceramics, glass, composites, and even concrete, we can do little to adjust or refine the wood's properties to suit our needs. At least in the past, we had to accept wood's natural strengths and limitations. But, as will be seen in Section 11.3, we are not totally at Nature's mercy any longer.

---

[1] Cutting logs from trees to size is not a matter of any pre-determined plan or design. Because each and every tree—and, thus, log—is different in size, straightness, number, size, and location of knots (which occur where branches grow from the main trunk) or other defects, what cross-sectional shapes and dimensions and lengths of specific product forms (e.g., boards, beams, such as 2×4's, and trim molding) can be extracted can only be determined once the log is stripped of its bark (i.e., "debarked") and rough-cut to be straight and "squared." Once done on-the-fly by highly experienced mill saw operators, this task is now largely left to a computer program that "reads" the end of the log and attempts to maximize the output from each log while trying to generate needed product either to inventory or order.

Wood is more than a utilitarian structural material. It is also generally the best loved of all materials that we use for building things. There are woods that are a delight to look at because of their widely varied grain patterns and wonderful natural colors. Wood seems to ask to be touched because of its varied textures, and when it is touched it has subtle but pleasing warmth. It even smells good when it is freshly cut or worked or when it is an aromatic variety, such as cedar.

More than the aesthetic properties of look and feel and smell, it is the properties of strength and stiffness combined with low density[2] that make wood especially attractive as a structural material. Those properties, the fact that wood is easy and relatively inexpensive to work and join, and, finally, that it is readily recyclable in a number of ways (e.g., re-use, chipping, burning for energy) lead to its importance and popularity across the ages.

But, all is not perfect with wood, or anything else, for that matter. The grain structure of wood, the origin of which will be described shortly, is related to the way the tree from which it comes grows and grew, and the grain structure, in turn, leads to anisotropy of properties that can be pronounced. As wood ages, it loses water and tends to shrink and/or warp across its grain structure and split and/or splinter along its grain structure. Like all living things, wood is prone to decomposition from moisture and from insects, bacteria, and fungi once it dies. Finally, it is combustible, which doesn't make it unique among structural materials, as many polymers and a few metals (e.g., Mg) are too, but it gives rise to another cause for caution, if not concern.

With all of its advantages and shortcomings, wood is a major material around the world. In the United States alone, in excess of 300,000,000 tons of wood is used annually, more than the combined tonnage of concrete and steel! More than 60 native woods are in common use, and another 30 types are imported. Besides this tremendous usage for lumber is the increasing use of wood for the production of engineered wood products (see Section 11.3).

Before beginning a description of what wood is and how and why it behaves the way it does, it needs to be said that wood is a very complex material. (In fact, only living tissue in higher animals—like humans—is more complex as a material.) It is complex because it is fundamentally a composite polymeric material, but also because its properties are difficult to describe. As a framework for discussing wood as a material, it must be recognized that, unlike most other materials, but not unlike some other composite materials (e.g., steel-reinforced concrete), wood exhibits a macrostructure and a microstructure, both of which give rise to its properties. In contrast to almost all other materials, the *macrostructure* is the more important feature of wood. The macrostructure can

---

[2] The combinations of strength and density, to give strength-to-weight, or stiffness and density, to give stiffness-to-weight, are known as "specific strength" or "specific stiffness," respectively. Both allow materials to be compared for their relative efficiency in a structure.

be seen with the unaided eye and is based entirely on the grain pattern[3] that arose from the particulars of the way in which the tree grew. Many properties of wood vary by a factor of 10–20 times, depending on the location of the wood in the log and the direction of testing. The *microstructure* in wood, as in other materials, is only observable with the aid of magnification, as the sizes of distinguishing features are too small to resolve otherwise. The microstructure of wood also leads to the anisotropy of properties, as it is the microstructure that gives rise to the macrostructure.

11.2 PROPERTIES OF WOOD THAT FACILITATE INTEGRAL MECHANICAL ATTACHMENT

As already stated, what makes wood different than all other engineering materials described previously in this book (i.e., metals, polymers, ceramics, and concrete) is that it is the product of growth in a living thing. Accordingly, the structure of wood and the properties that structure gives rise to are a response to the forces imposed on the tree as it grows, and not to the processing. By having to grow under the ever-present force of gravity, a tree needs to develop strength in the principal direction of growth that is able to resist that force. It must, after all, support its own weight against both gravity and wind. Not surprisingly, the structure in the direction of growth is quite different than in transverse directions. As a result, the properties in the growth direction are quite different than in transverse directions

The cross-section of a tree, like that shown in Figure 11.1a, reveals a key feature in the macrostructure at the interface between the wood within the tree and the protective bark on the tree's outer surface. This feature is the "cambium layer." Remarkably, the *cambium layer* is the source of the wood cells and the bark cells. As the tree grows, year by year, the cambium layer produces new wood cells just inside it, as a new layer, and new bark cells just outside it, as a new layer. The new wood cells immediately begin to grow in both length and diameter, and each cell—and the new "sapwood" these cells create—is very pliable. Growth occurs in bands or layers to create the rings we all know so well. The new sapwood consists of mostly living and some dead wood cells. The living cells carry fluids vital to the tree's nutrition and health. The inner rings of a tree make up what is called the "heartwood." The heartwood is composed entirely of dead wood cells. It is darker in color than the sapwood, as it contains greater deposits of tar-like materials and minerals that make it denser,

---

[3] It is important to recognize that the "grain pattern" or "grain structure" of wood and the "grain structure" of a crystalline metal or ceramic are quite different in many respects. In wood, the grain structure is visible to the unaided eye. In metals and ceramics, the grain structure is usually too small to be observed except under a microscope. Hence, the grain structure of a metal or ceramics is what we call the "microstructure." On the other hand, the grain structure in wood and in crystalline materials has its origin in growth processes, albeit the growth in metals and ceramics being based on nucleation of a crystal, but not so in wood.

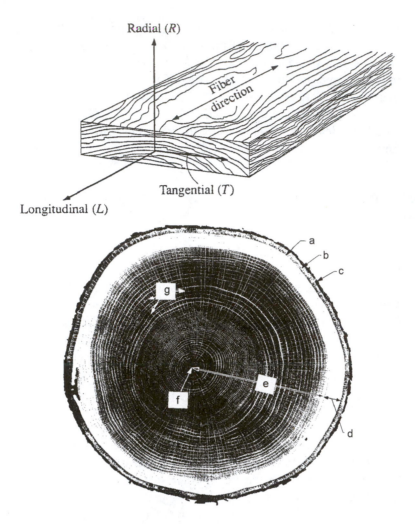

FIGURE 11.1 A photograph of the cross section of a typical tree showing key features and a schematic illustration of the axes used to specify directions in wood (as an insert). In this cross-section of a white oak tree trunk, the following typical structural features can be seen: (a) outer bark (dry dead tissue), (b) inner bark (living tissue), (c) cambium, (d) sapwood, (e) heartwood, (f) pith, and (g) wood rays. (Photograph courtesy of the U.S. Forest Products Laboratory, Madison, WI; used with permission.)

stronger, and more resistant to decay than the sapwood. Finally, old dead bark cells at the outermost surface tend to wear off; being replaced by newer, underlying bark cells, all to protect the underlying wood from disease and damage.

Figure 11.1b shows the axes used to specify directions in wood based on growth directions in the tree, as these will be important later when the properties of wood are discussed.

Because of the way they grow, trees tend to exhibit "yearly growth rings," the thickness of which varies with the amount of growth that occurs in the wood cells produced in that ring that year. More growth occurs with better growing

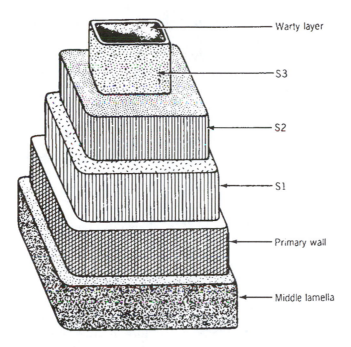

Warty layer

S3

S2

S1

Primary wall

Middle lamella

FIGURE 11.2 A diagram of a typical wood cell from a tracheid wall depicts the various layers and their microfibrillar organization. S refers to secondary wall, with S3 sometimes interpreted as a tertiary wall layer. (Courtesy of the U.S. Forest Products Laboratory, Madison, WI; used with permission.)

conditions, and vice versa. In tropical regions, where there are less seasonal differences, growth rings are less distinct.[4]

Two other features that appear in the macrostructure of a tree are "wood rays" and "pith." *Wood rays* are horizontal radial canals that connect the various layers from the center of the tree to its bark. They transfer and store water and food to keep the tree healthy. *Pith* is the original soft tissue around which the newly formed wood cells first grow. Since trees grow vertically, as well as radially, the pith is found along the entire length and across the entire section of the tree.

Woods are divided into two classes, each of which largely tends to arise from a fundamentally different class of trees. So-called "softwoods" tend to come from trees known as conifers or "evergreens," such as pines, spruce, and cedar. Trees in this class have needles or needle-like leaves, instead of broad leaves, and have seeds or seed-bearing cones. So-called "hardwoods" tend to come from what are known as deciduous trees, such as maples, oaks, cherry, walnut, ash, and so on. Trees in this class have broad leaves that they lose annually, they often

---

[4] There are also some fundamental differences in the macrostructure of so-called "palm trees" that are found only in tropical regions, although, for the purposes of this treatment, these differences compared to deciduous and conifer or evergreen trees will not be described.

have true flowers, and they tend to have seeds that are contained in a fruit or nut. Not surprisingly, hardwoods tend to be physically harder and stronger and more dense than softwoods, but there are exceptions (e.g., Douglas fir, a softwood tree, is harder than aspen, a hardwood tree). The difference in the properties of hardwoods and softwoods arises from the microstructure of these woods.

At the microstructural level, wood fundamentally develops its hardness and strength (and density) from details of the structure of the wood cell (or woody cell), a schematic of which is shown in Figure 11.2.

Like all composites, of which wood is a naturally-occurring example, wood has a hard, strong phase (lignin) and a tougher but softer phase (cellulose). Interestingly, in wood, the hard phase comprises the matrix, while the softer, tougher phase comprises the reinforcing fibers.[5] The wood cell consists of a primary wall and a secondary wall, as shown in Figure 11.2. In the primary wall, which is built up first by the cells in the cambium layer, the strong linear polymer molecules that make up cellulose (and, thus, cellulose fibers) are in a loose, irregular, flexible network. Following this, the secondary wall is created in three distinct layers: S_1, a crisscross network, S_2, a parallel, spiral-type network, and S_3, also irregular network. The S_2 layer typically makes up 70–90% of the cell wall, and the closer the fibers of cellulose in the S_2 layer approach the longitudinal direction, the stronger the cell—and, thus, the wood—is in this direction. All of this is in complete analogy with other fiber-reinforced composites, such as fiber-glass/epoxy.

In the transverse direction, strength in the cell—and in the wood—arises from lignin and hemicellulose deposits, which form a matrix that surrounds the cellulose fibers. The lignin is composed of phenol-propane network structures that are similar to thermosetting polymers, while hemicellulose is composed of much shorter, branched cellulose molecules. Two other important constituents are groups of oil-like hydrocarbons that help resist bacterial action and weathering and of minerals, like silica, which add hardness by the formation of a hard mineral network. In the tree, such hardness helps resist boring insects and insect larvae.

Woody cells, as shown in Figure 11.2, make up both softwoods and hardwoods. What differs between these two classes of wood are details of the larger-scale microstructure, which include long-cell fibers, parenchyma food cells, and vessels, which transport fluids. Softwoods tend to have much higher percentages of vessels and lower percentages of fiber than hardwoods. For example, the differences between soft basswood and hard hickory are: 56 versus 6% vessels; 36 versus 67% fiber; and 8 versus 22% food cells.

Figure 11.3 schematically illustrates the microstructure of softwood (left) versus hardwood (right).

---

[5] In virtually all synthetic composites, a softer, more ductile, tougher matrix phase is reinforced by a stronger, harder, but more brittle phase. The fact that Nature tends to do the opposite is curious but, perhaps, revealing and significant.

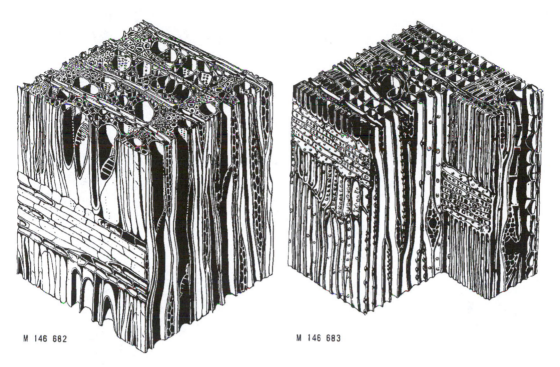

M 146 682   M 146 683

FIGURE 11.3 Illustrations of the typical microstructures of (left) softwood and (right) hardwood. (Courtesy of the U.S. Forest Products Laboratory, Madison, WI; used with permission.)

The properties of wood are, as for all materials, the result of the material's structure. However, given that the structure of wood is complex, the properties are more complex than for most other materials, but not much more complex than for other directionally-reinforced composite materials. Suffice to say here that the properties of wood are anisotropic,[6] since the structure of wood is anisotropic. Mechanical properties, in particular, differ in the longitudinal (L), radial (R), and tangential (T) directions (see Figure 11.1b).

Table 11.1 shows some important mechanical properties for some important woods native to the United States. All are similar from the standpoint of their response to applied stresses, in that they tend to be particularly strong in compression compared to tension, resist fatigue, and have reasonably good elasticity and toughness and resilience.[7] They can have their shape change permanently, for example, by deforming them while they are wet and letting them dry in the deformed state, but no atomic- or molecular-level flow occurs.

---

[6] Wood technologists tend to use the term "orthotropic" to indicate that properties are anisotropic in three orthogonal directions.

[7] "Resilience" in a material is a measure of its ability to absorb energy without deforming permanently and is determined by taking the area under a stress–strain curve out to the point where a permanent strain occurs. "Toughness" is a corresponding property, but all the way to fracture.

TABLE 11.1 Properties of Some Important Woods Native to the U.S. (Compiled from data from the U.S. Forest Products Laboratory, Madison, WI)

Species	Specific Gravity	Static Bending		Compression		Shear	Side Hardness Load (Perp. to Gr.) (lb_f)
		Rupture Mod. (Para. to Gr.) (lb/in^2)	Elastic Mod. (Perp. to Gr.) (10^6 lb/in^2)	Max. Crush Str. (Para. to Gr.) (lb/in^2)	Fiber PL Stress (Perp. to Gr.) (lb/in^2)	Max. Shear Str. (Para. to Gr.) (lb/in^2)	
Softwoods							
Eastern white pine	0.34 Gr	4,900	0.99	2,440	220	680	290
	0.35 KD	8,600	1.24	4,800	440	900	380
Douglas fir	0.45 Gr	7,700	1.56	3,780	380	900	500
	0.48 KD	12,400	1.95	7,240	800	1,130	710
Western red cedar	0.31 Gr	5,200	0.94	2,770	240	770	260
	0.32 KD	7,500	1.11	4,560	560	990	350
Redwood (young growth)	0.34 Gr	5,900	0.96	3,110	270	890	350
	0.35 KD	7,900	1.10	5,220	520	1,100	420
Hardwoods							
White ash	0.55 Gr	9,600	1.44	3,990	670	1,380	960
	0.60 KD	15,400	1.74	7,410	1,160	1,950	1,320
Yellow birch	0.55 Gr	8,300	1.50	3,380	430	1,110	780
	0.62 KD	16,600	2.01	8,180	970	1,880	1,260
Black cherry	0.47 Gr	8,000	1.31	3,540	360	1,130	660
	0.50 KD	12,300	1.49	7,110	690	1,700	950
Hickory (pecan)	0.60 Gr	9,800	1.37	3,990	780	1,480	1,310
	0.66 KD	13,700	1.73	7,850	1,720	2,080	1,820
Maple (big leaf)	0.44 Gr	7,400	1.10	3,240	450	1,110	620
	0.48 KD	10,700	1.45	5,950	750	1,730	850
Oak (white)	0.66 Gr	8,300	1.25	3,560	670	1,250	1,060
	0.68 KD	15,200	1.78	7,440	1,070	2,000	1,360

Gr = green state; KD = kiln dried to 12% moisture

These properties suggest that wood is an ideal candidate for joining mechanically, either using fasteners or using integral design features. When integral design features are used, those that operate by relying on the rigid behavior of the wood are most common, but features that rely on elastic behavior are also possible but much more unusual.

Before moving on to the joining of wood, it is important to briefly describe one more interesting development taking place in wood technology: the proliferation of engineered wood products.

11.3 ENGINEERED WOOD PRODUCTS

A major advancement in wood construction—and in ecology—began when naturally-occurring wood was first processed into what are now known as "engineered wood products." *Engineered wood products* were motivated by a need and desire to overcome various shortcomings of solid wood structural members. The idea almost certainly first began when the builders of wooden ships glue-laminated thin planks to create thick beams and small, precut arc-segments to form U-shaped timbers. The need was motivated by limitations in suitably large trees, but even more by limitations in the ability to form large planks into the complex-curved shapes needed to create the frames for hydrodynamically efficient hulls. It was also motivated by the desire to have the strength-enhancing grain of the wood run in the directions needed to carry loads and to create large, highly-loaded timbers from suffering from naturally-occurring defects such as knots.

Even early furniture makers used thin veneers of precious woods to cover less expensive, more readily available structural woods, creating a laminated wood composite as it were. Veneers are cut from logs soaked in hot water to soften the wood and that are then rotary sliced on large lathes against a stationary knife to peel off a continuous strip of thin wood, much as paper is unwound from a roll. The thin veneers are adhesively bonded to the underlying wood.

As veneer production advanced, one of the first "engineered wood products" emerged, that is, plywood. In plywood, an odd number of layers of veneers are bonded in stacks in which alternating plies are rotated 90 degrees to one another so that the direction of the grain alternates from ply to ply. This causes the plywood to have properties that are virtually the same in both orthogonal directions in the plane of the panel. Even more important than equal strength is the resistance to warpage, which normally occurs worse across the grain than along the grain in planks.

As time has passed, the variety of different types of engineered wood products has increased, and the motivation has broadened to include: (1) improved strength and/or stiffness (i.e., greater load-carrying capability); (2) better utilization of all of the wood from trees; (3) better quality (from both freedom from defects and enhancement by polymer additives); (4) lower cost; (5) larger size (than can be obtained from available trees); (6) greater versatility for producing complex shapes (e.g., curves and varying cross-sections); and (7) improved aesthetics. The basic categories of engineered wood products include:

- Laminated wood
- Structural composite lumber
- Wood panel products

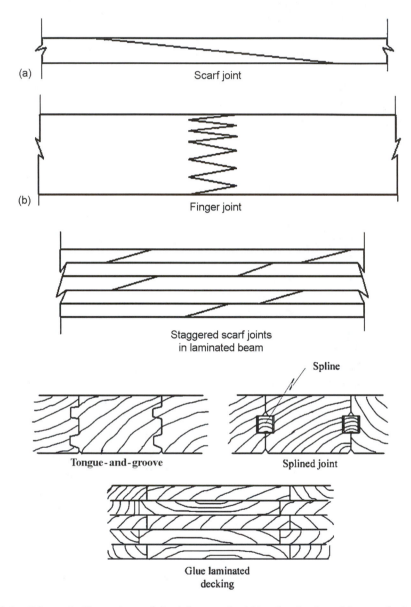

FIGURE 11.4 Schematic illustrations of the joints used within glue-laminated beams, including (a) scarf joints and (b) finger joints for end-to-end splices (top) and other types of joints, including glue laminated ones.

Within each there are sub-types. The following gives a very brief overview:

Laminated wood. In laminated wood, large structural members are produced by joining many small strips of wood together with glue. For this reason, laminated wood is commonly called "glulam" for short. By laminating smaller pieces, very large, high-load–carrying beams can be produced. It is also possible to produce angled, curved, or variable cross-sections and/or defect-free material. Within glue-laminated beams, strips of wood are adhesive bonded end-to-end, as well as layer-to-layer. Figure 11.4 (top) shows that either scarf or finger joints are used for end-to-end bonding, while Figure 11.4 (bottom) shows some other joints, including glue laminated ones.

Structural composite lumber. In structural composite lumber, veneers are laminated together, as in plywood (see "Wood panel products," next bulleted item), but all oriented with their grain running in the longitudinal direction of the piece of lumber to achieve maximum bending strength. Two sub-types are: (1) *laminated veneers lumber (LVL),* which uses sheets of veneer stacked to look like plywood without cross-bands, and (2) *parallel strand lumber (PSL),* which consists of narrow strips of veneers coated with adhesive and pressed and heat-cured into rectangular cross-sections. The advantage of PSL over LVL is that the former can be produced much more economically by using narrower strips, as opposed to sheets of veneer, and with a much higher degree of mechanization.

Wood panel products. Wood in the form of wood panel products offers advantages for many building applications. Panel dimensions, in the United States, are usually 4 × 8 feet (1.22 × 2.44 m), so less labor is needed for sheathing of walls, floors, ceilings, and so on. Panels also have much more comparable strength in both principal directions and, as will be seen for many sub-categories and types, make much more efficient use of forest resources than solid wood products.

The three general sub-categories of structural *wood panel products* are: (1) *plywood* (which is a cross-plied, bonded laminate, as described above); (2) *composite panels* (which have two parallel face veneers bonded to a compacted core of reconstituted wood fibers); and (3) *nonveneered panels* that include *oriented strand board* (OSB) (which consists of long, strand-like wood particles compressed and glued into three or five criss-crossed plies, like plywood), *waferboard* or *flakeboard* (which is composed of wafer-like flakes compressed and bonded into panels), and *particleboard* (which is composed, in several classes, from compacted and bonded smaller particles than waferboard or oriented strand board).

The joining of most wood panel products usually requires different techniques than solid wood or laminated wood (see Section 11.7).

11.4 GENERAL OPTIONS FOR JOINING WOOD

Wood can be joined chemically using adhesives or mechanically either using fasteners or by employing only integral design features. As will be seen from the brief discussion that follows, both general methods have their shortcomings.

Adhesive bonding generally works quite well for joining wood because, being inherently porous as a result of its cellular microstructure, wood offers the possibilities for chemical agents to develop an adhesive force from three of four mechanisms by which adhesives operate.[8] The porous nature of wood leads to openings at the surface that can allow suitably fluid adhesives to penetrate and develop a degree of mechanical locking. Furthermore, this same porosity allows some of the more fluid, lower viscosity components of adhesives to penetrate by absorption, taking advantage of the diffusion mechanism by which adhesion can develop. Finally, being polymeric, wood is amenable to the development of adhesive forces due to adsorption and secondary bond formation (especially due to water) at its surface. There may even be some contribution from electrostatic attraction arising from polarization within the adhesive and the wood. Phenolic-based adhesives are, by far, the most widely used adhesives for joining wood, although animal glues, fish glues, starch, casein, and other natural adhesives, including rubbers, have been used for hundreds of years. Various formaldehyde-based adhesives, epoxies, and cyanoacrylates are also used.

Table 11.2 lists adhesives commonly used with wood and in wood products.

The shortcoming of adhesive bonding of wood is that it is largely limited to factory installation, as critically-important control of the bonding process is difficult to impossible to achieve on-site and, especially, outdoors.

TABLE 11.2 List of Adhesives Used with Wood and in Engineered Wood Products

Natural Adhesives
• Animal or hide glues
• Blood glues
• Casein glues (from milk)
• Fish glues (widely used in early wood ship construction)
• Soybean glues
• Starches

Synthetic Adhesives
• Epoxies (used for wood-to-metal)
• Mastics of reclaimed rubber, neoprene, butadiene-styrene, polyurethane, and butyl rubber
• Phenol-formaldehyde (heat and weather resistant)
• Polyvinyl acetates (water-based, quick drying)
• Resorcinol-formaldehyde (cold setting)
• Rubber-based contact adhesives
• Urea-formaldehyde modified with melamine-formaldehyde (used for interior plywood and veneers)

Compiled from several sources, most notably Arthur H. Landrock's *Adhesive Technology Handbook,* Noyes Publications, Park Ridge, NJ, 1985.

---

[8] Adhesives are believed to operate by a combination of one or more of four possible mechanisms: surface adsorption, diffusion, electrostatic attraction, and mechanical interlocking (Landrock, 1985).

The fact that most woods are soft enough for holes to be cut or pierced into them explains the widespread use of fasteners for their joining. Nails and staples, self-tapping screws (known as "wood screws" and "lag screws"), and bolts have been and are widely used for fastening wood. Nails and staples are forced into wood using hammers or air- or gas-powered guns. They make their own hole as they are forced in and are held in place by the frictional gripping force of the wood as it tries to elastically recover from the penetration. Wood screws and lag screws are capable of producing their own holes using specially-designed threads, but they can also be installed more easily into pre-drilled holes with a diameter slightly smaller than the body of the screw. Bolts are placed through pre-drilled holes in wood and are held in place by nuts. To prevent the bearing loads that develop under bolt heads and nuts due to the clamping force from preload needed for their proper operation from digging into the surface of the wood, washers are normally advisable.

Figure 11.5 schematically illustrates a variety of the types of nails used for joining of, to, and through wood, as well as the three common ways in which nailing is done. Figure 11.6 shows wood screws, lag screws, and machine and carriage bolts and nuts used for fastening in, to, or through wood, as well as the use of special split rings for extra through-the-thickness strength.

An increasingly popular method of fastening wood structural members in factory-produced lightweight roof and floor trusses is to use "toothed plates" such as those shown in Figure 11.7. These steel plates are inserted into the wood at junctions between joint members using hydraulic, pneumatic, or mechanical presses, and result in an extremely effective connection because of their multiple, closely-spaced points, which interlock with the fibers of the wood.

The problem is that fasteners have always been the weak link in wood construction, as opposed to in steel or other metal construction, where they detract little from the load-carrying capability of the structural members. In modern wood construction, it is usually not possible to insert enough nails, screws, or bolts to develop the full strength of the structural members being joined.

Special hangers and brackets are often used to help support wood structures at joints, as shown schematically and in a photograph in Figure 11.8.

The fact that wood can be easily shaped leads to the old and still widespread use of designed-and-fabricated-in rigid interlocking features. Section 11.5 describes the most popular rigid interlocking integral attachment features used to join wood end-to-end, edge-to-edge, and at various corner and cross junctions.

11.5 WOOD JOINERY

The advent of bar-like metal fasteners (see "toothed plates" in the previous section) by the metalworking industry at the beginning of the 20th century has led

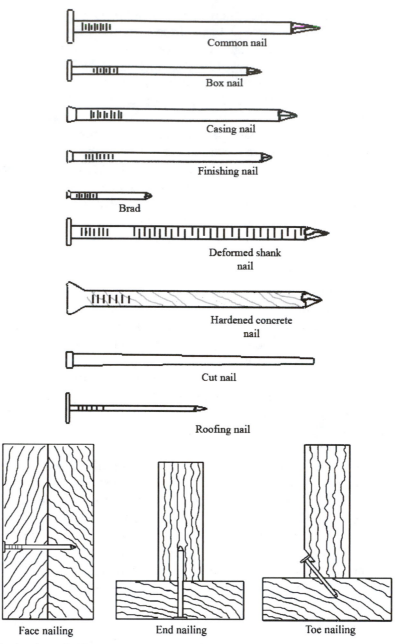

FIGURE 11.5 A schematic illustration showing various types of nails used with wood, as well as the three common ways in which nails are used.

to a diminished reliance, if not use, of traditional methods for wood-to-wood attachment or wood joinery. *Wood joinery* is defined as "the mating of two or more surfaces [in wood components] to form a solid unit that serves a specific function." Without question, metal fasteners (see Section 11.4) are easy to

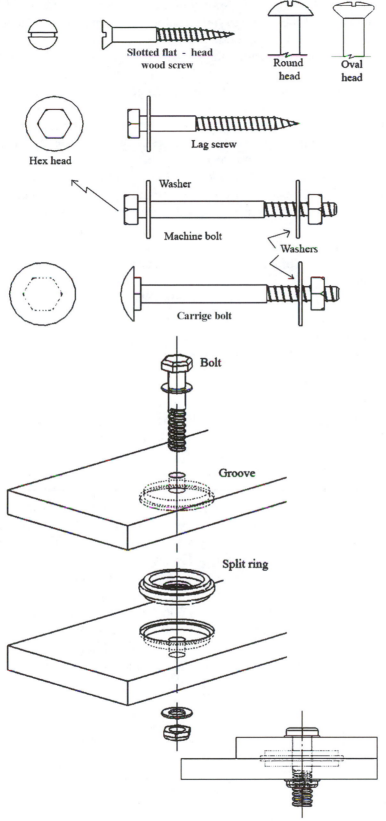

FIGURE 11.6 A schematic illustration showing wood screws, lag screws, machine and carriage bolts and nuts used with wood (top), as well as the use of special split-ring fasteners used to provide greater through-the-thickness strength (bottom).

install and allow easier assembly of wood than traditional wood joinery. Also, calculation of the load-carrying capability of fasteners is far more manageable than that of wood-to-wood integral joints. Extensive analyses and comprehensive tables of load-bearing capacity have been complied by the manufacturers of metal fasteners and are invaluable resources to engineers, architects, and planners. Comparable tables, no less comparable analyses, for wood-to-wood joinery methods simply do not exist and would be prohibitively time-consuming and expensive to generate today. In the past, such wood-to-wood joints developed with experience and success, and any design and analysis were largely, if not totally, empirical in nature.

The simple fact is that traditional wood joinery evolved over hundreds of years of experience, constantly encompassing and improving the best ways to use the unique quality of wood. However, it must be said again that the greatest disadvantage of wood as a structural material lies in the weakness inherent across the grain. But, this too can be offset by appropriate design of a joint, including joints created with geometric features cut into the wood.

The very oldest method of erecting shelters constructed entirely of wood[9] found in Japan used hollows dug into the ground that were covered with a roof of poles supported by posts sunk into the ground. The same "pole-and-post" method was used to construct wood buildings in Europe only shortly thereafter. This type of construction was followed in Europe by so-called "palisade" structures in which walls were constructed of vertical tree trunks or heavy logs set side-by-side in the ground and covered with a wood roof. Palisade construction was followed by so-called "stave construction" in both northern and southern Europe around 300 A.D. Here, long (up to 10 m) wood columns extended from foot purlins in the ground to a ridge purlin, making it possible to build purlin roof trusses with relatively large spans. This led to the "groundsill" design around 1000 A.D. in which a wood frame or sill was erected on the surface of the ground to support a wood frame for walls and roof, thereby preventing the decay of supports dug into the ground. Finally, timber-frame construction emerged around the 12th century A.D. It is basically this type of construction for wood buildings that survives to this day.

The key to wood construction from the earliest times until well into the 17th century and beyond is the fabrication of geometric features in the wood that allowed one piece to be interlocked with or into another. Joints were made—and are still made—where supporting members and supported members meet, where structural members must be spliced together, or where structural members need to be braced. Consequently, different joint forms for

---

[9] Older buildings are known that were constructed from stone or bricks, with or without mortar, and covered with a roof held up by wood beams set into notches in the walls, and further covered with smaller sticks.

FIGURE 11.7 The use of toothed plates for joining wood into pre-fabricated trusses, like the roof trusses shown here. (Photograph courtesy of the Wood Truss Council of America, Madison, WI; used with permission.)

wood were not developed exclusively for a particular function. The great majority of wood joints came about as generations of crafts-persons developed and adapted existing joints in response to changing conditions and demands at the time. Hence, simple lap joints evolved into more complex joint forms that could withstand loads imposed from several directions. Today, more than 600 wood joints are known. Without question, the earliest designs for wood joints were inspired by the branching form of trees. Recall our earliest ancestor wedging a rock into a forked stick, or resting one stick in the fork of another.

The essential concepts to be grasped here are (1) that wood is highly amenable to the creation of integral features for allowing attachment or joining because it is easy to work to a variety of fairly intricate shapes and (2) that the properties of wood allow such joints to operate as rigid interlocks, with some degree of elasticity.

The remainder of this section will describe the major types of joints used to join wood members to wood members end-to-end (i.e., splice joints), edge-to-edge (i.e., edge joints), at corners or T-intersections (i.e., corner joints), at crossing junctions (i.e., cross joints), and at angles (i.e., oblique joints). No effort

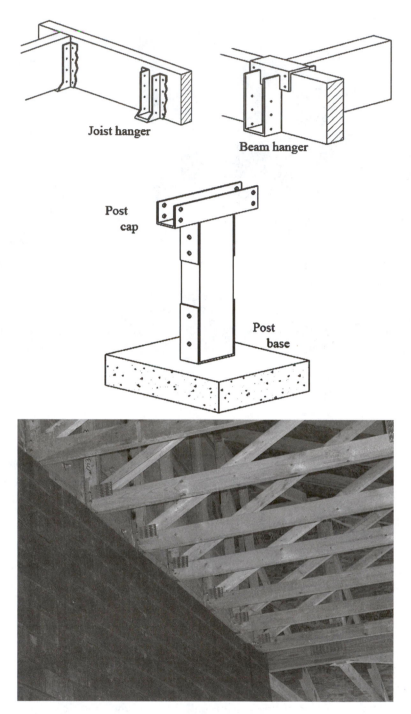

FIGURE 11.8 Schematic illustrations (top) of some metal (usually steel) hangers and brackets used to join wood structural members for increased support at the joint, and photograph (bottom) showing the use of metal hangers for supporting the joining of trusses (pre-assembled with toothed plates) to walls. (Photograph courtesy of the Wood Truss Council of America, Madison, WI; used with permission.)

will be made to describe every type of joint used, no less known.[10] This is left to dedicated references, of which there are several fine ones (e.g., Graubner, 1992; Zwerger, 1997).

11.5.1 Splice Joints

Splice joints in wood joinery are rigid interlocking integral design features intended to lengthen horizontal and vertical timbers. They are used in original construction to lengthen members and to allow replacement of portions of horizontal beams or vertical posts exposed to decay from weather, the ground, or other environmental conditions. They are also used by furniture or cabinet makers to lengthen available stock.

There are many different types of splice joints from simple straight or square butt joints that really are not joints at all, as they do not resist any loadings (except compression in the direction of the joint faces) without the assist of fasteners or adhesive, to very complex double gooseneck joints that resist separation loads and shear loads out of plane and laterally. Table 11.3 lists the major types, along with the directions of loading each resists without enhancement from fasteners, keys, dowels, or adhesives, and shows each with a small inset schematic.

As can be seen from Table 11.3, various end-designs on the wood members being spliced vary in angle, length of overlap, and number of load-bearing surfaces to increase the load-bearing capability either out-of-plane or, in some

TABLE 11.3 List of Major Types of Splicing or Splice Joints Used in Wood Joinery*

Butt Joints
• Simple butt joints (without and with a key)
• Offset butt joints (without and with a key)

Splayed Joints
• Splayed scarf joints (without and with a key)
• Halved splayed joint
• Offset splayed joints (without and with a key)
• Tapered finger joint
• Bird's mouth or V-joint

Lapped Scarf Joints
• Straight half lap joint
• Half lap joint with splayed shoulders

(Continued)

---

[10] The Chinese, and later the Japanese, evolved wood joinery to an art form beyond a remarkable structural engineering achievement. The interested reader is referred to the superb work by Zwerger (1997).

TABLE 11.3[*]   (*Continued*)

- Half-lapped V-joint
- Undercut half-lapped V-joint
- Beveled half lap joint
- Wedged half-lapped V-joint
- Keyed halved splayed joint
- Halved splayed joint with tongue and groove

Mortise-and-Tenon Joints

- Mortise-and-tenon joint
- Lapped mortise-and-tenon joint
- Through lapped rod mortise-and-tenon joints (wedged and haunched, and with diagonal keys)
- Slot mortise-and-tenon joints (without and with undercut shoulders)
- Double-tenon joint
- Loose tenon joints (without and with keys or pins)
- Offset double-tenon joint
- Upright mortise-and-tenon joint
- Bird's-mouth mortise-and-tenon joint
- Offset stub tenon joint
- Half-blind stub tenon joint
- Square- and fan-shaped stub tenon joints
- Crossed stub tenon joints

Dovetail Joints

- Through dovetail joints (single and double)
- Lapped dovetail joints (single and double)
- Tabled lap joint with two dovetails
- Gooseneck joints (simple and mitered; including 4-way)

Tabled Splayed Joints

- Tabled splayed scarf joint
- Hooked scarf joint
- Tabled splayed scarf joint with stub tenons

The Gerber Joint
Tabled Lap Scarf Joints

- Lipped tabled lap scarf joint
- Pegged tabled lap scarf joint
- Wedged arch clasp joint
- Cogged lap joint with straight shoulders
- Bridled cog joint
- Tabled joints with wedges
- Wedged locking joint or French joint
- Half-blind tabled joints (various types)

[*] Compiled from *Encyclopedia of Wood* by Wolfram Graubner, The Taunton Press, Newtown, CT, 1992.

designs, laterally, or both. Those designs with re-entrant angles in recesses that accept tangs are capable of resisting separating loads.

Figure 11.9 shows an example of a splice joint in wood construction.

11.5.2 Edge Joints

Edge joints in wood joinery are rigid interlocking integral design features intended for joining boards edge to edge to increase the width of the assembly. There are a variety of types as listed in Table 11.4 along with the directions of loading each resists, and as pictured schematically in small insets.

As can be seen, load-carrying capability can be made quite high by using multiple tongues and grooves, culminating in the so-called "toothed joint" used in very heavy timbers, such as in old wooden bridges.

Figure 11.10 shows an example of an edge joint in wood construction.

FIGURE 11.9   Wall detail of farmhouse in Wawrzenczyce (Poland) showing two splice joints; one in the vertical member at the center and one in the bottom cross-beam. (Photograph courtesy of Dr. Klaus Zwerger of the Institute for Artistic Design at Vienna Technical University, Vienna, Austria; used with permission.

TABLE 11.4   List of Major Types of Edge Joints Used in Wood Joinery[*]

Rabbeted and Grooved Joints

- Rabbeted joint or ship-lap joint
- Tongue-and-groove joint
- Spline-and-groove joint or loose spline joint
- Double spline-and-groove joint
- Toothed joint
- Mortises with glued loose tenons

Lap Joints

- Mitered lapped scarf joint
- Halved mitered lapped scarf joint

Carcase Joints

- Half-blind rabbet joint
- Fully housed dado joint
- Sliding dovetail joint
- Tongue-and-dado joints (various types)

Dovetail Battens or Sliding Dovetail Joints

- Dovetail batten
- Blind keyhole batten

Finger Joints and Dovetails

- Simple finger joints
- Multiple finger joints
- Dovetail finger joint
- Blind and half-blind dovetail joints
- Decorative dovetail joints (various types)

[*] Compiled from *Encyclopedia of Wood* by Wolfram Graubner, The Taunton Press, Newtown, CT, 1992.

11.5.3   Corner and Cross Joints

Corner joints in wood joinery are rigid interlocking integral design features intended for connecting timbers at right angles to form L- or T-shaped junctions. *Cross joints* in wood joinery are closely related to corner joints, in that they are rigid interlocking integral design features intended for joining timbers at right angles, but in both directions to form an X-shaped junction. There are a great variety of designs suited for creating corner or cross joints. In most cases, the same basic design can be used for a T-junction or, by re-creating the design on both sides of a member, an X-junction.

Table 11.5 lists the major types of corner and cross joints along with the directions of loading each resists, and as pictured schematically in small insets.

FIGURE 11.10 Roof shingles of church in Owczari (Poland) showing interlocking edge joints. (Photograph courtesy of Dr. Klaus Zwerger of the Institute for Artistic Design at Vienna Technical University, Vienna, Austria; used with permission.)

TABLE 11.5 List of Major Types of Corner and Cross Joints Used in Wood Joinery*

Bridle Joints (various types)

Butt Joints

- Housed butt joint
- Doweled butt joint

Mortise-and-Tenon Joints

- Sloped mortise-and-tenon joint (without and with pegs)
- Right-angled mortise-and-tenon joint
- Simple mortise-and-tenon joint
- Straight housed mortise-and-tenon joint
- Sloped housed mortise-and-tenon or tusk tenon joint
- Haunched mortise-and-tenon joint
- Grooved through mortise-and-tenon joint
- Open slot mortise-and-tenon joint
- Offset mortise-and-tenon joint
- Cogged mortise-and-tenon joint
- Wedged mortise-and-tenon joints (through and blind types)
- Draw-pin joints
- Multiple tenons (double, triple, and quadruple types)

(Continued)

TABLE 11.5 (*Continued*)

- Mitered mortise-and-tenon joints (blind mitered and half-blind mitered types)
- Dovetailed mortise-and-tenon joints (blind mitered dovetail and dovetail keyhole types)
- Wedged dovetails
- Gooseneck mortise-and-tenon joints (various types)

Lap Joints

- Corner lap joint
- Simple cross lap and T-shaped lap joints
- Oblique cross lap joints
- Beveled half lap (or French lock) and blind beveled half lap joints
- Tabled lap joints (including sloped types)
- Dovetail laps (full and half types)
- Lap joints used in log buildings (various types)
- Stop lap joints (simple, sloped, half-dovetail, and dovetail types)
- Cross-lap joints (simple, offset, tabled, dovetailed, and V-mitered types)

Cogged Joints (simple, double, housed, cross, half-cross, oblique, and dovetail types)

Tongue Joints (blind mitered double and oblique types)

*Compiled from *Encyclopedia of Wood* by Wolfram Graubner, The Taunton Press, Newtown, CT, 1992.

Figures 11.11 and 11.12 show examples of a corner and a cross joint used in wood construction.

It should be obvious by now that many basic designs used for making splice or edge joints are used, with appropriate modification, to create corner or cross joints. Also obvious should be that as the geometry of a joint becomes more elaborate, two penalties are incurred. First, more wood must usually have to be removed, which can actually reduce the load-carrying capability of the structural member. Second, as the details of the joint's geometry become more complex, the cost of producing the joint are increased. It also requires more skill to create the more complex joints, one reason integral mechanical attachment of wood is more art than science!

11.5.4 Oblique Joints

Oblique joints in wood joinery are rigid interlocking integral design features intended for joining timbers at angles less than 90 degrees. There are many different designs for oblique joints, as listed in Table 11.6, as for other joint types along with load-resisting capability and schematics.

Figure 11.13 shows an example of an oblique joint in wood construction.

The use of integral mechanical attachment in wood joinery in shipbuilding and in the construction of wood-framed homes and large wooden buildings, over the centuries, and around the world, reached a high level of sophistication.

FIGURE 11.11A Stabbur in Gjellerud (Norway) showing corner joints in base frame. (Photograph courtesy of Dr. Klaus Zwerger of the Institute for Artistic Design at Vienna Technical University, Vienna, Austria; used with permission.)

FIGURE 11.11B Corner detail in Tsumago Nagano (Japan). (Photograph courtesy of Dr. Klaus Zwerger of the Institute for Artistic Design at Vienna Technical University, Vienna, Austria; used with permission.)

(*Continued*)

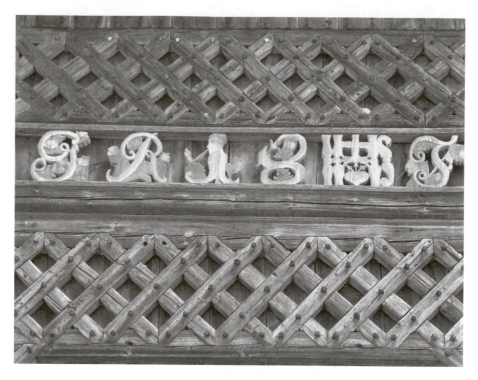

FIGURE 11.12 Bundwerk detail right angle cross joints in lattice. (Photograph courtesy of Dr. Klaus Zwerger of the Institute for Artistic Design at Vienna Technical University, Vienna, Austria; used with permission.)

TABLE 11.6   List of Major Types of Oblique Joints Used in Wood Joinery[*]

Notched Joints
• Notched heel joint
• Notched mortise-and-tenon joint (slot and beveled slot types)
• Shouldered heel joint (slope-shouldered and straight-shouldered types)
• Notched butt joint (single and double types)
• Double-notched toe and heel joints
• Triple-notched joints
• Multiple-notched joints for tooth and spline girders or beams
• Right-angled notched joints (beveled-shoulder notched mortise-and-tenon and offset-shouldered tenons)
• Bird's-mouth joints (various types)
• Angled lap joints (lap, half-dovetail lap, and angled housed half-lapped dovetail)

[*] Compiled from *Encyclopedia of Wood* by Wolfram Graubner, The Taunton Press, Newtown, CT, 1992.

Figure 11.14 shows a remarkable example of ancient temple construction using wood joinery brought to the level of an art form, while Figure 11.15 shows wood joinery techniques used in the construction of a modern-day church constructed using shipbuilding techniques.

FIGURE 11.13   Pillar detail of church in Tusa (Romania) showing oblique joints in curved arms and at vertical support. (Photograph courtesy of Dr. Klaus Zwerger of the Institute for Artistic Design at Vienna Technical University, Vienna, Austria; used with permission.)

11.6   ENHANCEMENTS FOR INTEGRAL RIGID INTERLOCKS

Sometimes to keep the design and fabrication of the integral features of a wood joint simple and other times because of the complex and/or severe loading to be resisted, integral rigid interlocks will not suffice without some assist or enhancement. Three basic options exist for providing such enhancement to joints in wood joinery: (1) adhesives; (2) conventional mechanical fasteners; and (3) mechanical assists in the form of supplemental parts also made of wood. Let's consider each briefly.

Adhesives have been used to strengthen and seal wood joints for hundreds of years. The earliest glues were of natural origin, coming from plants or animals. Fish glues[11] were widely used in shipbuilding, along with pine pitch. In

---

[11] "Fish glues" are natural adhesives made from gelatinous materials either secreted by or extracted from fish by processing.

FIGURE 11.14   Shanhoa in Datong_Shanxi (China) showing the remarkable level of detail to which wood joinery evolved in China. (Photograph courtesy of Dr. Klaus Zwerger of the Institute for Artistic Design at Vienna Technical University, Vienna, Austria; used with permission.)

FIGURE 11.15   The use of wood joinery techniques found in wood ship building can be seen throughout the interior of Our Lady of the Sacred Heart Catholic Church in Boothbay Harbor, ME. Here, the ends of timbers in trusses can be seen to be notch into vertical beams (left), while vertical supports tie into roof structure with mortise-and-tenon joints augmented with wood pegs (right). (Photographs by the author, Robert W. Messler, Jr.; used with permission.)

more recent times, excellent synthetic adhesives have been developed and used (see Section 11.4).

Because adhesives perform far better in shear than in peel, they are most effective when used with wood joints that have large surface areas that are loaded in shear.

Mechanical fasteners obviously are required to join wood if the geometry of the joint itself doesn't provide the needed interlocking in all of the directions in which motion is to be prevented (see Section 11.4). However, fasteners have been, and are still, also used to enhance the performance of wood joints primarily relying on rigid interlocking features of the joint members. If, for any reason, the joint members cannot be designed and prepared to resist all of the loading expected in service to prevent unwanted movement, particularly separation in the direction opposite that used to cause assembly, then fasteners may be needed. The first fasteners were wood *pegs* inserted into holes that passed through one joint member into or through the other(s). To ensure that such pegs remained in place, they were often tapered so that they had to be forced into place by wedging. Even greater tightness could be obtained by soaking the pegs in hot water, to soften them, prior to forcing them into place. Later, metal nails and, then, screws and bolts were developed. Metal fasteners tend to prevail today, although wood pegging is used for aesthetics, if nothing else.

Last, but by no means least, among methods for enhancing the security of wood-to-wood rigid integral mechanical attachments is the use of supplemental pieces of wood or wood parts. These come in a variety of types, many of which are designed to pin the integral features of the joint members to prevent their unwanted separation or simply tighten the action of these features bearing against one another using wedges. Hence, a list of such devices, without much description, includes:

- *Treenails* (which, in many ways, are pegs that either pin or wedge the joint elements)
- *Dowels* (which can, themselves, be integral to one of the joint members or be separate, normally-cylindrical pieces)
- *Wedges* (which are simply forced between the surfaces of interlocked joint members to tighten the fit and increase the friction that resists unwanted movement)
- *Biscuits* (which are small, thin, disk-like pieces that operate like keys in keyway slots in opposing joint members)
- *Spline inserts* (which operate like biscuits but are normally simple rectangular strips that fit into recesses in each joint member)
- *Butterfly, dog-bone,* and other *keys* (that operate as links between joint members containing suitably shaped recesses to accept the protruding features of the key)

Figure 11.16 schematically illustrates the various types of supplemental wood pieces used to enhance rigid interlock attachments.

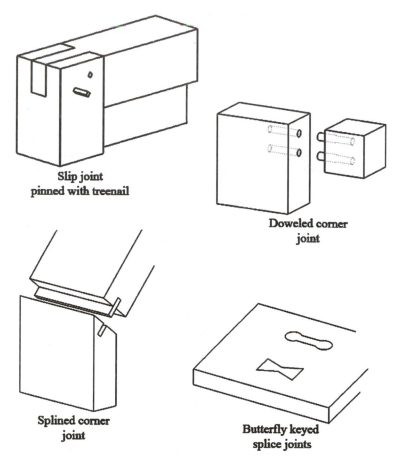

Slip joint
pinned with treenail

Doweled corner
joint

Splined corner
joint

Butterfly keyed
splice joints

FIGURE 11.16 A schematic illustration showing various types of supplemental wood pieces used to enhance rigid interlock attachments in wood joinery.

11.7 INTEGRAL FASTENERS USED WITH WOOD AND ENGINEERED WOOD PRODUCTS

Many engineered wood products (see Section 11.3) make it difficult or impossible to use conventional metal fasteners for joining. While most laminated wood and structural composite lumber, and some wood panels, can be nailed, screwed, or bolted, some cannot. Some newer laminated products contain layers of resin adhesive or even plies of polymer that make nailing difficult, if not impossible, although bolting is usually still possible. Some newer structural composite lumber contains so much strong but brittle thermosetting resin that nailing and screwing are impossible. In some of the wood panel products made up from wood flakes, fibers, chips, or sawdust, conventional nails and self-tapping screws will not work. The material simply lacks suitable through-the-thickness strength to allow these fasteners to operate properly. For all of these, as well as for solid wood for reasons of higher productivity (e.g., easier assembly), special integral fasteners are—or may be—the answer.

Various sleeves or bushings to line drilled holes, pre-placed (often bonded) anchor nuts, clinched fasteners, quick-release fasteners, and other special inserts are used with success. Many of these inserts are designed to facilitate the easy assembly of pre-fabricated furniture made from compressed wood panels with bonded veneer. One common type is a locking nut that is pushed into a pre-drilled hole in one part and is engaged by a quarter-turn screw inserted through another mating part.

11.8 SUMMARY

In many ways, wood is the most remarkable of all materials. First and foremost, while not the only material to come from Nature and be used in its natural state, it is the only material that is renewable. Wood comes from trees, and trees grow and can be grown in controlled and managed ways to improve their quality and abundance. In growing, however, trees—and the wood they consist of—exhibit patterns in their structure that reflect details of that growth. Wood cells produced by the cambium layer just under the bark (which the cambium layer also produces from bark cells) are present in both living and dead forms to produce lighter colored, more pliable or flexible sapwood. The deeper-lying, previously grown, heartwood consists of only dead wood cells and is stronger, harder, denser, and darker, but less flexible than the sapwood. Both the heartwood and the sapwood are composed of a complex structure of fibers, vessels, pith, rays, and so on, which differ in proportions between softwoods and hardwoods to give these their respective properties.

At the finest level, wood cells consist of long fibers of tough, strong, pliable cellulose aligned primarily in the longitudinal direction, that is, along the length of the tree's trunk and branches, and hard, very strong, but brittle lignin. Both cellulose and lignin are polymers, but with quite different molecular structures and, hence, properties. Together, the cellulose and the lignin form a polymeric composite that we know as wood.

Like its structure, which is complex, the properties of wood are also complex or, at least, are difficult to describe. Most importantly, the most important properties of strength and stiffness are highly directional as a direct result of the growth mechanism of and grain pattern in the tree. Properties differ in the longitudinal, radial, and transverse directions in lumber, which greatly influences the way in which wood should be used for optimum performance as a structural material.

The properties of solid wood, as well as the fuller utilization of the precious resource of our forests, have been augmented with the development of engineered wood products, including laminated wood, structural composite lumber, and various types of wood panel products.

Joining of wood can be done with adhesives or with mechanical fasteners or, oldest and, in many ways, most sophisticated of all, by employing integral design features. These features are almost always used to produce rigid inter-locks for joining wood-to-wood in what is known as "wood joinery." A large number of widely diverse types of designed- and fabricated-in geometric features are used for lengthening wood members through splice joints, widening wood members through edge joints, and for producing 90-degree junctions with corner or cross joints or less than 90-degree junctions with oblique joints. While able to resist a variety of different loads by themselves, through their own design features, rigid integral interlocking joints in wood joinery can—and often must—be enhanced through the use of adhesives, metal fasteners, or various wood pieces, parts, or devices.

Engineered wood products sometimes make joining with fasteners, as opposed to integral design features, necessary but difficult because of their unique properties compared to solid wood. For these situations, various inserts and integral fasteners are used.

Table 11.7 summarizes the various ways in which wood can be joined using integral mechanical attachment schemes and attachments.

TABLE 11.7   Summary of the Methods for Joining Wood Using Integral Mechanical Attachment

Integral Mechanical Attachment (Wood Joinery)
• Dado or rabbet joints
• Tongue-and-groove joints
• Mortice-and-tenon joints

Other Mechanical Enhancements in Wood Joinery
• Biscuits
• Dowels
• Keys (butterfly, dog-bone, etc.)
• Spline inserts
• Treenails
• Wedges

Mechanical Attachments
• Metal brackets
• Metal hangers
• Metal straps

Mechanical Fasteners
• Lag screws
• Machine and carriage bolts and nuts
• Nails
• Pegs
• Split rings with bolts and nuts
• Toothed plates or plate nails
• Wood screws

Adhesive Bonding

REFERENCES

Graubner, W., *Encyclopedia of Wood Joints,* The Taunton Press, Newtown, CT, 1992.

Landrock, A.H., *Adhesives Technology Handbook,* Noyes Publications, Park Ridge, NY, 1985.

Zwerger, K., *Wood and Wood Joints: Building Traditions of Europe and Japan,* Birkhauser, Boston, MA, 1997.

BIBLIOGRAPHY

Allen, E., F*undamentals of Building Construction: Materials and Methods,* 3rd edition, John Wiley & Sons, Inc., New York, NY, 1999.

Graubner, W., *Encyclopedia of Wood Joints,* The Taunton Press, Newtown, CT, 1992.

Smith, W.F., *Foundations of Materials Science and Engineering,* 3rd edition, McGraw-Hill Higher Education, New York, NY, 2004.

U.S. Department of Forestry, *Encyclopedia of Wood,* Washington, DC, 1980.

Zwerger, K., *Wood and Wood Joints: Building Traditions of Europe and Japan,* Birkhauser, Boston, MA, 1997.

12 MECHANICAL ELECTRICAL CONNECTIONS

12.1 DEMANDS ON ELECTRICAL CONNECTIONS THAT FAVOR MECHANICAL ATTACHMENT

A strong case can be made that no other technology has favorably changed the lives of humans or favorably impacted the economy of countries more than the ability to generate, control, and distribute electricity. In fact, it is hard to imagine the world without it! Electricity lights and often heats and cools our homes, offices, and factories. It enables the production of essential chemicals (e.g., chlorine), primary metals (e.g., aluminum), as well as many engineering ceramics, cement, and composites. It powers the machines[1] that, in turn, produce the automobiles, trucks, buses, trains, ships, and airplanes we rely on for local, interstate, and international transportation, trade, and commerce. It brings us entertainment via radio and iPod™, TV and CD, cinema and DVD, and PlayStation™ and Xbox™. And, the thought of life without cellphones and computers is, after only decades, almost unimaginable! The ready availability of electricity is one of the greatest differentiators between developed and underdeveloped countries. It "brings good things to life."[2]

To move electricity requires that it be conducted through suitable carriers, which are, quite literally, known as "conductors." Almost all conductors used in the transportation of electrical energy are made of copper or aluminum or alloys

---

[1] While it is true that machines were once powered by water, electricity largely freed manufacturers from having to locate their plants next to fast-moving rivers or streams. This, in turn, led to the spread of people and wealth across the landscapes of the world.

[2] The slogan "We bring good things to life!" is used and copyrighted by The General Electric Company, further making the point!

based on one or the other of these metals. The reason for this is simple: Both allow the electricity to move relatively unimpeded compared to all other metals.[3] To move electricity from where it is generated to all those places it is needed to be used requires that lengths of conductive copper or aluminum solid wires or bundled-wire cables be joined to one another. At the point of use, electricity must be brought from the utility's power line to a circuit box that distributes electricity at the proper voltages and currents needed to run lights, appliances, machines, and so on. Many joints are needed between different-size wires and different types of terminals. Such joints are known as "electrical connections," even though they are largely enabled by mechanical techniques.[4]

The making of sound electrical connections between electrical conductors requires careful consideration of the mechanical and chemical properties of the conductors, not just their electrical properties. Except in rare situations and instances where two metal conductors can be welded together, almost all conductors, particularly those used in transporting electrical energy, are joined by a mechanical process normally accomplished using pressure. Sometimes conventional mechanical fasteners (e.g., machine screws, washers, and nuts) are used, sometimes specially-designed fasteners are used (e.g., "wire nuts"), and sometimes only the conducting elements (wires, terminals, etc.) themselves are used. In the last case, the process involves attachment and not fastening, as used throughout this book, and natural, specially-designed-in, or formed-in features are used, making the process of joining that of integral mechanical attachment.

When any two conductors are to be joined mechanically, by any means, it is essential that the contacting surfaces be clean (i.e., free of at least heavy layers of naturally-occurring oxides and of contaminants of all sorts). The reason that cleanliness is so important is to keep the resistance at the contact (i.e., the "contact resistance") as low as possible. The reason for this is that any resistance, R (measured in ohms, which is really volts per ampere), causes heat to be generated as a loss to the electricity being transported. For a current flow of I (measured in amperes), the heat lost (in watts, which is volt-amperes) is given by I^2R. I^2R losses occur for every resistance in a circuit, whether in the conductor itself or at the interface or contact between conductors. If the contact resistance is much higher than the resistance of the conductor itself, excessive heat will be generated at the contact, with the possibility of melting the conductor at the contact and subsequently breaking the circuit.[5] The contact resistance will

---

[3] All metals are conductors of electricity. This important property arises from the arrangement of electrons in the outermost shells of atoms of what we call "metals" and from the way in which such atoms bond to one another to create solid aggregates (which is also based on their electron configurations). However, other than pure silver (Ag), copper (Cu) has the highest conductivity, followed by aluminum (Al). Other materials, such as ceramics and polymers, are, in general, such poor conductors of electricity that they are known as "electrical insulators."

[4] Joints can be made between electrical conductors using thermal techniques, such as welding and soldering, but, as will be seen in Section 12.2, such techniques pose serious restrictions.

[5] The use of pure aluminum for wiring homes and offices in lieu of using pure copper was largely stopped because of problems associated with electrical fires started when such wires overheated due to I^2R heating at contacts where ever-present aluminum oxide developed.

always be higher than the resistance of the bulk conductor materials being placed in contact for two reasons. The first reason is that the surfaces of all materials, regardless of how smooth they appear to have been made, are, in reality, quite rough on an atomic scale. As a consequence, the actual cross-sectional area of pathways for allowing current to pass across the interface is much smaller than the apparent or nominal area. In reality, only the high points on the surface of each mating conductor are in intimate contact. The smaller the cross-sectional area of a conductor is, the higher the resistance. The second reason for resistance being higher at contacts than in conductors is that any contamination on the surfaces in contact impedes the flow of current since such contaminants are virtually always non-conductive. Paint, grease, remnants of rubber or plastic insulation used on the conductor, and oxides are all non-conductors.

A further concern in making electrical connections arises if the conductors are different metals or alloys. Some dissimilar metal combinations lead to galvanic corrosion. Some combinations of metals, such as aluminum and copper, cause problems when one dissolves in the other, thereby increasing the previously pure copper's resistance,[6] for example.

A final consideration when electrical connections are made is the voltage level under which the conductor must operate. High voltage demands more and better protective insulation. Furthermore, high voltage leads to the distinct likelihood of arcing across gaps between conductors as they approach or separate from one another. Such arcing is potentially dangerous to an operator, as the electricity can jump to the operator causing electrocution, and often damages the contacts by spark erosion and/or melting.

The production of electrical connections must, absolutely (1) provide good electrical conduction; (2) allow for efficient dissipation of unwanted, but inevitable, heat from I^2R losses; and (3) provide needed mechanical/structural integrity for the service conditions. As will be seen in Section 12.2, mechanical integrity is often either overlooked or, at least, under-estimated by electrical engineers, so many electrical failures actually arise from the loss of mechanical integrity at a connection.

Beyond the just-mentioned property-based requirements of electrical connection, there are the practical considerations of ease of joining on-site as well as in-plant, for repairing connections or replacing parts quickly and easily, and to keep costs for required skill level and labor intensity to a minimum.

All of these requirements or desires can be met by mechanical methods of joining, some of which clearly involve the use of fasteners, some of which clearly involve the use of designed- and processed-in features, and some of which seem to lie somewhere between.

---

[6] Adding any other metal to a pure metal raises its resistance (actually, its resistivity) due to the strain-field such foreign or solute atoms produce in the crystalline lattice of the host or solvent. Such strain fields lead to increased scattering of the electrons that comprise a current flow.

All three major processes for joining—mechanical joining, adhesive bonding, and welding—can be used for making electrical connections, although adhesive bonding is, by far, the most rarely used and, even when it is used, only under very special circumstances.

The essential requirement of any process that is to be used to make electrical connections obviously is that the resulting joint be electrically conductive. Since conductors of electricity almost always also conduct heat,[7] a joint that allows good electrical conduction also allows good heat dissipation. Beyond these essential properties for electrical connections, there is also always a need for some degree of mechanical/structural integrity. How much obviously depends on how much stress is placed on the joint or connection. As a rule, for large-diameter conductors (e.g., thick solid-wire or bundled-wire high-power transmission lines), loads and stresses can be high, if for no other reason than their inherent stiffness, which transfers loads imparted on them to the joints. For very small conductors, as found in microelectronic devices, loads are small and stresses can be small or can be surprisingly high. If nothing else imposes mechanical forces on the very small-size contacts in microelectronic devices and circuit boards, thermally-induced stresses from thermally-induced strains from differential CTEs between metal conductors, ceramic chip packages, silicon wafers, and glass-reinforced polymer circuit boards can still be significant. Another source of stress on the fine electrical connections in microelectronic devices and circuit boards is mechanical vibration from the mechanical system of which they may be a part (e.g., an automobile engine). So, mechanical integrity—and a degree of structural strength—is almost always also essential.

Let's look at the three options for producing electrical connections.

12.2.1 Welding

For connections intended to be permanent, welding is the joining process of preference. The problems are: (1) rarely is anything truly needed to be permanent (either because it needs to be repaired or because it eventually becomes obsolete) and (2) usually joints need to be produced in environments unsuited to the types of welding processes preferred for making electrical joints. Often, either access to the point requiring connection is limited or obtaining the level

---

[7] In most materials, the major contributors to thermal conduction are so-called "free electrons" in either partially-filled valence bands or from filled valence bands that happen to overlap the next higher-energy, empty conduction bands. These are the same electrons that contribute to electrical conductivity. In materials with very stiff interatomic bonds, another contributor to thermal conduction is passage of the vibrations of one atom to an adjoining atom through such stiff bonds. This is known as "phonon conduction." Such materials can be electrical insulators but thermal conductors. The best-known example is carbon in the form of crystalline diamond, although some other covalent-bonded ceramics, such as aluminum nitride, also exhibit this unusual behavior.

of process control needed to ensure a high-quality joint is produced is difficult to impossible to achieve.

The types of connections that are most frequently welded are copper bars, large-diameter (over 1/4 inch or 6 mm) solid wires, and bundled-wire cables to one another, as well as copper conductors to steel rails for grounding. The process employed has been and continues to be thermit welding (TW). Also known as "aluminothermic welding," TW involves a highly exothermic reaction between finely-divided aluminum metal and fine, powdered copper oxide. Upon reaction, the Al reacts with the CuO/Cu_2O to produce molten Cu metal along with solid Al_2O_3 as a dross. The process can be performed outdoors, on-site, but the need for full access to the intended joint limits field utility. Variations on the basic class of exothermic chemical reactions involved allow the welding of copper to copper, copper to aluminum, or copper to steel, and of aluminum to aluminum. Examples are shown in Figure 12.1.

For smaller-diameter conductive elements, welds are made by percussion welding (PEW) and ultrasonic welding (USW). PEW is a resistance welding process also known, rather more descriptively, as "capacitor-discharge welding." In this process, electrical energy stored in a capacitor is suddenly discharged to flow through two pieces, usually a wire to a plate. Rapid and highly-localized I^2R heating at the high-resistance point of initial contact leads to localized melting and the formation of a weld. USW is a form of friction welding in which the high-frequency (over 25,000 cycles per second), small-amplitude movement of one piece in contact with another under pressure leads to sufficient heat to cause localized melting and the formation of a weld.

Beyond these processes, welding, as such, is not performed to produce electrical connections.

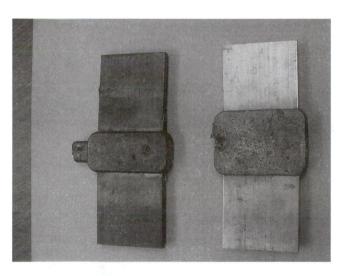

FIGURE 12.1 A photograph of Cu-Cu welds (left) and Al-Al welds (right) made by the thermit welding process for electrical connection of cables to buss bars for high current applications. (Photograph by Sam Chiappone for the author; used with permission.)

12.2.2 Soldering

In terms of the sheer number of electrical connections made, no process has made—or makes—more joints than soldering. Soldering, of course (see Chapter 2, Section 2.2), is a sub-process of welding in which a filler that melts below 425°C (840°F) is caused to flow by wetting and spreading on the surface of base metal that does not melt and the soldering temperature. A properly made solder joint produces a metallurgical bond between a metallic solder alloy and metallic base materials, although there is also almost always some contribution to the strength of the joint from mechanical locking of the solder into asperities on the surfaces of the base materials.

In the past, the predominant method of producing solder interconnections was known as "through-hole technology." In this approach, shown schematically in Figure 12.2a, the fine wire leads on what once were individual electrical components, like resistors and capacitors, and are now ceramic chip packages, were passed through holes drilled into circuit boards. These holes were lined with metal (usually copper) and solder was applied to the lead to be pulled into the hole to interconnect the Cu-metallized circuitry on the board with the device. While the solder definitely improved the electrical and thermal connection, tolerance of imposed mechanical or thermo-mechanical loads came from the combination of compliance provided by the leads themselves and the mechanical locking obtained when the portions of the leads that extended through the hole to the backside of the circuit board were crimped over. The more modern and increasingly prevalent approach is called "surface-mount technology." In this approach, shown schematically in Figure 12.2b, there are no actual wire leads. Instead, there are metal tabs or pads that run along the sides and onto the bottom surface of chip packages. These are soldered to similar pads metallized

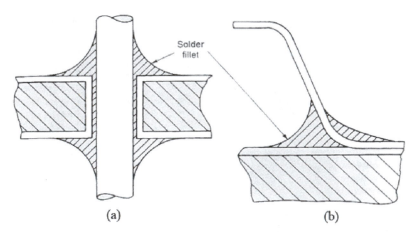

Solder fillet

(a) (b)

FIGURE 12.2 Schematic illustrations of (a) through-hole and (b) surface-mount technology for producing soldered joints in electronics.

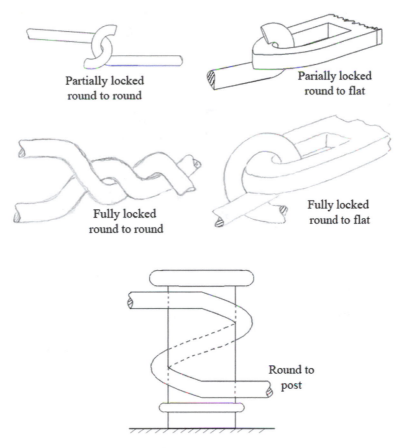

Partially locked
round to round

Parially locked
round to flat

Fully locked
round to round

Fully locked
round to flat

Round to
post

FIGURE 12.3 Schematic illustrations of the basic mechanical-joint configurations used with soldered electrical connections.

onto the surface of the printed circuit board. Without compliant leads, any thermally-induced strains from CTE mismatches between the leads, metallizations, ceramic package, and polymeric board are transferred directly to the solder joint. This leads to much greater demands on the solder interconnect to provide mechanical integrity. Also missing is any contribution from crimped-over leads on the backside of the circuit boards.

So, while soldered interconnects or connections are popular and profuse, they are perilous under the most severe conditions found under the hood of an automobile or in flying aircraft or spacecraft, as examples.

More will be said about how soldered electrical connections are, or should be, made to provide structural integrity in Section 12.3.

12.2.3 Adhesive Bonding

The synthetic polymeric adhesives that tend to prevail in electronics and elsewhere are electrical insulators, not electrical conductors. Hence, they cannot be used to make electrical connections. However, for a variety of reasons, including greater tolerance of strains arising from CTE mismatch, greater damping of potentially damaging vibrations, and for less potential for adverse impact on the health of workers and on the environment, adhesive bonding in lieu of soldering as a means of producing electrical interconnects in microelectronics is highly desirable. To achieve this goal, polymer chemists are attempting to develop polymers that are, at once, intrinsic electrical conductors and effective adhesives. What is meant by "intrinsic electrical conductors" as it pertains to polymers is that the ability to conduct electricity is the result of the polymer's molecular structure itself and not to the addition of finely-divided powders of copper or silver to the polymer resin in what are known as "metal-filled conductive adhesives." If such adhesives can be developed with conductivities closer to metals than can now be obtained, microelectronic interconnection could be revolutionized.

12.2.4 Mechanical Joining

Electrical connections can be produced using mechanical means using either mechanical fasteners or attachment features integral to the components being joined. Regardless of which approach is used, it is important that the non-metallurgical joint[8] have the most intimate contact possible. This means that pressure needs to be applied at least during the creation of the joint and possibly throughout the operation of the joint as well. Such pressure can be applied and sustained using a fastener that develops a clamping force. The most common examples in electrical connections are screws threaded into internally-threaded terminal strips or blocks. The conductive component to be connected to the terminal is literally clamped under the head of the screw, often with the help of a washer. Other designs use split cylinders that tightly clamp around a large-diameter wire or cable using screws between the half-cylinders to apply pressure.

Another common approach uses the spring action of some type of clamping device to apply a sustained pressure to the contacting components. Without going into details here, this is the basis for some elastic interlocks described in Chapter 4 and to be described for making electrical connections in Section 12.4.

By far the most common approach for achieving intimate contact between metal conductors is to plastically deform one up against or around the other. This is, of course, the basis for plastic interlocks described in Chapter 5 and to be described for making electrical connections in Section 12.4. Suffice to say here that the predominant method involves the creation of crimps, wraps, or twists.

Table 12.1 shows major types of pressure connections used for making electrical connections along with key requirements met by each and gives comparisons

---

[8] A "metallurgical joint" between metallic components refers to one that involves actual atomic-level metal-to-metal bond formation, often with a degree of interdiffusion. Any joint that does not involve the formation of metallic bonds across the interface is said to be "non-metallurgical" and relies solely on intimate metal-to-metal contact, point-to-point.

TABLE 12.1 Classification and Comparision of Pressure Connections Used in Electrical Applications

			Types of Pressure Connections			
	1	2	3	4	5	6
Requirements Wrapped	**Fahnestock Clip**	**Plug**	**Crimp**	**Wire Nut**	**Screw**	**Solderless**
1. Large contact area			X	X	X	X
2. High contact force			X	X	X	X
3. Long life			X	X	X	X
4. Small size		X	X			X
5. Mechanically stable			X		X	X
6. Easily disconnected	X	X		X	X	X
7. Low cost						X

(*Continued*)

TABLE 12.1 (*Continued*)

Requirements Wrapped	Types of Pressure Connections					
	1	2	3	4	5	6
	Fahnestock Clip	Plug	Crimp	Wire Nut	Screw	Solderless
Size	No. 15	0.040 in diam pin	0.120 in diam × 0.112 in long	0.350 in diam × 0.550 in long	No. 4–40 (0.112 in)	0.0148 × 0.062 in Terminal
Contact force, lb	1.4	2.2	22	Unknown	135	90
Contact area, sq in	0.000079	Unknown	Unknown	Unknown	0.0074	0.0031
Contact pressure, psi	18,000	Unknown	Unknown	Unknown	18,250	29,000
Space, 10^6 cubic mils	41	8.78	1.75	52.8	15.6	1.53
Elastic energy, mil-lb	21	Unknown	Unknown	Unknown	2.77	3.05

(Taken from Tables 12.2 and 12.3 on page 416 of the *Handbook of Fastening and Joining of Metal Parts* by Vallory H. Laughner and Augustus & Hargan, McGraw-Hill, New York, NY, 1956, for which a copyright seemingly no longer exists.)

of the types for making electrical connections for No. 24 (0.020 inch or 0.5 mm diameter) wire.

12.3 SOLDERED ELECTRICAL CONNECTIONS

While the purpose of this chapter is to consider joints made for electrical applications (i.e., electrical joints or electrical connections) using mechanical methods, it is important, in order to put such mechanical methods in the proper context, to also describe electrical connections made by soldering.

Soldering is used to join metal components for four general purposes: (1) electrical connections or interconnects, which are designed for their current-carrying capacity; (2) structural joints, which are designed for their mechanical strength; (3) hermetic joints or seals, which are designed for their leak tightness against positive or negative pressure in gases or liquids; and (4), occasionally, to provide intimate contact for good thermal conduction and for effective heat dissipation. The choice of soldering for producing the joint, in each case, is that a metallurgical bond can be achieved with minimal risk of adverse effects on the composition, microstructure, or properties of the base materials by heat. The compromise, however, is that low-melting metals and alloys have low mechanical strengths. This is a natural consequence of the fundamental strength of the interatomic bonds in low-melting metals. Lower bond strengths or binding energy means that atoms can be caused to separate more easily than for higher bond strengths or binding energy, whether the energy to cause separation comes from a mechanical or a thermal source. As a result, it must always be recognized that when metal components are soldered, the solder is almost always the weakest link in the assembly. The only exceptions would be when soldering is used to join low-melting metals or alloys, which could be even weaker than the solder alloy.

When soldering is used to make structural joints, as in plumbing, automobile radiator cores, heat exchangers, and so on, two basic approaches are employed. First, in virtually all cases, lap joint designs are used to provide the maximum possible area over which the solder will act for the smallest size or space and to force the joint to transfer loads via the solder in shear. Like adhesives, solders (and also braze fillers) perform better in shear than when loads cause stresses that act out of plane in other than pure tension, which is difficult to ensure. Solder joints produced without any mechanical assist try to maximize the joint area that places the solder in shear, by design.

The second basic approach, used when joints to be soldered must provide maximum mechanical security before soldering and maximum strength following soldering, uses various methods for causing interlocking of the joint members, independent of the adhesion provided by the solder itself. As shown schematically in Figure 12.3, interlocks created by plastic deformation (i.e., plastic interlocks) are created to carry the main proportion of stresses from applied loads and forces, with any solder helping to hold the interlocks together, even as it tends to yield or creep. All of the various methods shown in Figure 12.3 behave primarily as rigid joints.

The provision of some mechanical assist in soldered connections is also advisable for electrical connections. Table 12.2, originally from a superb paper by H.H. Manko (1961), shows a variety of designs for electrical connections to be soldered. The degree of mechanical assist increases from Group I, where no mechanical security is provided before soldering and, hence, no mechanical assist is provided to the finished soldered joint, through Group II, where there is partial security and assist, and, finally, to Group III, where there is full security and assist.

To understand the need for mechanical security before soldering and mechanical assist in completed solder joints, one has to recognize that soldering is occasionally used to make electrical connections in high-power electronics and electrical machines (motors and generators) and appliances, not simply solid-state microelectronic devices. Solid and multi-strand wires from small to large diameters, as well as larger cables, are often soldered to make connections. Such wires and cables are often subjected to loads from pulling, bending, twisting, flexing, and vibration, not simply thermal-expansion mismatch, and, so, connections with these wires or cables must be capable of tolerating the resulting stresses. As mentioned earlier, in microelectronics, particularly when soldered joints are made using surface-mount technology, the degree of thermally-induced loading from CTE mismatch and of mechanical loading from vibration can be much more severe than electrical engineers seem to recognize or account for.

12.4 MECHANICAL PRESSURE ELECTRICAL CONNECTIONS

Some references state that the methods for joining wires or cables to terminals on an apparatus for the purpose of electrical conduction can be broadly divided into two groups: (1) solder connections and (2) pressure connections (Parmley, 1989). As discussed in Sections 12.2 and 12.3, *solder connections* provide good electrical connection by presenting lower resistance to the passage of electric current than purely mechanical pressure connections present due to ever-present contact resistance. However, solder alone does not provide much strength to the connection. Because of this, and because of the high cost of manual soldering, the pressure connection is of great importance to the electrical industry.

Purely mechanical pressure connections use only a mechanical force to create and maintain physical—and, therefore, electrical—contact. In addition, by also creating physical interlocking between one metal conducting element and the other, a degree of mechanical strength is also provided. The level of strength achievable in a pressure connection depends on the specific method used, the contact area over which pressure is applied and sustained, and the magnitude of the contact force. Different methods for producing pressure connections lead to different degrees of interlocking, with some giving rise to surprisingly high

TABLE 12.2 Data for Electrical-Connection Design.

A. GROUP I—NO MECHANICAL SECURITY PRIOR TO SOLDERING						
BUTT CONNECTIONS						
NO.	TYPE	DIAGRAM	CONTROLLING FORMULA	CONDITIONS	FIXTURES	CURRENT
1	ROUND TO ROUND		$D_s = \sqrt{\delta}\, D_{c_1}$	$\rho_{c_1} \geq \rho_{c_2}$ $D_{c_1} \leq D_{c_2}$	YES	SMALL
2	SQUARE TO SQUARE		$D_s = \sqrt{\frac{4}{\pi}\delta}\, T_{c_1}$	$\rho_{c_1} \geq \rho_{c_2}$ $T_{c_1} \leq T_{c_2}$	YES	SMALL
3	RECTANGLE TO RECTANGLE		$T_s = \delta T_{c_1}$	$\rho_{c_1} \geq \rho_{c_2}$ $W_1 = W_2 = W_s$ $T_{c_1} \leq T_{c_2} \neq T_s$	YES	SMALL
LAP CONNECTIONS						
1	ROUND* TO ROUND		$L_j = \frac{\pi}{2}\delta D_{c_1}$	$\rho_{c_1} \geq \rho_{c_2}$ $D_{c_1} \leq D_{c_2}$ $W_s \geq \frac{D_{c_1}}{2}$	YES	LARGE
2	ROUND TO FLAT		$L_j = \frac{\pi}{4}\delta D_{c_1}$	$\rho_{c_1} \geq \rho_{c_2}$ $A_{c_1} \leq A_{c_2}$	OPTIONAL	LARGE
3	FLAT TO FLAT		$L_j = \delta T_{c_1}$	$\rho_{c_1} \geq \rho_{c_2}$ $W_1 = W_2 = W_s$ $T_{c_1} \leq T_{c_2}$	OPTIONAL	LARGE
4	WIRE TO POST		$L_j = \frac{1}{2}\delta D_{c_1}$	$\rho_{c_1} \geq \rho_{c_2}$ SOLDER FILLET $\geq \frac{D_{c_1}}{2}$	NO	MEDIUM
5	WIRE TO CUP		$L_j = \frac{1}{4}(\delta - 1) D_{c_1}$	$\rho_{c_1} \geq \rho_{c_2}$	NO	LARGE
6	WIRE TO HOLE		$L_j = \frac{1}{4}\delta D_{c_1}$	$\rho_{c_1} \geq \rho_{c_2}$	OPTIONAL	MEDIUM

(Reprinted from *Soldering Manual,* 2nd edition, Table 4.2 and 4.3, pages 30 and 31, American Welding Society, Miami, FL, 1978, with permission. Originally appeared in a paper by H.H. Manko in 1961.)

(Continued)

TABLE 12.2 *(Continued)*

GROUP II—PARTIAL MECHANICAL SECURITY PRIOR TO SOLDERING						
HOOK CONNECTIONS						
NO.	TYPE	DIAGRAM	CONTROLLING FORMULA	CONDITIONS	FIXTURES	CURRENT
1	ROUND TO ROUND		$D_{c_1} = \dfrac{2}{\delta} D_{c_2}$	$\rho_{c_1} \geq \rho_{c_2}$ $D_{c_1} \leq D_{c_2}$ HOOK $\geq 180°$	NO	LARGE
2	ROUND TO FLAT		$D_{c_1} = \dfrac{1}{\pi\delta}(8L_j + 4T_{c_2})$	$\rho_{c_1} \geq \rho_{c_2}$ $A_{c_1} \leq A_{c_2}$ HOOK $\geq 180°$	NO	MEDIUM

GROUP III—FULL MECHANICAL SECURITY PRIOR TO SOLDERING [†]						
WRAP CONNECTIONS						
NO.	TYPE	DIAGRAM	CONTROLLING FORMULA	CONDITIONS	FIXTURES	CURRENT
1	ROUND TO ROUND		$L_j = \dfrac{\pi}{2}\delta D_{c_1}$	$\rho_{c_1} \geq \rho_{c_2}$ $D_{c_1} \leq D_{c_2}$ $n > 1$	NO	LARGE
2	ROUND TO FLAT		$D_{c_1} = \dfrac{8}{\pi\delta}(L_j + T_{c_2})$	$\rho_{c_1} \geq \rho_{c_2}$ $A_{c_1} \leq A_{c_2}$ $n = 1$	NO	MEDIUM
3	ROUND TO POST		$D_{c_1} = \dfrac{4n}{\delta} D_{c_2}$	$\rho_{c_1} \geq \rho_{c_2}$ $D_{c_1} < D_{c_2}$ $n \geq 1$	NO	LARGE

D_{c_1} – DIAMETER OF SMALLER CONDUCTOR

A_{c_1} – AREA OF SMALLER CONDUCTOR

S – SOLDER

W – WIDTH

L_j – LENGTH OF JOINT

T – THICKNESS

N – NUMBER OF TURNS

δ – RESISTIVITY RATIO $\dfrac{\rho_s}{\rho_{c_1}}$

ρ – RESISTIVITY (MICROHM–CM)

*USE ONLY WHEN LARGE CONDUCTOR DIAMETER IS 3 TO 4 TIMES LARGER THAN SMALL DIAMETER; OTHERWISE USE ROUND–TO–FLAT LAP–JOINT FORMULA.

[†] IN CASES WHERE LOOSENING OR BREAKING OF THE JOINT WOULD RESULT IN A HAZARDOUS CONDITION, MECHANICAL SECURITY SHOULD BE SPECIFIED.

strengths. The contact area also depends on the method used but is particularly important for methods that create interlocking strictly—or predominantly—at the microscopic as opposed to macroscopic level. That is, contact area is important to the degree that friction is the predominant force holding the joint elements together, with greater contact area giving rise to a greater friction force, all else being equal. Methods that produce pressure connections using cold (versus hot) flow of the metal result in greater contact forces due to the creation of residual stresses that help keep the connection tight, if done properly. Of course, once cold-formed, pressure connections cannot be allowed to become too hot from electrical heating (say, from high currents or especially high contact resistance from poor fit or from contamination) or the residual stress in the joint is relieved and the pressure creating contact is relaxed.

Mechanical pressure contacts can be produced in a number of ways without employing mechanical fasteners (see Section 12.2.4). Such contacts can reasonably be considered integral mechanical attachments, as they rely on only mechanical force arising from the physical interference between geometric features designed into or processed into the joint components. There are, in fact, methods that are reasonably considered rigid interlocks, ones that are absolutely elastic interlocks, and ones that are absolutely plastic interlocks. Basic methods identified by the author include the following:

- Wrap-around electrical connections
- Spring (electrical) connections
- Crimped (electrical) connections

12.4.1 Wrap-Around Electrical Connections

Wrap-around electrical connections, also known as "solderless wrapped connections," produce rigid interlocks between solid wires and terminal lugs. The method was developed to a high level in the 1950's to produce potentially gastight joints, if needed. The pressure connection is formed by wrapping about six turns of the wire element tightly around a terminal lug, for instance. The special tool used is a hand-held gun that has a rotating spindle that is powered by compressed air or an electric motor. The spindle has an axial opening that receives the terminal lug and a second opening radially separated from the axial opening into which the stripped end of the conductor wire is inserted. Rotation of the spindle causes the stripped wire to wrap around the lug in a tight helix, creating a firm metal-to-metal mechanical joint. Figure 12.4 shows the tool (right) and process and finished connection (left).

By maintaining a pulling force (i.e., tension) on the wire as it is being wrapped, it is possible to obtain very tight wraps with high contact pressures against the lug. Contact pressure in the finished assembly can be a remarkable 15,000 psi (approximately 100 MPa) minimum, and, under proper use, that pressure is sustained over the 24 contact areas produced when a rectangular terminal lug is wrapped with six full turns.

Data on the wrap-around electrical connection is shown in Table 12.1 under the heading "solderless wrapped."

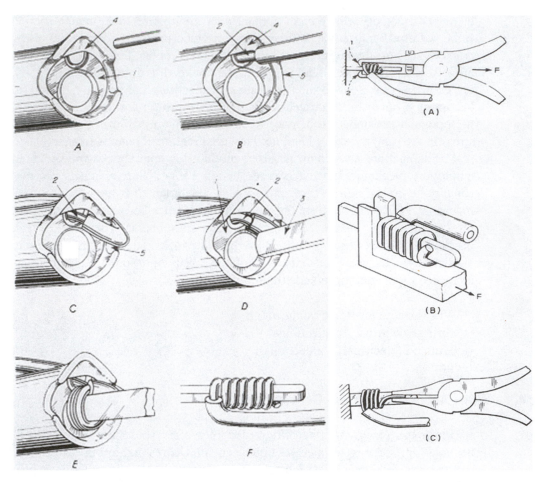

FIGURE 12.4 Schematic illustrations showing the tool and process steps used for making wrap-around electrical connections. On the left, Position A shows the tool tip; B the base wire 2, inserted into the lead slot 4; C the anchoring of the wire by bending it into the notch 5; D the terminal insertion; and E the wrapping of the wire 2 by rotating the spindle 1, around the terminal 3. Position F is the finished connection. On the right, the wire may be removed from its terminal by two methods: the most convenient method is stripping. Two types of tools may be used: the specially formed jaws of a pair of pliers (A), or, (B), a formed block. Another method (C) is by unwinding the helix by using a pair of pliers. (From page 413 of the *Handbook of Fastening and Joining* by Laughner and Hargan, which is long out of print and for which no copyright seemingly exists.)

12.4.2 Spring (Electrical) Connections

Spring (electrical) connections create elastic interlocks. Small schematic illustrations and data are provided on two obvious and one not-so-obvious example in Table 12.1. The "Fahnestock clip" and "plug connector" (or simply "plug") clearly rely on the spring action of the clip when a wire is inserted through the opening created by the crossing right-side-up and inverted V's and of the thin sheet-metal conical collar when a wire is inserted into the opening in the plug.

In each, the elastic recovery or "spring-back" of the deflected features of the connectors provides a force that grips the wire at its surface and holds it in place by friction. Less obvious is the elastic interlock by the plastic cap of the "wire nut" on the twisted wires inserted into it. After the wires are twisted together— using a clockwise twisting motion—an internally-threaded metal insert in the plastic cap tends to further tighten the twist, while pulling the wires tightly into the cap and developing a squeezing force that holds everything together.

There are other spring-type connectors, some of which are actually terminal blocks that employ the same basic concept as the clip and the plug.

12.4.3 Crimped (Electrical) Connections

The method of crimping to create plastic interlocks was described in Chapter 5, Section 5.6. There are a great variety of electrical connectors that rely on crimping to create a sound joint. Some use crimping to hold the connector onto a solid or multi-strand wire so that the connector can be fastened to some apparatus using a screw. Others use crimping to splice two wires together using various soft-metal collars.

Figure 12.5 schematically illustrates a variety of such *crimped electrical connectors.*

12.5 MECHANICAL ATTACHMENTS IN MICROELECTRONICS

The predominant process for producing electrical connections in microelectronics is soldering. Many different specific process embodiments are employed from fully automated processes like wave soldering that produce interconnections en masse to manual soldering with a tiny soldering iron for touch-up repairs. Some more recent methods begin to border on integral attachment in that features designed and pre-processed into (actually, onto) each of the components being joined are used to accomplish the interconnections. Most of these methods still, ultimately, involve soldering to actually produce a joint. Stenciled or otherwise pre-placed solder at points to be joined are simply reflowed by subjecting mechanically assembled components on boards, for example, to heating in ovens, using the heat of condensation of vapors of organic solvents (i.e., vapor condensation soldering), or other means.

In one fairly recent method known as "flip chip," the soldering step almost becomes incidental, as adhesive bonding is sometimes used to secure the components and provide impressive mechanical strength.[9] Flip chip

---

[9] Flip chip microelectronic assembly has been used successfully to produce electronics used on rockets, and, in 2004/2005, Delphi Delco placed 300,000 flip chip die per day into automotive electronics, much of which must survive the relatively harsh environment under the hood in the engine compartment. Most electronic watches and a growing percentage of mobile cellular phones and pagers are also assembled using flip chip technology.

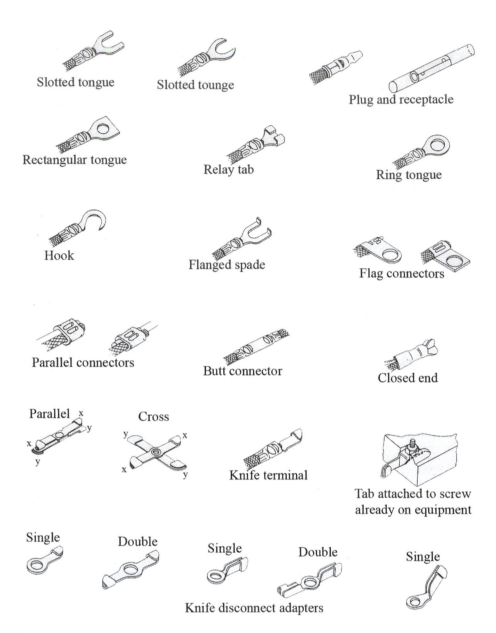

FIGURE 12.5 A schematic illustration showing a variety of different crimped electrical connectors. (From Supplement 72, page 569 of the *Handbook of Fastening and Joining* by Laughner and Hargan, which is long out of print and for which no copyright seemingly exists.)

microelectronic assembly is the direct electrical connection of face-down (hence, "flipped") electronic components onto substrates, printed circuit boards, or various carriers, by means of conductive bumps on the chip bond pads (Lau, 1995). The making of flip chip assemblies involves three steps: (1) bumping the die or wafer, (2) attaching the bumped die to the substrate or board, and, in most cases, (3) filling the remaining gap under the die with an electrically non-conductive material, including adhesives. Flip chip is also called "Direct Chip Attach" (DCA), a more descriptive term, since the chip is directly attached to the substrate, board, or carrier by the conductive bumps.

Figure 12.6 schematically illustrates a flip chip prior to, during, and after assembly. Figure 12.7 shows an example of some actual flip chip bumps and flip chip assemblies.

The bumps used in flip chip assembly can be produced by a number of different methods including sputtering, plating, or jet-spraying solder or other conductive metals (e.g., gold or copper) or even conductive adhesives, for that matter. While bumps made from solder are almost always reflowed to create an electrical connection, pressure alone could be used with bumps made from either solder or other metals. If only pressure is used, whether obtained by some mechanical means or from an adhesive when it cures and shrinks, flip chip assembly could easily be considered integral mechanical attachment.

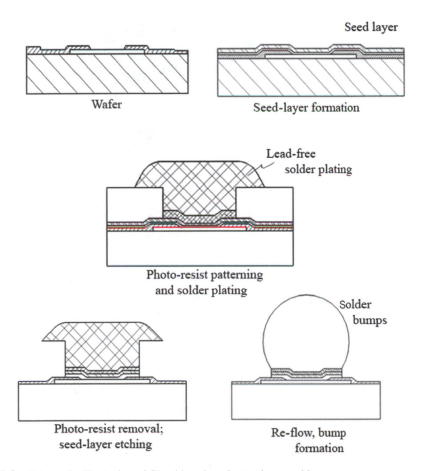

FIGURE 12.6   Schematic illustration of flip chip microelectronic assembly.

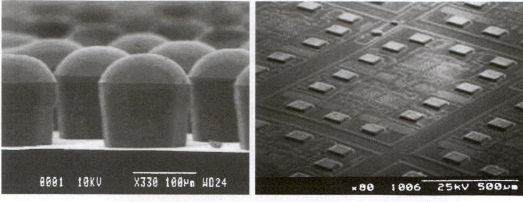

FIGURE 12.7 Photographs showing examples of the use of flip chip assembly in modern microelectronics (all as seen on www.flipchips.com). At the top left, a high-magnification view of copper bumps (courtesy of TMCI Corporation). At the top right, NiAu pads on a device (courtesy of PacTech GmbH, Nauen, Germany). At the bottom, a three-dimensional RAM with four 64 Meg chips on a folded flex substrate (rear) and a single 64 Meg packaged ram (foreground) (courtesy of Toshiba Corporation, Tokyo, Japan).

As the scale of features in microelectronics gets smaller and smaller (as it has since the beginning of the technology in accordance with what is known as "Moore's Law[10]), the utility of solder becomes more impractical, more difficult, and, eventually, technically impossible.[11]

---

[10] "Moore's Law" was named after Gordon Moore who, in 1965, made the observation that there had been an exponential growth in the number of transistors per integrated circuit over the 4 years since the first planar circuit was produced, and he predicted that this trend would continue. Through relentless technological advances, the trend has continued.

[11] The technological limit for soldering occurs when features to be soldered are so close together that molten solder at each location pulls together from individual bumps, or the molten solder simply doesn't separate to individual sites in methods relying on capillary flow, due to surface tension forces. That limit has been set at near 0.3 μ.

Thus, the likelihood of mechanical pressure contacts playing a greater role seems high. As will be seen in Chapter 13, Section 13.3, self-assembly using designed-in keying and/or interlocking features is already occurring in micro-electro-mechanical systems (MEMS) technology.

12.6 HIGH-VOLTAGE, HIGH-CURRENT, OR HIGH-POWER CONNECTIONS

Voltage, also known as "potential," is the force with which electrons are pushed to produce a flow of current; the higher the voltage, the greater the push. If the resistance of the path for current flow is low, the current will be higher as the voltage becomes higher. The combination of high voltage and high current gives rise to high power. High voltage is used to help move electricity from the site of its generation to the sites where it is to be used. High voltage (albeit at relatively low currents, of say hundreds of milliamperes) is also needed to operate some electrical or electronic devices, such as electron microscopes (e.g., 50,000 V with 100–200 milliamperes). Some devices require very high current, even though the voltage may be relatively low, such as resistance welding machines (typically 10,000–100,000 amperes at 10–20 volts). And some devices require very high power, which is the product of the voltage and the current. One example is arc melting furnaces used in primary metal refining and vacuum re-melting.

The risk with high voltage is that extra insulation is required to prevent the electricity from going where it shouldn't go (i.e., outside the conductor). The risk with high current and high power is usually the heating that can arise in the conductor from I^2R losses.

Beyond the extra care needed to preserve as much as possible of the insulation provided on conductors intended for high-voltage applications *or* the restoration of any insulation that had to be removed (i.e., stripped) to allow electric connections to be made, there is little different about making electrical connections for high voltage than for low voltage (i.e., typically, 110, 220, 440 V). As for making electrical connections for high current or high power, what needs to be different is that all conducting elements need to be more substantial (i.e., massive) and contact resistances need to be kept as low as possible.

A wide variety of specially-designed splice connectors and terminating connectors are available for high-voltage, high-current, or high-power applications.

Figure 12.8 schematically illustrates some compression (pressure) connectors used with line power from utilities, while Figure 12.9 shows a typical molded splice for use on high-voltage power lines.

12.7 SUMMARY

Producing connections that allow the efficient passage of electric current, whether at low or high voltages, is extremely important in a world dependent

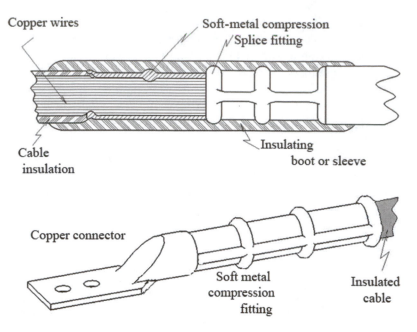

FIGURE 12.8 Schematic illustrations of some compression (pressure) connections used with common house-power electric lines. (Adapted from Figures 11-11 and 11-8, pages 11-9 and 11-8, in *Standard Handbook of Fastening & Joining,* 2nd edition, by Robert O. Parmley, McGraw-Hill, New York, NY, 1989; used with permission.)

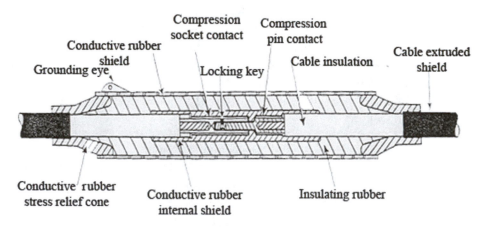

FIGURE 12.9 A schematic illustration of a molded splice used with high-voltage power lines. (Taken from Figures 11-40, pages 11–21, in *Standard Handbook of Fastening & Joining,* 2nd edition, by Robert O. Parmley, McGraw-Hill, New York, NY, 1989; used with permission.)

on electricity for light, heat, air conditioning, transportation, communication, entertainment, and computation, to name only a few.

When it is deemed necessary, for some small fraction of joints between electrical conductors, the electrical connection must be a permanent one, welding is the process of choice. The most common specific processes are thermit welding (for producing joints between substantial wires and buss bars or terminal lugs that are copper to copper, or aluminum, or steel, or aluminum to aluminum), percussion or capacitor-discharge resistance welding for smaller-diameter wire connections to plates, or ultrasonic welding for very fine wires to conductive pads on chips.

Adhesive bonding is only an option if the adhesive is made conductive by the addition of fine-particle silver or copper to the polymer adhesive resin or if the polymer used in the adhesive is inherently conductive. This is rare but possible, although the conductivity of the best intrinsically conductive polymers is orders of magnitude less than that of most metals. Adhesives are used, however, to provide structural support to electrical connections made by other means, including mechanically and, particularly, by what is known as flip chip assembly in microelectronics.

In sheer number, more connections, by far, are made by soldering than by any other joining process. Metal solder alloys provide excellent electrical conductivity and very good thermal conductivity to allow for the dissipation of unwanted, but inevitable, I^2R heat losses. Soldered joints can also provide hermeticity, if needed. What solders have difficulty providing is good mechanical strength, integrity, and durability under vibration or thermally-induced stresses from any CTE mismatch in the system. To help improve the strength of soldered electrical connections, some mechanical interlocking is used, as it was, without being obvious, when through-hole soldering technology predominated over current surface-mount technology. Mechanical interlocking using formed-in features that act rigidly is used when mechanical security is needed in an assembled joint before soldering and/or when structural integrity is needed in the soldered joint. Not surprisingly, when soldering is used for purely mechanical applications, such as in plumbing, mechanical interlocks tend to always be present in some form.

Purely mechanical methods of producing electrical connections not only exist but are widely used. Such methods offer convenience outdoors and in difficult-to-access areas. While mechanical fasteners are used, including screws (often with washers) in terminal connections and in some split-cylinder splicing collars, integral mechanical attachment is also used. In fact, all three methods of producing integral mechanical attachments are employed: rigid interlocks, elastic interlocks, and plastic interlocks. Respective examples include the categories of wrap-around connections, spring connections, and crimped or compression connections.

Mechanical methods relying on designed- and formed-in features even find use in microelectronics in flip chip assemblies and in some pressure contacts.

Finally, mechanical connections tend to prevail in high-voltage, high-current, and high-power electronics, where insulation becomes paramount and ease of use can be a matter of life and death.

Table 12.3 summarizes mechanical electrical connections, along with other methods of producing electrical connections.

TABLE 12.3 Summary of Mechanical and Other Methods of Producing Electrical Connections

Mechanical Pressure Electrical Connections

- Crimped (electrical) connections
- Spring (electrical) connections (including plugs)
- Wrap-around electrical connections/solderless wrapped connections

Microelectronic Interconnection

- Flip chip connections
- Soldered connections
- Wafer die bonding

Welding

- Percussion (resistance) welding/capacitor-discharge welding
- Thermit welding

Soldering

- Surface-mount technology
- Through-hole soldering (offers some mechanical support)

Adhesive Bonding

- Intrinsically-conductive polymer adhesives
- Metal-filled polymer adhesives

REFERENCES

Lau, J.H., editor, *Flip Chip Technologies,* McGraw-Hill, New York, NY, 1995.

Manko, H.H., "How to design the soldered electrical connection," *Production Engineering,* June 1961, 57.

Parmley, R.O., *Standard Handbook of Fastening & Joining,* 2nd edition, McGraw-Hill, New York, NY, 1989.

BIBLIOGRAPHY

Manko, H.H., *Solders and Soldering,* 4th edition, McGraw-Hill, New York, NY, 2001.

13 THE FUTURE OF INTEGRAL MECHANICAL ATTACHMENT: WHERE FROM HERE?

13.1 WHAT HAS CHANGED? WHAT IS CHANGING?

We've come a long way as a species, at least technologically. We can warm ourselves or cool ourselves at will. We can cross states, countries, continents, and oceans on land, on or beneath the seas, or through the skies. We can have light at night. We can talk to our friends from anywhere on the planet, anytime we wish. We can access more information than exists in any library faster than at any time in history, from anywhere. We can live into our 80's, hike or ski into our 70's, and routinely cure diseases that once kept most people from living into their 40's. Some everyday people have even traveled into space and returned safely, and many more seem so anxious to go that a multi-billionaire from England is taking reservations for a trip on a reusable spacecraft he's bankrolling with a private company.

Technology has advanced tremendously, so what we have learned in the last century exceeded all that we had learned since we first walked erect as a species. And, the rate of continued technological advancement is almost startling. It makes one wonder what could possibly change. Where could we possibly go from here?

Without a crystal ball, of course, no one knows. But, this is certain: we are a curious and capable species that seems predestined to go further, climb higher, dive deeper, move faster, and live longer and better. Can it be? Can we do it all?

It seems we already are! The Chinese are 10 years into a 20-year, $29 billion project to build the largest hydroelectric dam in the world, ever. It will stretch nearly a mile across and tower 575 feet above the third largest river on Earth, the Yangtze. The reservoir of the Three Gorges Dam will stretch over 350 miles upstream and force the displacement of 1.9 million people. But, the power it will generate will help energize the fastest growing economy on the planet. Rock, concrete, and steel are being moved, mixed, poured, erected, and joined to create an unprecedented structure. The amount of joining being done boggles the mind.

Figure 13.1 (left) reproduces a satellite photograph that shows that this megalith will join the Great Wall as one of only a few visible proofs from space of the existence of intelligent life on the Earth. Figure 13.1 (right) shows the Three Gorges dam complex from a high-altitude aerial view. Figure 13.2 shows a model of the planned dam and lock system, while Figures 13.3 and 13.4 show progress on the actual dam and lock, respectively, as of mid-2004.

FIGURE 13.1 A satellite photograph (left) that indicates that the Three Gorges Dam will join the Great Wall of China as one of only a few edifices made by human beings that can be seen from outer space. (Photograph courtesy of the National Aeronautics & Space Administration from its www.visibleearth.nasa.com website.) The scale of the complex can be obtained from the high-altitude aerial photograph (right) taken in late 2004. (From www.esamultimedia.esa.int/images/EarthObservation; for which no contact could be made, despite several attempts, for permission to use what might be a copyrighted image.)

FIGURE 13.2 A photograph of a model of the Three Gorges Dam complex shows the ship-lift locks in the foreground and the dam in the background. (Photograph obtained from the website www.northern.wvnet.com.edu, for which no contact for permission to use the image could be made, but with whom copyright may exist.)

FIGURE 13.3 Photograph showing that progress on a major portion of the dam at Three Gorges as of late 2004 had progressed to the point that the spillways were operational. (Photograph obtained from the website www.yangtzeriver.org/three_gorges_dam, for which no contact for permission to use the image could be made, but with whom copyright may exist.)

FIGURE 13.4 Photograph showing progress on the locks at the Three Gorges Dam as of the end of 2004. Note the suspension bridge that was also built above the locks and dam, which is to the right of the locks in this view. (Photograph by Karl Ernest Roehl from www.weter-roehl.de; used with his kind permission.)

Plans are also well along for the construction of the longest and tallest bridge ever built.[1] The Gibraltar Bridge would span 9 miles over the Straits of Gibraltar at the entryway to the Mediterranean Sea. It would connect cultures of Christianity and Islam and potentially increase ties between the economies of Europe and Africa.

Table 13.1 gives some statistics, while Figure 13.5 shows a satellite view of the location for this huge bridge. An artist's concept for the imagined bridge is shown in Figure 13.6. The imagined bridge, if superimposed over New York City, would span from New Jersey on the west to Long Island on the east, dwarfing existing bridges like the George Washington Bridge to New Jersey and the Triborough Bridge to Long Island. A sense of the massiveness of the bridge can be obtained from the artist's rendering shown in Figure 13.7.

---

[1] Renowned bridge designer T.Y. Lin, professor emeritus at the University of California at Berkeley and chairman of Lin Tung-Yen China, Inc., has a design that contains an architectural breakthrough that would enable the support needed for the massive spans his design has derived (http://idol.union.edu/~ferrerf/project/text.htm).

Location: Strait of Gibraltar; linking Spain and Morocco

Length: 9 miles, two spans of 4.5 miles each.

Height: Each tower is 3,000 feet tall; twice as high as the tallest skyscraper.

Width: 5 traffic lanes, 2 breakdown lanes in each direction.

Road Deck Material: Fiberglass-epoxy composite.

Length of Wire Cables: 1,000,000 miles.
  Enough to circle the Earth 30 times!

Closest Existing Relative: Akashi Bridge in Japan; currently the world's longest at 12,828 feet.

Estimated Cost: $15 billion.

Dangers: Wind speeds of 80 mph at tops of towers; ship collisions, strong ocean currents, traffic, Sahara
  Desert dust storms.

Concrete, steel, and fiberglass composite for the decking will be used in huge quantities, and joining will be a key to success.

Airbus of Europe and Boeing of the United States are each well along with designs for commercial airliners that will carry 600–1,000 people anywhere on Earth. A prototype of the Airbus A380 is shown in flight in Figure 13.8.

FIGURE 13.5   Satellite photograph showing the proposed location of the Gibraltar Bridge (as a bright drawn line), spanning the Strait of Gibraltar at the entrance of the Mediterranean Sea from the Atlantic Ocean. (Photograph from www.acrobat-service.com using a photograph taken by NASA; no permission needed.)

FIGURE 13.6 An artist's rendering of the planned Gibraltar Bridge showing the combined use of classic suspension and modern cable-stayed sections. (Image obtained from www.radio-bridge.net/www/links/STRAITS.html; for which contact for permission to use the image could not be made, but for which copyright of the image may exist.)

Design studies are still underway for hypersonic[2] aircraft-spacecraft hybrids that could fly anywhere in the world in less than 2 hours, whether for military or business purposes. Business-persons could have breakfast at home in New York, have a lunch meeting in Tokyo, and be home for dinner that evening! Two concepts being studied, "HyperSoar" and the "Aurora Project," are depicted in Figure 13.9.

FIGURE 13.7 A realistic artist's rendering of the proposed Gibraltar Bridge showing its massive size compared to ships passing beneath it. (Image from www.idol.union.edu/ferrerf/project; for which contact could not be made for permission to use the image, which may be copyrighted.)

---

[2] "Hypersonic" vehicles travel at speeds well exceeding the speed of sound (about 700 mph or 1,100 kph), which is designated Mach 1 or M1. Planned transports would travel over Mach 5 up to Mach 10.

FIGURE 13.8 A prototype of Airbus S.A.S.'s A380 super-jumbo jet capable of carrying nearly 600 passengers is seen in flight. (Photographs courtesy of Airbus S.A.S., Blagnac Cedex, France; used with permission.)

The Japanese and French have each independently built and are operating trains capable of traveling at over 200 mph (320 kph), with potential for even higher speeds. Figure 13.10 shows these trains.

The quest for more oil to feed the hungry engines of such a highly-mobile world is pushing oil exploration and recovery more and more and deeper and deeper into the seas using monstrous structures that rival cities on stilts (Figure 13.11). Such huge structures are currently built on-shore, towed to location, and erected on-site. Joining is a key at both sites.

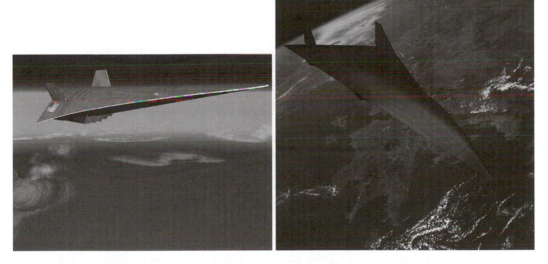

FIGURE 13.9 Concept drawings for two different hypersonic aircraft-spacecraft hybrids capable of flying anywhere in the world in less than 2 hours. HyperSoar (left) is intended for commercial air travel, while the secret "Aurora" (right) is intended as a next-generation spy-plane to replace the SR71 Blackbird used for more than two decades. (Photographs from www.beachbrowser.com and www.abovetopsecret.com, respectively; for which no contact could be made for permission to use images that might be copyrighted.)

FIGURE 13.10 The Japanese bullet trains and French TGV's are capable of traveling at over 200 mph (320 kph). The French TGV Atlantique' is shown speeding along a track in France (left), while the 300 (to the left) and 700 Series Shinkansen is shown at Tokyo Station (right). (Photographs from the website http://en.wikipedia.org; used without need for permission per website.)

And, as a final example, there are further plans for building great orbiting laboratories-"factories"-living quarters in Earth orbit, such as the several-year-old Orbiting International Space Station (Figure 13.12), which is still a work in progress. Such structures have to be pre-fabricated on Earth and assembled in space from modules delivered by the Space Shuttle and other rockets launched from Europe, Russia, and elsewhere. To facilitate assembly, modules are designed to literally snap together once properly engaged.

Longer, higher, bigger, faster, further out in space, deeper under the seas, we're already headed there! But, that's not all. We're also moving to a new world at the nano-scale, where there will be a need for structures to self-assemble from extremely small, uniquely shaped mechanical parts or from uniquely shaped molecules. Figure 13.13 shows how self-assembly is used at several levels to create a nano-scale high-coercive force magnetic material and device.

The world of the future—even our future, not only our children's and grandchildren's future—may seem driven by information technology, microelectronics and photonics, nanotechnology, and biotechnology, as it surely is, but it is also more dependent than ever on new technologies to enable the construction

FIGURE 13.11 A typical offshore drilling platform, built on land and towed to location and erected in-place at sea. (Photograph courtesy of Marathon-Ashland Petroleum LLC, Findlay, OH; used with permission.)

of mega-scale structures too. This inevitable expansion of our world to both larger and smaller scales will pose new challenges to materials and, perhaps particularly, to new processes and processing. Surely, as it has before, joining will have to be ready again to meet these new challenges.

13.2 JOINING METHODS MUST ADVANCE AS MATERIALS AND STRUCTURES ADVANCE

Joining has been as responsible as any process (and more responsible than most) for getting us where we are as a species (Messler, 2003). From the dawn of humankind, the ability to join materials into parts and devices, and parts and devices into assemblies and structures, has enabled, if not driven, the advancement of civilizations and the growth of economies. The Stone Age, the Bronze Age, and the Iron Age were more about the ability to work and join these materials than they were about the materials themselves.

As surely as joining has brought us to where we are, it will bring us to where we want to go. Information technology (IT), nanotechnology (NT), microelectronics and photonics, and biotechnology (BT) will surely dominate our immediate, if not long-term, future. Bill Joy, co-founder of Sun Microsystems and a computer scientist who, as much as anyone, revolutionized computer programming to the point that some programs are capable of generating other programs,

FIGURE 13.12 The Orbiting International Space Station must be assembled from sub-assemblies or modules ferried from Earth. To facilitate ease of assembly, sub-assemblies or modules literally snap together once they are properly engaged. (Image from www.space-technology.com; for which no contact could be made for permission to use the image.)

wrote in his thought-provoking article "Why the Future Doesn't Need Us" (Joy, 1999) that IT, NT, and BT, as well as robotics, will dominate and shape our future, with the near-equal potential for either unprecedented progress or the annihilation of our species. And, let there be no doubt, joining will not only be at the heart of each of these technologies, but it will remain the backbone of our infrastructure, our transportation, and our manufacturing-, agricultural-, and mining-based economies.[3]

In 2003, Intel produced its Itanium® 2 processor containing 410,000,000 transistors (slightly higher than predicted by Moore's Law). The industry has nearly reached the lower size limit of $0.03\,\mu$ predicted for solder-based joining. MEMS devices are already being built that employ what is known as "self-assembly" (see Section 13.5) and nano-scale versions are being talked about. Nanotechnology is announcing new discoveries, producing amazing new structures, and making even more amazing predictions for the future. Molecular self-

---

[3] Historically, and many would argue that it is still true, that manufacturing, agriculture, and mining are the only things that add to the true wealth of a society, as each takes things of little intrinsic value and creates much greater real value. Think of an acorn and an oak tree, wood and a house, or iron ore and steel. Service, many would say, simply moves wealth from one place to another but creates nothing.

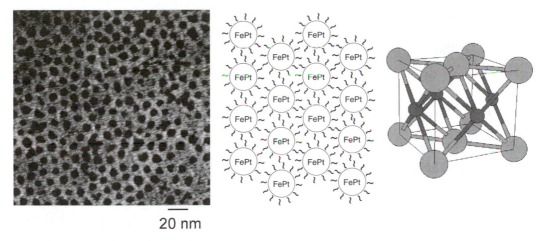

20 nm

FIGURE 13.13 Self-assembly and self-organization of nano-structures with disparate dimensions across different length scales. The micrograph (left) shows an array of Sb-doped FePt nano-particles assembled to form a thin film with a nominal height equal to the average particle size. The second level of self-organization is exemplified by the self-assembled layer of surfactant molecules (see center schematic) on the surface of each particle serve as spacers and inhibit the agglomeration of particles. Each particle has a chemically ordered face centered tetragonal LI_0 phase where all the Pt atoms occupy the face center positions of the (002) plane (right schematic taken from http://cst-www.nrl.navy.mil), constituting the third level of self-organization—within the unit cell. Forming the LI_0 phase requires annealing of about 300°C through Sb doping, which effectively increases the mobility of the Fe and Pt atoms at lower temperatures. The ordered nano-particles have high magnetic coercivity (e.g., ~500 mT), and may open up new possibilities for nano-magnetic information storage media. (Courtesy of Alex Yan and G. Ramanath; also see Q.Y. Yan, T. Kim, A. Purkayastha, P.G. Ganesan, M. Shima, G. Ramanath, *Advanced Materials* 17, 2005, 2233; used with permission.)

assembly has become a way of life in the NT community. And, as final examples, with the human genome nearly fully mapped, genetic engineering and the tremendous promises of tissue engineering are making new demands on joining the most precious structures of all—that is, those of life.

But, even as these marvels of modern engineering begin to materialize, we are faced with innumerable day-to-day problems. Highways and bridges need repair and replacement across the United States and elsewhere, and new highways are being built everyday where none existed before. The world still needs to deal with flood control, provision of water for irrigation and potable uses, and power generation, so new dams of less than world class are needed. It is a stark reality that no great concrete dam on the planet is over 100 years old, so we simply don't know what we may be facing in terms of repair or replacement of what we have already built more than half a century ago. In many places, railroads have fallen into disrepair even as fuel costs and highway congestion (and the associated pollution) are making trucking intolerably expensive economically, socially, and, perhaps, environmentally. Airline fleets have aged past the design life of many of their aircraft. It seems more than time to build replacements, but fuel costs again place new demands on greater efficiency.

And, finally, more people in more places need a place to live, to call their home, and no more so than in the already densely-populated cities of the world.

So, as in the past, we must look again to joining to solve problems and realize promises.

13.3 WHAT ARE THE LIKELY CHANGES TO BE EXPECTED OR THAT NEED TO OCCUR?

Several things are likely to change if joining is to keep pace with the advances in materials and structures described in Section 13.2, as examples. Other things will almost surely not change, and that will be okay!

Without going into great detail here, as there are other references that provide such detail (Messler, 2000a,b, 2002, 2004), the following are six changes that either can be expected or must occur in joining:

1. *Joining will continue to shift from a "secondary" to a "primary" process.* In the past, joining was often an afterthought, even though the need was or should have been obvious. Instead of having joining be a secondary process, performed as a last, discrete step in the material synthesis/material processing/detail part fabrication/assembly/inspection sequence common to most manufacturing or construction of the past, it will, increasingly, become a primary process that occurs at the same time as one or more of these other steps. If nothing else, the means for accomplishing joining must be considered in the design of details and included as part of their manufacture.

2. *Past successes must not be forgotten but must be extended.* We have to learn to look at what has worked somewhere else before and push it to new scales of size and/or performance, up or down. We have to learn to go with winners, but without hindering innovation. This is the basis of reverse engineering (Otto and Wood, 2001).

3. *Joining processes will be automated or will take place automatically.* Social pressures for higher education, the nature of humans to want what we want quickly, and the well-meaning desire to allow every person to use his or her brain more than his or her brawn, seem to be whittling away at the pool of those in the skilled crafts, particularly in the more affluent and technologically advanced countries of the world. This and the desire to maximize productivity and quality, simultaneously, are pushing for more and higher degrees of automation. This trend has not missed joining. Ways will have to be found to automate old processes for joining, and new processes will either have to be designed for automation or will have to be capable of taking place automatically. An example of the latter approach is "self-assembly" to be discussed later in this chapter (see Section 13.5).

4. *Quality assurance must be embedded and what needs to be of high quality needs to be monitored.* If it's not already happening for a process, assurance of joints of suitable quality will have to be embedded in the process, and not be expected to occur after the fact. To do this will require in situ sensing, continuous monitoring, and real-time adaptive control. Furthermore, we will have to learn to monitor what needs to be of high quality and not simply what is convenient to monitor.[4]

5. *High costs must be accepted for high value.* Some things just take time and are inherently costly to achieve. The desire to contain costs doesn't mean to do things cheaply. We have to be willing to pay a penalty in labor intensity, required skill level, or both for applications where the payoff will be truly high. Joining composite materials may never be inexpensive if it is to be done properly, and not the way it's done now (see Section 13.9).

6. *Thinking must broaden.* While we should look at past successes to see how those might help with new problems (see no. 1 above), we must also learn to look in places and directions where we haven't looked before to find answers to problems we haven't found where we've been looking. Materials have changed (and will likely continue to change) so radically that our methods for processing them—including joining—may have to change equally radically. We probably have to match the revolution that has been and is still occurring in materials (Easterling, 1990; Forester, 1988) with a comparable revolution in our approach to joining.

What is especially interesting, with a little reflection, is that with the many and profound changes that have taken place in materials, the changes that have taken place in joining are sometimes equally profound and other times are surprisingly absent and unnecessary. Examples of the former are that joining is accomplished at the lowest levels in microelectronics through the techniques of "self-forming" and "self-limiting" joint formation (Frederick and Ramanath, 2004; Suwwan de Felipe, 1998; Kirchner, 1996) and in some very advanced composites through what are known as "self-healing" materials (Chen et al, 2002). The example of the latter is not only the persistence over the millennia but the repeated historical coming to the fore and current ongoing resurgence of integral mechanical attachment. Such persistence and re-formation over time as new materials emerged (e.g., stone, bronze, iron, plastics, and, now, carbon nano-tubes) strongly suggests that something is fundamentally sound and economically robust about the use of integral features of physical things to allow and enable their joining.

---

[4] If in a weld what is meant by quality is obtaining the right microstructure free of defects, what should be monitored is the state of the microstructure, not the voltage, current, and travel speed at which welding was done—and hope for the best!

With the suggested changes for joining as guiding principles, and the fact that integral mechanical attachment started before, has persisted amidst and may be experiencing a resurgence among all other methods for joining, the remainder of this chapter looks at where the opportunities for integral mechanical attachment, in particular, could—or should—be headed.

13.4   MOVING TO LARGER SCALE APPLICATIONS OF INTEGRAL MECHANICAL ATTACHMENT

One of the major reasons for joining is to produce truly large structures, ones too large to produce anywhere except on site and/or ones for which the components of the structure exceed the size or shape-complexity limits of the primary processes used for producing them (Messler, 2004). Examples include large metal alloy reaction or storage vessels made from castings, forgings, and/or rolled plates, ships and submarines made from rolled plates, beams, and/or forgings, and bridges made from pre-cast, reinforced (often pre-stressed) concrete or steel or aluminum alloys. While the processes used to join the components of large structures are welding (for containment vessels, ships, and submarines), adhesive bonding (for concrete bridges), or mechanical fastening (for steel or aluminum alloy bridges), it is rare that there is not some provision for the rigid integral mechanical interlocking of joints. Joints are often shaped to allow at least some degree of keying to ensure fit fix alignment and help carry major service loads in certain directions. In fact, as structures get bigger and the parts to be joined get thicker, the only option becomes integral mechanical attachment.

It is also highly advantageous to fabricate details (especially very large ones) in a controlled environment, such as a plant or dedicated facility or site. With greater control over processes and procedures, closer tolerances and, hence, better fit and function can be achieved. Many modern shipyards fabricate both the hull and the superstructure in modules indoors, under close process control, and assemble these modules outdoors in drydock. Interestingly, such an approach is a necessity in certain shipyards because of the climate (e.g., the Bath Iron Works in Bath, ME) and has been proven to lead to far superior quality and productivity, so that other shipyards, even those located where climate would not be an issue, have adopted the approach. To facilitate the bringing together of such large modules, joints that self-key help greatly with fit-up.

More and more, the benefits of modular construction and the use of self-keying interlocking joints are being extended to new applications.

In Hong Kong, an extensive system of new tunnels were constructed around and under Victoria Harbour (one of the busiest in the world!) to carry vehicular traffic to ease congestion in this very densely-populated city. For the underwater sections, engineers elected to construct huge pre-cast reinforced-concrete rectangular cross-sectioned multi-channeled tubes large enough to carry six lanes of traffic and services for electricity, ventilation, and rescue. For production logistics, efficiency, economics, and process control, hundreds of these soccer field–sized tubular sections were fabricated in special lower-than-sea-level basins located along the shore of one of the large islands in the harbor. Once cast, the open ends of the huge tubes were capped, the basin was flooded with seawater, and the floating sections were towed to the proper location and sunk into position. Once in position on the bottom, huge hydraulic jacks forced the newest section tightly against the previously-set sections at rigid interlocking joints. The tight-fitting joints were subsequently sealed to be watertight. Structural loads, however, and easy fit-up were handled by designed-in rigid interlocks.

Similar interlocking, pre-cast, reinforced-concrete arch segments were used to construct the train-carrying tubes of the "chunnel" that links England and France under the English Channel.

Rigid integral mechanical attachment features are also used in modern building construction. Throughout Europe and Asia, many modern buildings for use as offices, hotels, or residential apartments are increasingly being erected on site from wall and floor-ceiling units (see Figures 13.14 and 13.15) or entire

FIGURE 13.14 Pre-fabricated reinforced concrete wall, floor, or ceiling panels used for the modular construction of buildings, as shown here. (Photograph courtesy of Christian Prilhofer Consulting and New Building Systems, Freilassing, Germany; used with permission.)

FIGURE 13.15   Here a double-wall unit is shown. (Photograph courtesy of Christian Prilhofer Consulting and New Building Systems, Freilassing, Germany; used with permission.)

modules that are pre-fabricated from reinforced concrete in highly automated factories (see Figure 13.16). The majority of these pre-fabricated units or modules interlock together using rigid integral attachment features. Once assembled, the mechanical interlocks may be augmented by bolting, but more to prevent movement and separation than to carry primary loads. This design approach dramatically speeds construction and provides a better structure at lower cost than conventional poured-concrete construction.

Similar modular construction is occurring in wood structures using pre-fabricated trusses and even entire wall, floor, and ceiling or roof modules (see Figure 3.17).

Earlier, in Chapter 10, Section 10.3, the use of integral rigid interlocking features for joining large pre-fabricated sections of reinforced, pre-stressed concrete sections for the erection of overhead roadways was described. Figure 13.18 shows an example.

FIGURE 13.16 Concrete modules for constructing buildings are fabricated in a modern, fully automated production facility to control quality to a high standard and to minimize cost. (Photograph courtesy of Christian Prilhofer Consulting and New Building Systems, Freilassing, Germany; used with permission.)

FIGURE 13.17 Pre-fabricated wood wall, ceiling/roof, and floor modules are used for shortening the time needed for on-site erection of new homes and buildings. Here, pre-fabricated floor/ceiling (left) and wall (right) units are being hoisted into place for installation. (Photograph courtesy of the Wood Truss Council of America; used with permission.)

FIGURE 13.18 Photograph showing how large, pre-fabricated sections of pre-cast, pre-stressed concrete are interlocked to create an overhead roadway. (Photograph courtesy of the Portland Cement Association, Skokie, IL; used with permission.)

In the United States and elsewhere, steel and aluminum-alloy space-frame designs are also finding increasing use in buildings because they speed construction on-site, reducing costs, and provide more open and aesthetically-pleasing architectural structures. The joints in such tubular-truss structures often use proprietary but standardized designs at junctions that utilize rigid integral mechanical interlocks. One system by Delta Structures, Inc. (Wood Dale, IL), uses forged ball-and-tube and other systems that allow easy assembly and high degrees of architectural design versatility. In some joints, tubular truss elements are simply locked in place at fittings by a strong friction-fit, while others may be bolted or adhesive bonded just to prevent separation, but primary loads are carried by the mechanical interlock.

Figure 13.19 shows two examples of the use of interlocked space frames in the construction, while Figure 3.20 shows some of the designs for the joining systems at the intersections of tubular members of the space frame.

The more that integral mechanical attachment is used for erecting large structures of metal or concrete or both, the more confidence grows within the design community for more bold designs. In the future, integral design features alone may be used to facilitate assembly, accomplish locking, and provide needed sealing of pre-fabricated details produced under controlled, largely automated conditions.

As interesting and impressive as the preceding examples are, the idea is far from new. The Egyptians sealed the inclined corridors that led to the sacred burial chamber of the pharaoh by sliding huge pre-shaped blocks of granite down the incline to lock in place by dropping into a shallow recess at the chamber end of the corridor. As the block slid down the corridor, it broke clay caps extending from the ceiling of the corridor, releasing sand from chambers above the corridor and forever sealing the corridor against intruders.

FIGURE 13.19 Various designs for joining tubular members, including systems that rely on only interlocking, are used in the construction of modern space frame for buildings and interiors. Here, the use of space-frame assembly is shown for an exterior, roof-support structure at the entrance to an airport terminal (left) and for an interior structure at a hair salon (right). (Photographs courtesy of Delta Structures, Inc., Wood Dale, IL; used with permission.)

13.5 MOVING TO SMALLER SCALE APPLICATIONS OF INTEGRAL MECHANICAL ATTACHMENT

There seems to be little doubt that the "hot" technologies are increasingly dependent on smaller and smaller scales in materials and structures. Microelectronics, which underlies and enables information technology,

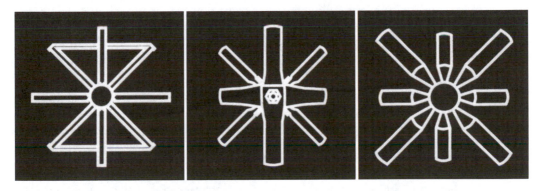

FIGURE 13.20 Various designs for the attachments of structural members of a space frame. (Designs courtesy of Delta Structures, Inc., Wood Dale, IL; used with permission.)

nanotechnology, and even biotechnology delve at the micro (10^{-6} m) scale, the nano (10^{-9} m) scale, and the molecular (10^{-10} m) scale. Not surprisingly, joining is still necessary, and obviously any joint formation must somehow form itself. "Self-assembly" and "self-joining" have become more than catch-phrases or goals; they've become realities.

Let's consider what is occurring:

Microelectronics. Shortly after the birth of microelectronics, Gordon Moore, in the early 1960's, recognized that the number of transistor devices[5] created on chips had increased exponentially. Furthermore, he predicted, such growth would continue until some fundamental technical limit was reached or when some processing obstacle blocked progress. Figure 13.21 shows in a plot that the growth in the number of transistors on chips has, indeed, followed what became known as "Moore's Law" for 40 years. Relentless pursuit of break-through processing methods for in situ doping, lithographic and beam-writing techniques, metallization and joining methods have, so far, kept chip-makers on track. But soon, serious, potentially insurmountable technological barriers are being approached, many say at a spacing of 0.1–0.3 m (about 10^{-7} m).

As described in Chapter 12, Section 12.3, one approach for producing electrical connections at the micron scale involves a variation of soldering that creates solder bumps at the sites where contacts are to be made, sometimes by reflowing the solder with heat and sometimes just by pressure. "Flip-chip," and also "ball-grid array," techniques for making solder interconnects overcome problems with solder bridging (and short-circuiting) between metallized pathways when conventional methods relying on capillary flow are used. Further densification of silicon-chip–based devices may indeed pose insurmountable problems, however. Achieving nano-scale features will almost certainly depend on the development of nano-scale joining techniques. One promising approach could be to employ conductive carbon nano-tubes or nano-wires in which connections are made through chemical activation by functionalizing the conductive carbon nano-tube to form joints where they are needed or wanted.

What is interesting is that functionalization of molecules is used by Nature to "self-assemble" tissue, for example. More will be said about this approach later, but, first, let's look at what else microelectronics gave rise to.

Micro-Electro-Mechanical Systems (MEMS). MEMS are integrated micron-level devices or systems combining electrical and mechanical components. They are fabricated using integrated circuit (IC) batch processing techniques

---

[5] Transistors in solid-state electronics are current amplifiers that operate by placing p-type and n-type extrinsic (doped) semiconductors together at p-n-p or n-p-n junctions. Diodes, which limit current flow to one direction only, are created when p-n or n-p junctions are created.

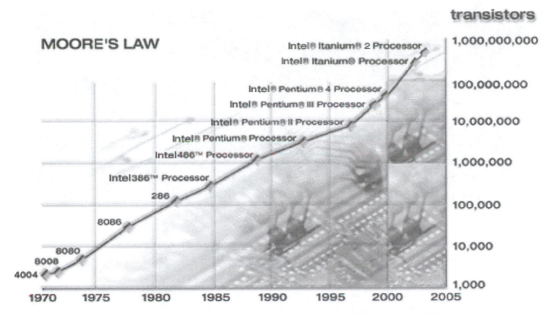

FIGURE 13.21 Plot showing that Moore's Law predicting the exponential increase in the number of transistors on a chip has been followed for over 40 years. (Plot from the website www.intel.com/research/silicon/mooreslaw.htm; for which contact for permission to use the image could not be made.)

adopted and adapted from microelectronics[6] and can range in size from microns to millimeters (10^{-6} to 10^{-3} m). These systems can sense, control, and actuate on the micro-scale and function individually or in arrays to generate effects on the macro-scale.

Figure 13.22 shows two examples of MEMS devices.

MEMS is an enabling technology, and current applications include accelerometers, pressure, chemical, and flow sensors, micro-optics, optical scanners, and fluid pumps, to name a few. By 2001, the industry reached $8 billion and has grown by 10–20% annually since. The potential of MEMS technology excites both the military and commercial sectors.

A key to past—and to future—success is what is known as "self-assembly." In fact, while the term always means what it says (i.e., that sub-millimeter-, micron- and sub-micron–level details assemble themselves into two-dimensional [2-D] or even three-dimensional [3-D] arrangements needed to produce a functioning system), there are two related approaches that differ in some, albeit important, details.

In the simpler-to-understand approach to self-assembly, sub-millimeter–size parts are fabricated and placed close to how they need to be arranged in a final assembly. Actual assembly takes place as each part is caused to self-align itself

---

[6] Device features can be created by etching away photosensitized resists, by depositing material, or both.

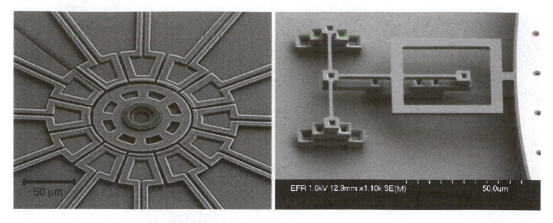

FIGURE 13.22 Scanning electron micrographs showing two examples of MEMS devices. The one on the left is a salient-pole micro-motor. The central portion rotates within the surrounding array of electromagnets. The one on the right is a unique hinged structure for mounting and tilting a mirror. (Photographs courtesy of MEMX at www.memx.com; used with permission.)

with other parts with which it must interact via fabricated-in geometric details. It is completely analogous to shaking uniquely-shaped and specially-keyed jig-saw puzzle pieces until they arrange themselves into the only assembly that is geometrically stable. This approach is limited to 2-D arrays. For the assembly of 3-D structures, other techniques relying on integral mechanical attachment have been proposed, fabricated, and demonstrated. One, shown in Figure 13.23, uses features integral to the design to allow a micron-scale box to be self-assembled much like "pop-up" greeting cards and children's books do.

Self-assembling Designer Molecules. Self-assembling designer molecules are, of course, the ultimate in integrated feature-based assembly, even though the mechanism is not entirely mechanical. The approach mimics Nature's most fundamental system of organizing cells to create living tissue. Biomolecules, which inevitably are complex, have geometric shapes that allow them to fit together with one another or other molecules in only one way. Their shapes are "keyed." Beyond this, most such mating molecules also have chemically active portions at these mating points that allow bonding only when the proper alignment is achieved between the proper molecules. It's Nature's way of preventing construction errors.

This new scientific approach designs macro-molecules that can be pro-grammed to encode information about the desired self-assembly behavior into their molecular architecture. Some molecules are cylindrical or arrays of cylinders, some look like a double sandwich, and some are continuous 3-D cubic structures. But all can self-assemble.

An example of self-assembly at the nano-scale involving both extremely small particles of long-range ordered Fe-Pt and special organic molecules is shown in Figure 13.13.

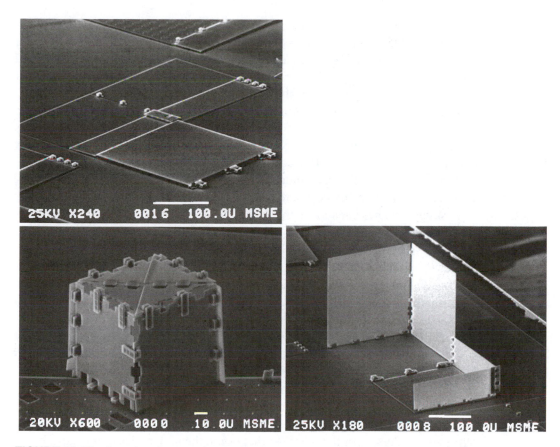

FIGURE 13.23 Scanning electron micrographs showing how a micron-scale box can be assembled much like a "pop-up" greeting card or children's book. Note the many integral features that allow interlocking once the box is folded up to its three-dimensional shape. (Photographs courtesy of Elliot Hui while a doctoral candidate under Professor Roger Howe at the University of California at Berkeley, Berkeley, CA; used with permission.)

While the body uses self-assembly for things like wound healing and tissue engineers are attempting to replicate the basic approach to permit the growth of new tissue to replace severely diseased or traumatized tissue, other scientists and engineers are researching the possibility of using certain complex molecules as the basic building-blocks for digital computers, to reduce the scale of devices to the nano-scale. Complex molecules that can, as a consequence of some unique structure they possess, have two distinctly different polar states—one oriented one way, the other the other way—could, conceivably, be used to store and process binary data bits the way present-day digital computers do. However, besides selecting or designing and creating the right molecule, methods have to be found for placing the molecules where they are needed and making secure electrical connections, as it were. The answer, while still a subject of research and development, is to use self-assembly (Yu, 2005).

Self-assembly for electrical interconnects in microelectronics, for production of MEMS devices, for the advancement of nanotechnology, and for biotechnology, including tissue engineering, will grow in sophistication and application over the next decades. Once again, the use of integral attachment features will be a key to successful joining.

Integral Micro-mechanical Interlocks. The trend in hook-and-loop fasteners or, more properly, elastic integral attachments (see Chapter 4, Section 4.7) has been toward smaller-scale features. As an example, some of 3M's (Minneapolis/St. Paul, MN) Dual Lock™ recloseable fastening systems (see Chapter 4, Figure 4.20) have interlocking features under 0.010 inch (0.25 mm). This may not be considered microscopic, but it's getting there.

A suggestion has been made that hook-and-loop elastic interlocks might provide impressive holding forces if features could be reduced in size, arranged in properly spaced and ordered 2-D arrays, and produced with proper feature shapes (Messler and Genc, 1998). Depending on the specific shape of the interlocking feature, the assembly force is proportional to some power of G, the shear modulus of the material from which the features are made, normally, thermoplastic polymers. The retention force, which is the force that resists separation of the interlocked features, is usually higher because elastic snap-fits, of which hook-and-loops are an example, are typically designed with inclined faces to allow easy insertion and flat or even re-entrant-angle faces on the opposite side of the catch and latch features to intentionally make separation difficult. The author has studied mushroom-shaped features for which the caps push past one another until they clear to elastically recover once the underside of one passes the underside of the other. At that point, the undersides of the caps, being either flat or having a slight re-entrant angle, prevent disengagement until the cap is either very severely deformed or fractures off. The steps of assembly are shown in Figure 13.24. By making individual interlocking features smaller, the load-carrying capability decreases approximately as the area of the cross-section of the feature decreases, but the number of features per unit area increases to offset that loss. Thus, very small, densely-packed elastic interlocks might yield surprisingly high retaining forces.

The current limit for reducing the size of elastic interlocking features in polymers seems to be one of how to process such features and arrays. The basic technology for creating tiny elastic hook-and-loop features, as well as small-scale cantilever hooks, comes from the textile industry. But, there is nothing to suggest a lower limit has been reached at approximately 0.010 inch (0.25 mm). Hence, there is almost certainly some downside potential.

New developments in the growth of carbon nano-tubes that protrude from surfaces like trees in a forest (see Figure 13.25) suggest that another possibility for nano-scale friction interlocks may be within the realm and reach of nano-technology. Imagine the gripping power of a densely packed 2-D array of up-standing nano-tubes, perhaps with roughened surfaces. Opposing arrays could be pressed together to interlock by surface friction just as the bristles on two

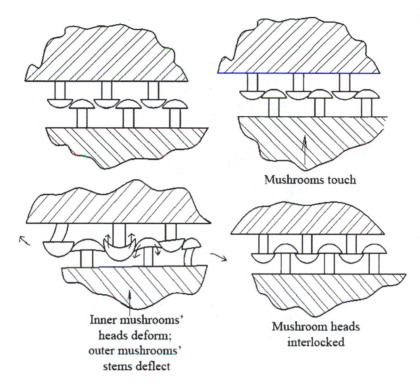

Mushrooms touch

Inner mushrooms'
heads deform;
outer mushrooms'
stems deflect

Mushroom heads
interlocked

FIGURE 13.24 Schematic illustration of the steps involved in assembling mushroom-shaped elastic interconnects.

hairbrushes do when forced together. If the surfaces of the nano-tubes are chemically activated in some way, the frictional gripping force might be greatly enhanced by chemical surface forces.

Is there value to such micro- or nano-scale hook-and-loop or other elastic interlocks? Only time will tell.

13.6 MOVING TO HIGHER PERFORMANCE INTEGRAL MECHANICAL ATTACHMENTS

It is not at all unusual that once a new technology has proven itself, it tends to be extended to higher performance applications. At least there's an attempt. This has already begun to occur for integral mechanical attachments in their latest resurgence. After all, many ancient peoples used the technique as the principal means for erecting great stone or wood fortresses, bridges, ships, and cathedrals or temples. In such structures, no other method of joining was known that could carry the huge loads. But, within the last decade, rigid, and to a slightly lesser extent, elastic integral mechanical interlocks, have appeared in

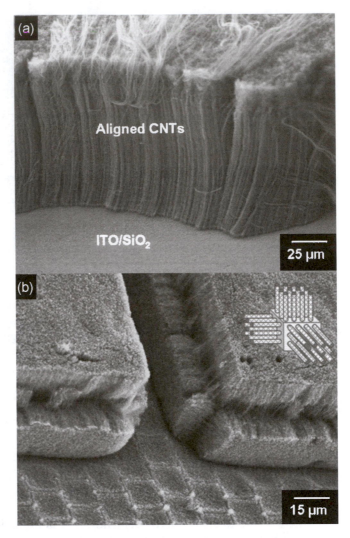

FIGURE 13.25 Micrographs depicting the self-organized growth of carbon nano-tubes in highly-oriented configurations on Si-based substrates coated with indium tin oxide—an optically transparent electrical conductor. Figure (a) shows vertically oriented bundles of aligned carbon nano-tubes, while (b) illustrates multi-directionally–oriented bundles through inheritance of the underlying topographical pattern on the substrate. Such oriented growth of electrically active nano-structures on conducting substrates are of interest for fabricating novel nano-device architectures. (Courtesy of S. Agrawal and G. Ramanath; also see S. Agrawal, M.J. Frederick, F. Lupo, P. Victor, O. Nalamasu, G. Ramanath, "Direct growth and electrical-transport properties of carbon nanotube architectures in indium tin oxide films on silicon based substrates." Vol. 15, pp. 1922–1926, *Advanced Functional Materials* [2005], used with permission.)

aluminum-alloy automobiles, and they have been used many times, for decades, in manned orbiting space stations.

Today, dozens of models of automobiles are considered to contain significant percentages of aluminum alloys, and a few are constructed predominantly of parts made from aluminum. The driving force, of course, is lighter weight, as a means of obtaining better fuel economy. Some of these employ a space-frame understructure that protects the occupants and provides the structure that allows the vehicle to operate under the variety of loads and forces it experiences. The space-frame approach to vehicle assembly also provides greater versatility of making modest changes to the exterior design, by simply changing the exterior body panels, and interior design, by simply changing interior décor panels. The space-frame design also provides greater interior room for the same exterior size or volume, since all structural stiffening is handled by the space frame and not under-chassis and structural body panels.

The design of automobile space frames generically consists of stiff, often tubular trusses and struts and net-shaped cast or machined fittings that allow these members to be connected at junctions. Sometimes the tubular members are simply fusion welded to one another at prepared transition joints, and other times they are welded to the fittings. But one major manufacturer, teamed with a major aluminum producer, elected to employ rigid and elastic interlocks. Some joints lock by interlocking mating features in extruded truss members and the fittings and others lock together from the frictional force provided by tapers. To secure friction-type joints, either pins or adhesives are used.

As design experience with integral mechanical attachments in aluminum structures grows, one can only wonder when the aerospace industry will consider using these easy and quick-to-assemble joints for airframes. Properly designed, integral interlocked joints would still provide the desirable frictional-damping and crack-arresting qualities afforded by riveted structure but would dramatically reduce the labor intensity required with hole-drilling, rivet insertion, rivet upsetting and head-shaving, and inspection during each of these steps. When one realizes that a modern airliner the size of a Boeing 767 contains over a million rivets, each of which takes upwards of 2 minutes apiece to fully install, the idea of snapping together an airframe becomes very intriguing. And for those concerned about snap-fit–assembled airframes falling apart during only slightly-jarring landings or in turbulence in flight, one only has to think about how secure those wood burr puzzles are that employ a "secret" keying piece to lock the entire assembly together until it is removed.

13.7   INTEGRAL ATTACHMENT IN BIOTECHNOLOGY

The human skeleton (as well as the skeletons of other higher animals, particularly mammals) is largely held together at numerous joints by integral features of opposing bones. Some, like ball-and-socket joints found in hips and shoulders (see Figure 13.26), more obviously employ a higher degree of mechanical interlocking, while others, such as the complex gliding or sliding joints found in wrists and ankles, rely to a greater degree on assistance from suspensor ligaments. Despite the essential assistance provided by elastic ligaments for holding joints from disengaging, the major loads and forces for which the joints are "designed" are carried, resisted, and transferred by the integral rigid features of the mating bones.

Nature's approach has not gone unnoticed by orthopedic surgeons or dentists; both of whom also rely on integral mechanical attachments for installation of artificial replacement joints, spine-straightening Harrington rods, and a variety of dental implants (e.g., posts, crowns, and even fillings). Inevitably, biotechnology will also emulate Nature in tissue engineering, as well as in the development of more advanced limb replacements, and use integral design features as a key approach for joining.

Figure 13.27 shows one example in a total hip replacement device.

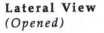

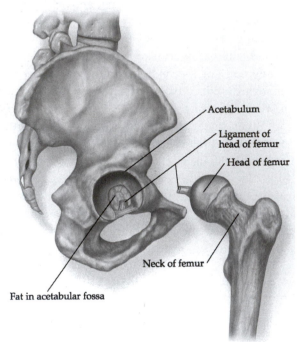

FIGURE 13.26 Illustration of hip of a human skeleton, wherein the use of integral mechanical attachments in the form of a ball-and-socket joint can be seen. Other joints in the skeletons of animals, including humans, use other types of integral mechanical attachments, always augmented by supporting ligaments. (Image taken from a chart of the hip and knee purchased by the author from the Anatomical Chart Company, Skokie, IL.)

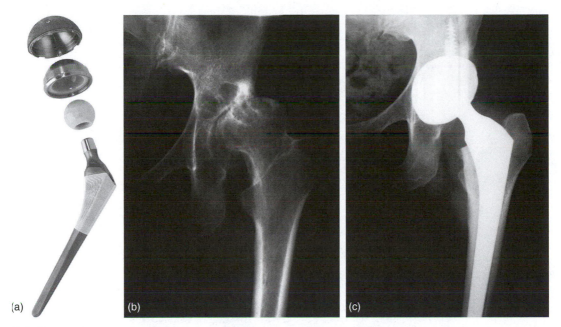

FIGURE 13.27 A total hip replacement device that employs a ball-and-socket design fully analogous to the human body's natural joint. Here, a hip prosthetic is shown in (a), while a diagram showing how such a prosthetic is placed is shown in (b) and an x-ray of such a prosthetic in place is shown in (c). (Photographs courtesy of Stryker Orthopaedics, Mahwah, NJ; used with permission.)

13.8 INTEGRAL MECHANICAL ATTACHMENTS IN HOSTILE ENVIRONMENTS

Many hostile environments once not even dreamed of are now being faced by creations of human beings. We seem drawn to some of these environments, such as the micro-gravity and infinite vacuum of outer space and the crushing pressures in the depths of the oceans, because they are there and because we are a curious specie, and we are forced into others by our need for energy and mineral resources. We also need to service or repair nuclear reactors made dangerously and permanently "hot" by radiation and increasingly are forced to seek and recover oil and natural gas in the forbidding cold of the Arctic tundra and inhospitable and unpredictably violent seas. The only way joints can be created, in the first place, and disassembled for repair or replacement, in the second place, is to design them so that they virtually assemble themselves or are so easy to assemble or disassemble that it can be done remotely. The solution, in many cases, is to use integral mechanical attachments.

One example, shown in Figure 13.28, is the Solar Thruster Sailor solar sail spacecraft being designed and planned in Europe. Powered by the inexhaustible "solar wind," which comes from the momentum transferred by photons moving

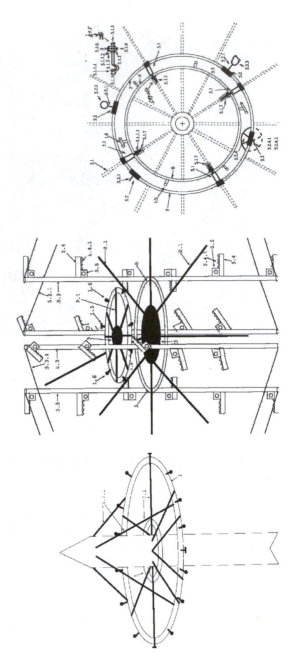

FIGURE 13.28 Line drawings of the German-proposed Solar Thruster Sailor solar sail spacecraft design showing the use of integral mechanical attachments to allow assembly in space with minimal, if any human intervention. (Images from the website http://solar-thruster-sailor.info; for which no contact could be made for permission to use the images.).

through space at the speed of light as they strike a surface, this large sail-craft must be assembled in space, preferably without the need for any assistance from astronauts during extra-vehicular activity. To allow such assembly, and to resist the forces of the solar wind as well as severe thermal-gradient–induced stresses from the sun-lit to the dark side of structures, joints will employ a combination of rigid and elastic integral interlocks. The technology for accomplishing such in situ assembly and of using such self-keying and self-locking joints comes from the experience of the current Orbiting International Space Station (see Figure 13.12), as well as earlier Russian Soyez and Mir and U.S. Mercury, Gemini, and Apollo spacecraft.

Another proposed vehicle that will have to be assembled in outer space is the Asteroid Tug (see Figure 13.29), which would use its nuclear propulsion system to nudge giant space rocks out of harm's way to the Earth.

As we are forced to plunge deeper under the seas to anchor, if not construct, next-generation off-shore drilling platforms, to build 3,000-foot (1,000-m) towers for a bridge that spans the Gibraltar Straits in 1,000-foot (350-m) fast-moving water and, inevitably, it seems, construct new power-generation plants employing nuclear reactors, we will increasingly rely on integral mechanical attachment for producing needed joints.

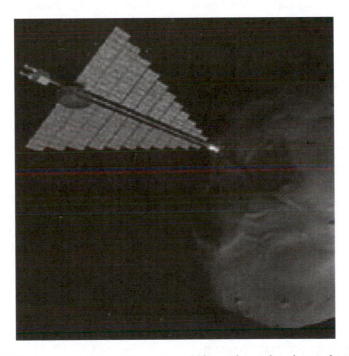

FIGURE 13.29 Proposed Asteroid Tug that would be used to nudge giant rocks out of harm's way to the Earth. Such a large vehicle would have to be assembled in outer space and would, inevitably, have to employ integral mechanical interlocks. (Image by Rick Sternbach, Copyright 2003; used with permission.)

13.9 WHAT ABOUT COMPOSITES?

Reinforced composite materials pose the greatest challenges to joining of all materials. Even more frustrating, the more effective the reinforcement is for providing the desired structural properties, the more difficult the joining becomes. The reasons for this are twofold: first, composite materials employ a reinforcing phase to obtain the desired enhancements in strength or stiffness or both (often at reduced weight), and where there are joints, there is no reinforcement. Even the best joints (i.e., those that are capable of matching base material properties in monolithic [non-reinforced] polymers, metals, ceramics, and carbonaceous materials) cannot hope to match those properties across joints. The second reason for difficulty in joining composites is the anisotropy that accompanies directional reinforcement. Worst of all is unidirectional reinforcement obtained from aligned, unidirectional continuous reinforcing fibers, followed by cross-plied, laminated, continuous-reinforcement composites. The former lack strength in both orthogonal directions to the direction of reinforcement, while the latter lack strength through their thickness. In either case, no known method of joining can match the properties found in the base materials.

Fusion welding almost always obliterates or irreparably degrades the reinforcements in the base composites and, of course, doesn't provide them in the weld proper. Non-fusion welding (e.g., using friction or diffusion) precludes or minimizes damage to the reinforcements but doesn't provide them in the weld proper. Adhesives can match the properties of a non-reinforced resin, so provide strength in the bond as good as that found in the matrix between reinforcements, but not through the thickness of the joint. Mechanical fasteners, unless special provisions are made in the design of the planned structure, force the destruction of reinforcements wherever fastener holes are drilled or exclude reinforcements in the area of pre-placed holes. The best option, therefore, may be to use some form of integral mechanical attachment, although there is no way to continue reinforcement across the mechanical joint—that is, physical interface—with this method either.

So, what to do? The answer, for the time being is "the best you can." Longer term, the answer probably lies in joining the most sophisticated composites for the most demanding applications the way surgeons reattach severed limbs. Structural reinforcements should probably be joined first (as bones are), followed by joining of the less important matrix (analogous to skin). If one thinks about self-assembling molecules, this may not be that far fetched. Not if reinforcements are made from materials that could be caused to grow in situ!

13.10 CLOSING THOUGHTS

We have, indeed, come a long way as a species, at least technologically. We have access to materials beyond those provided by Nature, and we have developed processes to shape and condition these materials to meet our many and diverse needs. Among the processes we have developed for shaping and conditioning materials, nothing may seem to have advanced more than the process of joining. However, as remarkable as our progress has been, have we really advanced that far?

We still wedge "stones" into "forked sticks" every time we plug an electric cord into a wall receptacle, snap the battery cover back onto the body of our wireless cellular telephone, and dock a space craft with an orbiting space station.

It seems that nothing works better than something that has worked since time immemorial!

Only time will tell how far integral mechanical attachment can take us.

REFERENCES

Chen, X., Dam, M.A., Ono, K., Mai, A., Shen, H., Nutt, S.R., Sheran, K., and Wudi, F., "A thermally re-mendable cross-linked polymeric material," *Science,* 295, 2002, 1698–1702.

Easterling, K., *Tomorrow's Materials,* 2nd edition, The Institute of Metals, 1990.

Forester, T., editor, *The Materials Revolution,* The MIT Press, Cambridge, MA, 1988.

Frederick, M.J., and Ramanath, G., "Interfacial phase formation in Cu-Mg alloy thin films on SiO$_2$," *Journal of Applied Physics*, 95(4), 2004, 3202–3205.

Joy, W., "Why the future doesn't need us," *Wired,* 1999, 238–262.

Kirchner, E., "Investigations of ultra-thin aluminum as an adhesion promoter and electrically stable diffusion barrier for copper metallization on SiO$_2$," Ph.D. thesis, Rensselaer Polytechnic Institute, Troy, NY, 1996.

Messler, R.W., Jr., "Joining composite materials and structures: some thought-provoking possibilities," *Journal of Thermoplastic Composite Materials,* 17(1), 2004, 51–75.

Messler, R.W., Jr., "Joining comes of age: from pragmatic process to enabling technology," *Assembly Automation,* 23(2), 2003, 130–142.

Messler, R.W., Jr., "Materials today, joining tomorrow: will the process be ready?" *Materials Today,* September 2002, 48.

Messler, R.W., Jr., "Joining technologies: where to go from here not to be left behind," *Assembly Automation,* Editorial Viewpoint, Special Issue on Joining Technologies, 20(2), 2nd Quarter 2000a, 99–100.

Messler, R.W., Jr., "Trends in key joining technologies for the 21st century," *Assembly Automation,* Special Issue on Joining Technologies, 20(2), 2nd Quarter 2000b, 118–127.

Messler, R.W., Jr., and Genc, S., "Integral micro-mechanical interlock (IMMI) joints for composite structures," *Journal of Thermoplastic Composite Materials,* 11(5), 1998, 200–215.

Moore, G.E., "Cramming more components on to integrated circuit boards," *Electronics,* 1965.

Otto, K, and Wood, K., *Product Design: Techniques in Reverse Engineering and New Product Development,* Prentice Hall, Upper Saddle River, NJ, 2001.

Suwwan de Felipe, T., "Electrical stability and microstructural evolution in thin films of high-conductivity copper alloys," Ph.D. thesis, Rensselaer Polytechnic Institute, Troy, NY, 1998.

Yu, Hongbin, Post-doctoral research on the self-assembly process and characterization of organic monolayers on Si surfaces and nano-wires for molecular electronic and biomolecule sensing applications, California Institute of Technology and, previously, the University of California at Los Angeles, 2005.

APPENDIX A: PERIODIC TABLE OF THE ELEMENTS

PERIODIC TABLE
Atomic Properties of the Elements

Courtesy of NIST, the National Institute of Standards & Technology, Washington, DC; from its website at physics.nist.gov.

APPENDIX B: SELECTED PROPERTIES OF IMPORTANT ENGINEERING MATERIALS

Material	Melting Point °C	Density at 20°C g/cm³	Coefficient Thermal Exp. ×10⁻⁶ °C	Th. Cond. W/m-°K	Yield Str. MPa	Tensile Str. MPa	Tensile Modulus GPa	Hardness VHN
Metals and Alloys								
Aluminum (Al)	660.2	2.7	23.6	247	40–100	90–140	74	50
Al alloys	–	2.6–2.9	18.0–24.5	88–243	80–450	120–580	69–75	50–160
Beryllium (Be)	1,278	1.85	11.5	188	–	–	300	–
Cast iron (gray)	115–1180	7.3	11.1–12.6	46.0–57.3	220–350	300–400	140–170	120–300
Copper (Cu)	1,084	8.92	17.0	398	60–80	160–240	120	50
Cu alloys	–	7.45–9.40	15.8–22.0	12–391	60–1000	200–1400	120–150	50–300
Gold (Au)	1,063	19.3	14.2	80	nil	130	77	<50
Hafnium (Hf)	2,230	13.28	5.9	–	–	450	138	190 DPH
Invar	–	8.05	1.6	10	276	517	141	120–150
Lead (Pb)	327.4	11.34	29.3	34	–	–	30	<50
Magnesium (Mg)	650	1.74	25.2	418	80	120	46	50
Mg alloys	–	1.75–1.9	25–27	51–138	80–200	165–265	41–45	50–120
Molybdenum (Mo)	2,620	10.2	4.8	142	500	630	320	500–600
Nickel (Ni)	1,455	8.90	13	70	148	462	204	75
Ni alloys	1,400–1,455	8.0–8.4	12.3–13.9	9.8–70	150–520	460–980	180–230	75–350
Niobium (Nb)	2,468	8.57	7.3	52.7	200	275	105	160
Platinum (Pt)	1,769	21.5	9.5	71.1	<13.8	124	160	100
Silver (Ag)	960.5	10.5	19.7	428	nil	170–400	74	<50
SS 316L*	–	7.8–8.0	15.9	16.2	205–310	515–600	190	250
Stainless steel, Aust.	1505–1550	8.0	15.0–17.5	14.4–17.2	205–310	515–620	190–195	200–25
Stainless steel, Ferritic	–	7.9–8.0	10.8–11.2	25.0–27.0	170–190	400–450	200	200–400
Stainless steel, Martensitic	–	7.8–7.9	10.2–10.8	24.0–24.8	415–1650	725–1790	200	450–600
Stainless steel, PH	–	7.65	11.0–11.5	16.0–17.0	1210–1310	1380–1450	204	300–450
Steel, HSLA	1350–1540	7.85	11.1–12,8	37.5–48.9	420–1620	650–1760	200	500–1000
Steel, Plain carbon	1350–1540	7.85	11.0–12.0	50–65.3	210–500	380–600	200	200–500

Material								
Tantalum (Ta)	2,996	16.6	6.5	54.4	165	205	185	400–500
Titanium (Ti)	1,668	4.5	8.4	17.0	170	240	110	200
Ti alloys	1,550–1,660	4.3–4.8	8.6–9.4	6.7–16.0	400–1100	700–1200	110–130	200–400
Ti, CP (Grade 4)*	~1,660	4.5	8.4	15.0–17.0	485	760	110	200
Ti-6Al-4V*	–	4.43	8.6	6.7	896–1034	965–1103	116	300–350
Tin (Sn)	231.9	7.30	23.8	17	11	15–20	44.3	<50
Tungsten (W)	3,390	19.3	4.5	155	760	900–1000	400	450–550
Vitallium (Co-Cr-Mo)*	–	–	–	–	448–1606	655–1896	210–253	–
Zinc (Zn)	419.5	7.14	39.7	113	–	135–160	104.5	50–100
Zirconium (Zr)	1,852	6.5	5.9	21.0	200–220	340–380	101	180 DPH
Intermetallics								
FeAl	1,232–1,310	5.7–5.8	–	–	300–600	–	260–280	–
MoSi	1,870	6.1	–	–	–	–	1,290	–
NiAl	1,600–1,640	5.5–5.8	–	–	200–600	–	220	–
TiAl (γ-Ti)	1,390–1,480	4.0	–	–	250–850	–	200	–
TiB$_2$	2,980	4.5	–	–	–	–	3,400	–
Ceramics[1] and Glasses								
Alumina*	2,050	3.9–4.0	8.8	30–40	–	280–550	280–380	1900
Bioglass-ceramics*	–	2.6	6.5	3.3	–	300–500	30–120	–
Calcium phosphate*	–	–	–	–	–	69–193T/510–896C	40–117	–
Concrete	2.25–2.5	10.0–13.6	1.25–1.75	–	37.3–41.3	25–50	1000+	–
Diamond	3,550	2.25	1.2	1450–4650	–	800–1400	800–1200	5000–8000
Glsss, Soda lime	~700	2.2	9.0	1.7	–	63–69	69	600
Graphite	3,550	1.30–2.25	~1.0	38	–	143–69	10–20	<100
Graphite, Pyrolitic*	–	2.2	2.7	110–190	–	280–560	100–200	100–200
Magnesia	2,800	3.58	–	–	–	–	225	–
Pyrex (borate) glass	850	2.23	3.3	1.4	–	70–90	64–70	600–700

[1] Strength is Flexural Strength

continued

Material	Glass Transition °C	Density at 20°C g/cm³	Coefficient Thermal Exp. ×10⁻⁶ °C	Th. Cond. W/m·°K	Yield Str. MPa	Tensile Str. MPa	Tensile Modulus GPa	Hardness VHN
Silica, Fused	1,700	2.2	0.5	2.0	–	104	73	800
Silicon (Si)	1,410	2.4	2.5	141	–	81.8–130	129–187	500–500
Silicon carbide	2,700	2.2–2.5	4.7	70–80	–	100–825	220–480	2550
Silicon nitride	1,900	3.2	2.1	10–35	–	250–1000	305–390	1750
Zirconia (PSZ)	2,690	6.0	9.6	2.0–4.5	–	800–1500	205	1000
Polymers								
Epoxy	–	1.11–1.40	81–117	0.19	–	27.6–90.0	2.41	<30
Nitrile rubber	–90	0.98	235	0.25	–	6.9–24.1	0.0034	<30
Nylon 6,6*	57	1.14	144	0.24	30	44.8–82.8	1.59–3.79	<30
PEEK	143	1.31	72–85	–	91	70.3–103	1.10	–
PET*	69	1.35	117	015	59.3	48.3–72.4	2.76–4.14	–
Phenolic	–	1.14	122	0.15	–	34.5–62.1	2.76–4.83	–
PMMA*	3	1.19	90–162	0.17–0.25	53.8–73.1	48.3–72.4	2.24–3.24	–
Polycarbonate (PC)	530–550	1.34	122	0.20	62.1	62.8–72.4	2.38	<30
Polyethylene (HDPE)	–90	0.959	106–198	0.48	26.2–33.1	22.1–31.0	1.08	<30
Poly(lactic acid)*	–	–	–	–	28–50	–	1.2–3.0	–
Polypropylene*(PP)	–18/–10	0.905	90–162	0.12	31.0–37.2	31.0–41.4	1.1–1.6	–
Polystyrene (PS)	100	1.05	146–180	0.13	–	35.9–51.7	2.28–3.28	–
Polytetrafluoroethylene*	–97	2.17	126–216	0.25	–	20.7–34.5	0.40–0.55	–
Polyvinyl chloride (PVC)	87	1.30–1.58	90–180	0.15–0.21	40.7–44.8	40.7–51.7	2.41–4.14	–
Silicone rubber*	–123	1.1–1.6	270	0.23	2.0–8.0	10.3	1.0–5.0	<30
Styrene-butadiene rubber	–	0.94	220	0.25	–	12.4–20.7	0.002–0.010	<30
UHMW Polyethylene	–	0.94	234–360	0.33	21.4–27.6	38.6–48.3	0.69	–

Material	Melting Point °C	Density at 20°C g/cm³	Coefficient Thermal Exp. ×10⁻⁶ °C	Th. Cond. W/m-°K	Yield Str. MPa	Tensile Str. MPa	Tensile Modulus GPa	Hardness VHN
Composites								
Boron fiber	~2,300	2.57	–	–	–	400	–	~2500
Carbon-carbon (generic)	–	1.4–1.9	–0.06–10.0	11–70	–	100–850	105–250	–
C fiber	3,367**	1.78–2.15	7.0–8.5	7–10	–	2500–6350	230–400	–
Concrete	–	2.4–2.5	–	–	–	37.3–41.3 Flexural	25–50	1000+
E-glass fiber	–	2.50	1.3	5.0	–	3450	72.5	–
Epoxy	–	1.15	35	0.19–0.34	–	27.5–90	2.4	–
Epoxy/B fiber (70 v/o)	–	2.3–2.5	–	–	–	1400–2100	210–280	–
Epoxy/E-glass (73.3 v/o)	–	2.1–2.3	6.6–30	–	–	1020–1640	45–60	–
Epoxy/Gr fiber (54 v/o)	–	1.3–1.7	–0.5–30	–	–	700–1,400	180–220	–
Epoxy/Kevlar (82 v/o)	–	1.2–1.4	–4.0–60	–	–	1350–1500	76–86	–
Kevlar fiber	375***	1.44	–	60	–	131	3600–4100	–
Nylon 6,6	57***	1.10–1.15	50	0.17	–	75.9–91.5	1.58–3.79	–
Nylon 6,6/glass (43 v/o)	–	1.51	–	–	–	140–160	0.85	–
PEEK/glass (20 v/o)	–	1.37	–	–	–	149	–	–
PEEK/C fiber (20 v/o)	–	1.40	–	–	–	165	–	–
Wood (Hardwood)	–	0.35–1.1	–	–	–	46.9–139	7.0–15.7	–
Wood (Softwood)	–	0.31–0.59	–	–	–	44.8–112	5.5–13.7	–
Natural Materials								
Bone, Cortica	–	–	–	–	30–70	70–105	15–30	–
Wood (see Table 11.1)	–	–	–	–	–	–	–	–

* Biocompatible
** Sublimes/*** T_{glass}

INDEX